Excavation Planning Reference Guide

The McGraw-Hill Engineering Reference Guide Series

This series makes available to professionals and students a wide variety of engineering information and data available in McGraw-Hill's library of highly acclaimed books and publications. The books in the series are drawn directly from this vast resource of titles. Each one is either a condensation of a single title or a collection of sections culled from several titles. The Project Editors responsible for the books in the series are highly respected professionals in the engineering areas covered. Each Editor selected only the most relevant and current information available in the McGraw-Hill library, adding further details and commentary where necessary.

Church · EXCAVATION PLANNING REFERENCE GUIDE

Gaylord and Gaylord · CONCRETE STRUCTURES REFERENCE GUIDE

Hicks · BUILDING SYSTEMS REFERENCE GUIDE

Hicks · CIVIL ENGINEERING CALCULATIONS REFERENCE GUIDE

Hicks · MACHINE DESIGN CALCULATIONS REFERENCE GUIDE

Hicks · PLUMBING DESIGN AND INSTALLATION REFERENCE GUIDE

Hicks · POWER GENERATION CALCULATIONS REFERENCE GUIDE

Hicks · POWER PLANT EVALUATION AND DESIGN REFERENCE GUIDE

Higgins · PRACTICAL CONSTRUCTION EQUIPMENT MAINTENANCE REFERENCE GUIDE

Johnson and Jasik · ANTENNA APPLICATIONS REFERENCE GUIDE

Markus and Weston · ESSENTIAL CIRCUITS REFERENCE GUIDE

Merritt · CIVIL ENGINEERING REFERENCE GUIDE

Ross · HIGHWAY DESIGN REFERENCE GUIDE

Rothbart · MECHANICAL ENGINEERING ESSENTIALS REFERENCE GUIDE

Woodson · HUMAN FACTORS REFERENCE GUIDE FOR ELECTRONICS AND COMPUTER PROFESSIONALS

Woodson · HUMAN FACTORS REFERENCE GUIDE FOR PROCESS PLANTS

Excavation Planning Reference Guide

HORACE K. CHURCH (Deceased)
Consulting Engineer

JEREMY ROBINSON ***Project Editor***

McGraw-Hill Book Company

New York St. Louis San Francisco Auckland
Bogotá Hamburg London Madrid Mexico
Milan Montreal New Delhi Panama
Paris São Paulo Singapore
Sydney Tokyo Toronto

Library of Congress Cataloging-in-Publication Data

Church, Horace K.
Excavation planning reference guide.

(The McGraw-Hill engineering reference guide series)
"The material in this volume has been published previously in Excavation handbook by Horace K. Church."
—Verso t.p.
Bibliography: p.
Includes index.
1. Excavation. I. Robinson, Jeremy, date.
II. Church, Horace K. Excavation handbook. III. Title.
IV. Series.
TA730.C484 1988 624.1′52 88-6835
ISBN 0-07-010904-4

1234567890 DOC/DOC 8921098

ISBN 0-07-010904-4

Printed and bound by R. R. Donnelley & Sons Company.

A visionary example of open-pit mining is the spectacular workings of the United States Borax and Chemical Corporation in the midst of the Mojave Desert of southern California.

This awesome amphitheater is 4700 ft west to east, 3300 ft north to south, and 550 ft deep. It is carved out of the desert alluvium and clays and the borate ore. Overburden and the blasted ore are loaded by outsize electric shovels and hauled by monstrous diesel trucks over the boulevardlike haul roads. Overburden is lifted hundreds of feet to waste pile areas beyond the rim of the pit and ore is hauled to the primary crusher at the foot of the conveyor, whence it is raised to the stockpiles of the processing plant.

In 20 yr some 168 million tons, or about 96 million yd^3, of combined overburden and ore have been wasted and processed. Such colossal quantities are enough to fill a freight train of hopper cars stretching more than around the world, or some 32,000 mi. *(United States Borax and Chemical Corporation.)*

Contents

Preface

The *Excavation Planning Reference Guide* discusses the methods and costs for excavating the materials of the Earth's mantle, or regolith, from their initial locations in situ to their final deposition in construction or their final or near final deposition in mining. These materials range from the earthy to the rocky.

Soft earths are really finely divided particles of rock and they are, petrologically, little rocks. Frequently in the text the material is called rock, rock-earth, earth-rock, or earth, the two intermediate designations meaning that either rock, rock-earth, or earth, earth-rock, predominates in the material.

The word "excavation" means work in an open cut, exposed to the sky, and it does not include underground excavation, as in mining and tunneling, except when tunneling is supplementary to open-cut excavation.

Purpose of the Book

The *Excavation Planning Reference Guide* may be used for general and specific information, for selecting methods for work, for estimating productions and costs of machinery for a given task, and for estimates of the total cost of excavation in construction and mining in open cuts.

The book is for the use of all college students in engineering, construction, and geology, for private engineering companies, for public engineering agencies, for construction and mining companies, and for machinery, explosives, and supplies companies. It may also be entertaining to people who have a general interest in the art and science of moving the Earth.

Coverage of the Book

The *Excavation Planning Reference Guide* includes these aspects and phases of many excavations:

1. Geology of excavation as it relates to rocks, ores, minerals, rock formations, rock weathering, and landforms.
2. Kinds of open-cut excavations, which are indicative of the diversity and size of these cuttings and fillings.
3. Examination and exploration of excavations, which are requisite to a well-considered estimate of costs and to a later successful prosecution of the work.
4. The costs of ownership and operation of the machinery and facilities, into which the productions of machinery are divided in order to secure unit costs for the work.
5. Preparation of a typical bid for excavation and a typical means for scheduling excavation as a part of the complete project.
6. Appendixes, a glossary, and a bibliography covering the science and practices of excavation.
7. A complete, detailed index.

Special Note

This book consists of sections covering the excavation planning process derived from McGraw-Hill's 1,024 page *Excavation Handbook* published in 1981. The material included here continues to be relevant for the solution of a wide variety of planning problems.

JEREMY ROBINSON
Project Editor

CHAPTER 1

Earth and Geology of Excavation

PLATE 1-1

Aggradation or uplift of the Sierra Nevada of California. The Sierra Nevada are a classical example of Nature's aggradation or uplift of the Earth's surface. Mount Whitney towers at an elevation of 14,495 ft along with the eastern edge of the batholith, which is some 350 mi from north to south and some 80 mi from west to east. This orogeny commenced some 180 to 135 million yr ago in the Jurassic period and, although intermittent, the continuous raising of the westward dipping fault block is still in progress. Along with the aggradation is the relentless degradation of the land surface by wind, rain, snow, and ice. Throughout the aeons the uplifting of the massive granite has triumphed over the opposing weathering of the many kinds of surficial rocks, and Mount Whitney is slowly increasing in elevation. *(National Park Service.)*

Excavation is the act of removing, moving, and depositing the veneerlike surface of the Earth's outermost crust or regolith. The material excavated may be in the solid or semisolid state, that is, rock; in the weathered state, usually a mixture of rock and earth; or in the loose state, earth.

In many cases excavation, because of the different degrees of weathering, may be called rock, rock-earth, earth-rock, or earth. The distinction depends upon the relative amounts of rock and earth. Lithologically, all excavated material may be called rock, the rock differing only in the size of the individual particles. Extremes are found in the material basalt. In the solid state in a lava flow it is basalt rock. In the loose state in a cultivated field it is red clay or earth. Between these extremes is the weathered state, a mixture of rock-earth or earth-rock, depending upon the relative amounts of rock and earth.

In construction of public works such as dams and highways and in the mining of quarries and open pits, the finished grade is rarely at a depth greater than 400 ft below natural or original ground. There are some notable exceptions such as the iron mines of Minnesota and the copper mines of Utah and Arizona, as well as some other large open-pit works.

Such deep excavations, spectacular as they may seem, are a mere 50 thousandths of the Earth's radius. They are in a weathered and solid zone wherein nature's destructive forces have been toiling for millions of years. These ruinous energies are the most important ally of human excavators, because if all excavation were in the rock or solid state much more time and money would have been expended for the trillions of cubic yards of excavation since the beginning of recorded history some 6000 yr ago.

The veneerlike surface of the Earth's crust is a variable and fascinating earth-rock mixture and it is presumptious and dangerous for the excavator to remove it without preliminary study. The greater the understanding of geology and its kindred sciences, the greater the mastery of the complexities of excavation.

Some 5 billion yr ago the planet Earth was formed from a gaseous substance which, through cooling and related processes, became a slowly consolidating molten mass. Some 4 billion yr ago the igneous rock basement complex of Earth had been formed, to be followed by systematic degradation by nature's forces. This relentless debasement resulted in the deposition of both marine and nonmarine or continental sedimentary rocks.

Simultaneously with the formation of the sedimentary rocks, the igneous rocks did not remain dormant but rather continued the orogeny by their intrusions and extrusions. From these rocks both metamorphic and additional sedimentary rocks were derived, not only in the past but in the geologic present.

Nature's powerful processes of building up and tearing down are continuously in evidence. An example of building up or aggradation is provided by the frequent lava flows of the volcanoes of the Hawaiian Islands. The six islands are the result of volcanic eruptions from the bottom of the sea during a period of some 8 million yr. Judging by the degrees of rock weathering, the oldest island is the westernmost Kauai and the youngest is the easternmost Hawaii. Mauna Loa is the perennial active volcano of the island chain, the accumulative lava of Hawaii towering 13,800 ft above the Pacific Ocean. The sea bottom is about the same distance below the ocean surface. The sea-bottom floor area of Hawaii is some 16,000 mi^2, and so Hawaii Island represents about 28,000 mi^3 of lava flows.

Turning to more understandable present-day volcanic aggradation, Figure 1-1 shows the 1955 flow of the less pretentious Kilauea Crater of Hawaii.

The area of the picture is 3400 ft north to south and 3400 ft west to east. The black basalt flow within the limits of the picture is 1.5 million ft^2 with an average 8-ft depth. The flow of 440,000 yd^3 took out the north-to-south road, isolating the farmers and cattlemen. The road was restored five days after the flow. Bulldozers worked atop the flow when temperatures had cooled from 1000°C (1830°F) to 700°C (1290°F), the incandescent molten lava being visible through the cooling cracks of the solidifying lava.

Every decade or so there is a sizable, perhaps 1-million-yd^3 lava flow on Hawaii, a sizable contribution to the land but infinitesimal when contrasted with the 153 trillion (153,000,000,000,000) yd^3 of lava of the island Hawaii.

Aggradation, even in the form of breathtaking molten lava flows, is not as spectacular as degradation in the form of landslides. One such massive downward movement of rock

Figure 1-1 **Aggradation by flow of lava.** Basalt lava flow of Kilauea Volcano, Hawaii, in 1955. The area of the picture is 0.41 mi^2, 3400 ft by 3400 ft. The volume of the black igneous extrusive rock within the picture is 440,000 yd^3. *(Hawaiian Volcano Observatory.)*

was the Wood's Gulch slide of 1974 on the abandoned old California Highway 1 of San Mateo County.

Figure 1-2 illustrates this slide of an estimated 110,000 yd^3, which weakens the lateral support for many new homes located on an artificial terrace above the shore.

There were several mutually contributing causes of this slide. First, the Pacific Ocean is constantly battering the shoreline foundation of the old Coast highway. Second, the rock structure is unstable, as the moisture-laden Merced formation of sedimentary sandstones, siltstones, and claystones is poorly indurated and rather steeply dipped. Third, earthquakes along the Wood's Gulch fault and the nearby parent San Andreas fault have further weakened the rock formation over the millenia. And fourth, the highway was excavated within a cliff in a location fraught with natural and artificial hazards such as encroaching home-building sites destructive of natural drainage of the steep slopes.

The slide-prone shoreline extends for 12 mi from Daly City southward to Montara and, wherever possible, the Coast highway is being relocated inland away from the treacherous cliffs of the sea.

In the Wood's Gulch slide only an unused highway was immediately affected. This was not the case in the Orinda slide of 1950 in Contra Costa County, California.

Figure 1-3 emphasizes the tragic possibilities of this 250,000-yd^3 slump slide of 800-ft height and 300-ft width. After several days of heavy rains the hillside with its cover of shrubs and trees began to slip at 10:30 A.M. on December 9. Within 15 min the four-lane arterial California Highway 75 was covered and the earth flow continued for about 2 h.

In spite of normal Saturday morning traffic in the San Francisco Bay area, averaging 35,000 vehicles daily, there were miraculously no fatalities and no injuries.

The view shows the emergency excavation for the four-lane detour around the toe of the slide. By Saturday night 9000 yd^3 of near-liquid-state sediments had been removed, the detour was paved, and the highway was ready for the Monday morning rush of commuters.

Figure 1-2 Degradation by slide of coastline highway. Massive coastline slide of 110,000 yd^3 at Wood's Gulch, San Mateo County, California. This degradation occurred in 1974 on the old Coast highway, previously abandoned because of ill-considered inroads on the instability of nature and the subsequent weakening of a rock structure by excavations for a highway and a huge home site along the coast line. *(California Division of Mines and Geology.)*

The causes of the slide are portrayed graphically in Figure 1-3. First, the area is slide-prone, as the rather steep natural slope of the young 10-million-year-old Orinda formation is made up of soft sandstones, shales, and conglomerates, sometimes steeply dipped. The wet shale beddings act as a lubricant for the sliding actions of the adjacent sandstones and conglomerates. Second, the area is within 3 mi of the Pinole and Wildcat faults, subparallel to the notoriously active Hayward fault, and, accordingly, is weakened structurally. Third, excavating to the back slope of the existing highway had reduced the stability of the natural ground by removing the buttress. Fourth, percolation of the winter rains provided lubricated slip planes in the layered sedimentary rocks.

The regrading of the slide area provided for two important future slide preventatives. The slope of the cut was reduced to a ratio of 2:1, considerably increasing the stability. Some 20,000 lineal feet of horizontal drains were installed in the area of the slide and a total flow of some 135,000 gal/day of water was withdrawn to dry out the rock formation. The slope setback at a ratio below the angle of repose and the dewatering have resulted in stable conditions.

Although degradation is more noticeable than aggradation, the two different and opposing natural processes are in delicate balance so that the mountains and the valleys are eternally in equilibrium.

An example of these adversary forces occurs in the Black Hills of the Wyoming–South Dakota border. Degradation is obvious in the daily weathering of the limestones and redbeds and in the stream erosion. Not apparent is the slow, inexorable uplifting of the sedimentary rock dome by igneous rock intrusion, which commenced some 80 million yr ago. After initial uplifting, ending some 35 million yr ago when the dome towered 7500 ft above the plains, weathering reduced the high elevation to the present 4000 ft. Thus,

Figure 1-3 **Degradation by slide of mountain highway.** The beginning of emergency slide removal of the Orinda slump slide of 1950 in Contra Costa County, California. During the winter rainy season 250,000 yd^3 of liquidlike soft rock avalanched over the arterial highway in 15 min. Four-lane traffic was restored in 2 days by emergency crews of the California Division of Highways. Phenomenally, there were no injuries or deaths to travelers and there was little inconvenience to San Francisco Bay area commuters, as the slide and the construction detouring took place on a weekend. *(San Francisco Chronicle, Dec. 10, 1950.)*

building up has exceeded tearing down by 3500 ft. The differential rate is only ⅝ in every 1000 yr, but it has been of a positive or growth nature.

Another striking example of the triumph of aggradation over degradation is the story of the Sierra Nevada, as described in Plate 1-1.

It is well for the excavator to have a good understanding of geologic time because it will be advantageous to correlate the same rock formations in terms of the geologic time scale in diverse locations around the world.

The Principle of Uniformitarianism was established brilliantly by the father of modern geology, James Hutton (1726–1779). Some 200 yr later the earthmover finds it safely practical to conclude that he will have blasting in the Hornbrook sandstone formation near Hilt, northern California, because he blasted the similar Rosario sandstone formation at Point Loma, southern California. Both formations are Upper Cretaceous marine sedimentary rocks of thick beddings. They are 700 mi apart but of the same geologic series of rocks, 63 to 99 million yr of age.

The early studies of James Hutton and the subsequent laborings of many geologists have produced the main divisions of geologic time for North America, as set forth in Table 1-1.

The rocks of the Cenozoic, Mesozoic, and Paleozoic groups are fairly accurately dated by fossil correlations in the sedimentary rocks and by isotypes in the igneous and metamorphic rocks.

The rocks of the Proterozoic, Archeozoic, and Azoic groups are likewise precisely dated by isotypes, as fossils do not exist in these groups except for rare instances of algae fossils.

TABLE 1-1 Main Divisions of Geologic Time for North America

Era or group of rocks		Period or system of rocks		Epoch or series of rocks	Scale of time (millions of years) Interval	Age
Cenozoic	63	Quaternary	2	Holocene		
(age of				Pleistocene		
mammals,				or Glacial	2	2
grass, and		Tertiary	61	Pliocene	11	13
land forests)				Miocene	12	25
				Oligocene	11	36
				Eocene	22	58
				Paleocene	5	63
Mesozoic	167	Cretaceous	72	Upper Cretaceous	36	99
(age of				Lower Cretaceous	36	135
reptiles)		Jurassic	45		45	180
		Triassic	50		50	230
Paleozoic	340	Carboniferous	110	Permian	50	280
(age of				Pennsylvanian	30	310
fishes,				Mississippian	30	340
amphibia,		Devonian	60	Upper Devonian	20	360
and swamp				Middle Devonian	20	380
forests)				Lower Devonian	20	400
		Silurian	30	Cayugan	10	410
				Niagaran	10	420
				Oswegan	10	430
		Ordovician	70	Cincinnatian	23	453
				Mohawkian	24	477
				Canadian	23	500
		Cambrian	70	Saratogan	23	523
				Acadian	24	547
				Georgian	23	570
Proterozoic	630	Keweenawan	250		250	820
(Algonkian		Huronian	250		250	1070
age of scum,		Temiskamian	130		130	1200
algae, and						
jellyfish)						
Archeozoic	800	Keewatinian	800		800	2000
(Archean;						
age of life)						
Azoic (no life)	2500				2500	4500

NOTE: This is a composite table, representing the thoughts of several geologists of these times. The ages are in millions of years, and the age of the earth is thought to be about 4.5 billion yr.

The oldest rocks have been dated at 3.3 billion years of age in Rhodesia, or about three-quarters of the age of the Earth.

SUMMARY

The continual and balancing processes of aggradation and degradation have produced the infinitely complex surface of the Earth. It is this regolith that the excavator must remove. Accordingly, it is fundamental that a prudent excavator must be acquainted with the uniformity and the irregularity of the regolith.

CHAPTER 2

Rocks, Ores, Minerals, and Formations

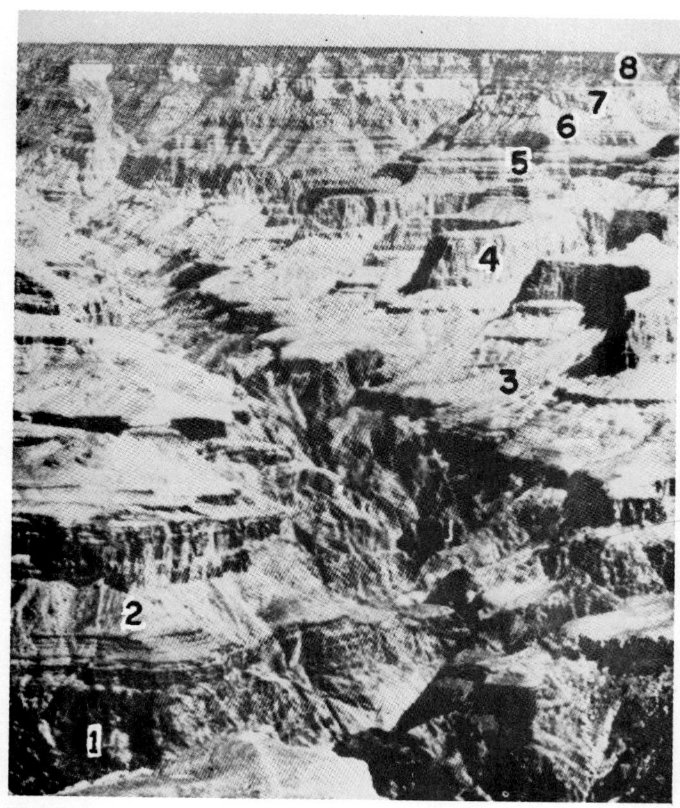

PLATE 2-1

Aged rocks of the Grand Canyon of the Colorado River. The majestic Grand Canyon of the Colorado River in Arizona is Earth's greatest display of rocks over the longest time. On the Kaibab Trail from the South Rim down to the Colorado River and up to the North Rim is a 1.2-billion-yr-old geologic record of igneous, sedimentary, and metamorphic rocks which encompasses about one-fourth the age of Earth.

Plate 2-1 shows the panorama of this assemblage of rocks, ranging from the oldest black metamorphic rocks of the Inner Gorge of the Colorado River up to the white sedimentary rocks of the South Rim, some 1 mi above the Colorado River. One may cross-refer this picture of the stratigraphy of the Canyon to Figure 2-1 and Table 2-1 of this chapter. The numbers of Plate 2-1 approximate the zones of the principal rock formations of the Canyon.

1. Archeozoic group of metamorphic and igneous rocks. Vishnu schist with granite intrusions. Perhaps 1200 to 2000 million yr ago.
2. Proterozoic group of metamorphic, igneous, and sedimentary rocks. Vishnu schist with granite dikes. Bass limestone and Hakati shale. Perhaps 570 million to 1200 million yr ago.
3. Cambrian system of metamorphic and sedimentary rocks. Shinume quartzite, Bright Angel shale, and Muav limestone. 500 to 570 million yr ago.
4. Devonian system, Temple Butte limestone, and Mississippian series, Redwall limestone, of sedimentary rocks—310 to 400 million yr ago. Between the Cambrian and Devonian periods there was a long interval of erosion or unconformity lasting some 100 million yr.
5. Pennsylvanian series of sedimentary rocks. Supai limestone and shale—280 to 310 million yr ago. Between the Mississippian and the Pennsylvanian series there is another unconformity of unknown time interval in the sequence of rocks. Consequently no time for the unconformity is hazarded.
6. Lower Permian series of sedimentary rocks. Hermit shale. Perhaps 263 to 280 million yr ago. Between the Pennsylvanian and the Lower Permian series of rocks there is another erosion surface of unknown length and so no time for this unconformity is speculated.
7. Middle Permian series of sedimentary rocks. Coconino sandstone. Perhaps 247 to 263 million yr ago.
8. Upper Permian series of sedimentary rocks. Toroweap limestone and sandstone in lower zone and Kaibab limestone and sandstone in upper zone, reaching the surface of the land. Perhaps 230 to 247 million yr ago.

Between 230 million yr ago and the present there are no rocks, except for the surficial residual sands of the desert country. Mesozoic and Cenozoic groups of rocks are missing. The Grand Canyon is in the midst of another great period of unconformity and another immense erosion surface is being formed, encompassing much of northern Arizona.

If geologic history customarily repeats itself, some millions of years in the future the sea will again invade the land and it will deposit another series, system, or group of sedimentary rocks. Then the land will rise again and future geologists will mark the sands of the South Rim as the fourth unconformity of the Grand Canyon of the Colorado River. *(U.S. National Park Service.)*

THE GRAND CANYON—THE CLASSICAL RECORD OF ROCKS

The Grand Canyon of the Colorado River in northern Arizona features the Earth's hugest excavation by weathering and river erosion and probably the most complete display of rock types of the last 1.2 billion yr. In the 1-mi-deep and 15-mi-wide Canyon the walls show a variety of igneous, sedimentary, and metamorphic rocks.

Let us go down a mile into the Earth and go back into the aeon of 1.2 billion yr ago.

The scene is the Kaibab Trail, commencing at Yaki Point on the South Rim of the Canyon at elevation 7260 ft and ending at the suspension bridge across the river at elevation 2420 ft. The distance is about 8 mi and the downward grade is about 10 percent throughout the 4860-ft depth.

If one wishes to examine the multicolored rocks in detail and to appreciate the plants, the animals, and the magnificent scenery, a 3-day hike is in order.

Figure 2-1 and Table 2-1 show the rock and rock-formation relationships. Table 2-1 gives the log of the probe into the earth's crust and it may be cross-referred to Table 1-1.

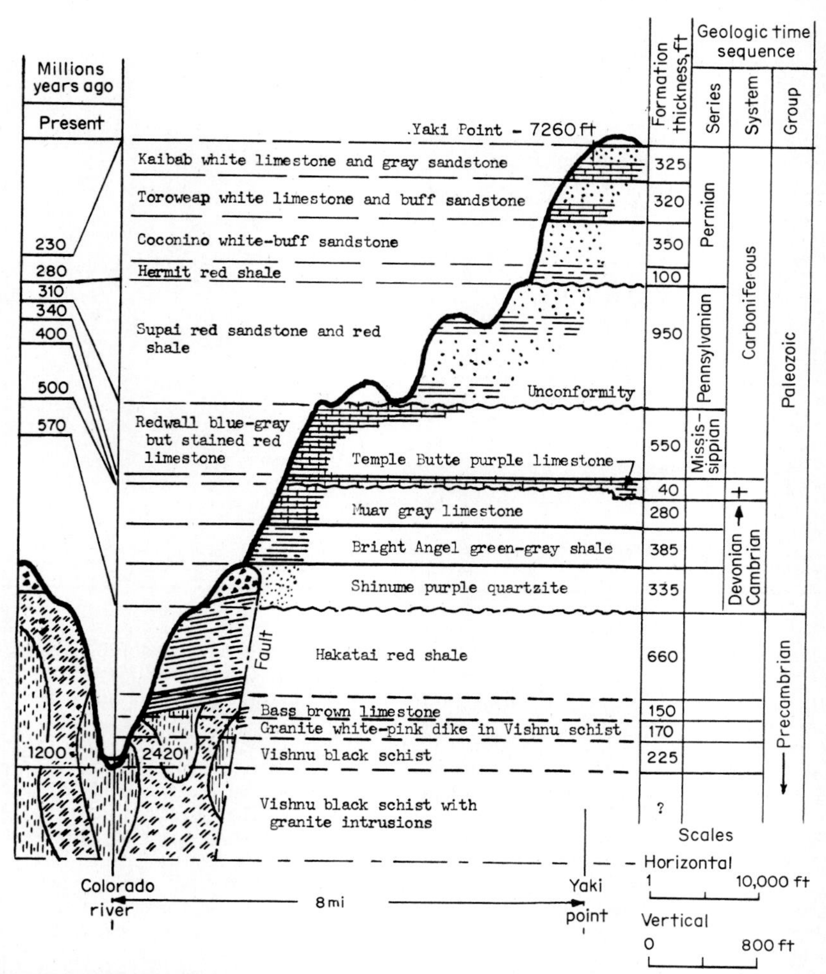

Figure 2-1 Geologic column of the south wall of the Grand Canyon along the Kaibab Trail from Yaki Point to the Colorado River.

TABLE 2-1 Geologic Column of the South Wall of the Grand Canyon along the Kaibab Trail from Yaki Point to Colorado River

	Formation thickness	Elevation (ft)
Yaki Point on the south rim, about 230 million yr ago		7260
Paleozoic group of rocks		
Permian series, 230–280 million yr ago		
Kaibab white limestone and gray sandstone	325	
Toroweap white limestone and red-buff sandstone	320	
Coconino white-buff sandstone	350	
Hermit red shale	100	
	1095	
Erosion surface or unconformity of unknown time interval		6165
Pennsylvanian series, 280–310 million yr ago		
Supai red sandstone and red shale	950	
Erosion surface or unconformity of unknown time interval		5215
Mississippian series, 310–340 million yr ago		
Redwall blue-gray but stained red limestone	550	
Devonian system, 340–400 million yr ago		
Temple Butte purple limestone	40	
Erosion surface or unconformity of 400–500 million yr ago, representing missing Silurian and Ordovician systems		4625
Cambrian system, 500–570 million yr ago		
Muav gray limestone	280	
Bright Angel green-gray shale	385	
Shinume purple quartzite	335	
	1000	
Erosion surface or great unconformity of unknown time interval		3625
Proterozoic and Archeozoic or Precambrian group of rocks, 570–2000 million yr ago		
Hakatai red shale	660	
Bass brown limestone	150	
Granite white-pink dike in Vishnu schist	170	
Vishnu black schist	225	
	1205	
Colorado River at the Inner Gorge, about 1200 million yr ago		2420

It is said that the 8-mi hike gives a white-to-black spectrumlike picture of rocks equivalent to a 2400-mi journey from Mexico's Sonora Desert to Canada's Hudson Bay. The range is from white sedimentary Kaibab limestone through pink igneous granite to black metamorphic Vishnu schist.

Down the trail to the Colorado River we pass through at least 970 million yr of many kinds of rocks in about 10 h at a rate of 97 million yr/h. However, if one wished to analyze these same rocks on the surface of the land from the Sonora Desert to Hudson Bay, it would probably require 3 months.

Sixteen hours or so should be devoted to the long uphill hike so as to observe the geology, the landforms, the formations, the rocks, the animals, and the plants. Literally we emerge from the depths of time of the earliest traces of life in the Bass limestone, 470 ft above the river, to the tracks of reptiles in the dune sands of the Toroweap sandstone, 485 ft below Yaki Point. At Yaki Point let us consider Nature's prodigious excavation job during the assumed 10 million yr of erosion of the Canyon.

Let the Canyon be confined to Grand Canyon National Park with a length of 100 mi, an average width of 15 mi from rim to rim, and an average depth of 1 mi. The following statistics may be derived from these dimensions.

Rock volume, eroded and removed ultimately by the river	750 mi^3
or	4,095 billion yd^3
Yearly excavation	410,000 yd^3

Excavating 410,000 yd^3 of rock annually is not impressive to the average earthmover, but so awesome is geologic time that 4.1 trillion yd^3 of rock excavation is mind-boggling. It is equivalent to a trench from Seattle, Washington, to Miami, Florida, with a length of 2600 mi, a width of 1 mi, and a depth of 1/4 mi.

After our return to Yaki Point we stand on the 230-million-yr-old Kaibab formation and we have come some 4840 ft vertically upward from the Colorado River. Our upgrade trail has taken us through almost 1 billion yr of rock formations and from a time 1.2 billion yr ago.

Such great time is astounding; let us visualize it by means of a descriptive book commencing 1.2 billion yr ago, each page representing the average threescore years and ten of our lifetimes. The book has some 17,142,857 pages and it is about 1429 ft in thickness. Page 1 describes the Vishnu schist and the last page describes our last view from Yaki Point.

Along about page 3,510,375 we find a description of an algae fossil in the Bass limestone, the beginning of the chapter of life, Genesis. Near page 9,527,000 there is a reference to trilobite fossils in the Bright Angel shale, a definite form of life. An earliest recognizable fish is discussed on page 11,857,143 or thereabouts. This fossil is in the Temple Butte limestone. In the windblown sands of the Coconino sandstone are the fossil tracks of reptiles and amphibians, as described on or about page 13,322,244. The record of the rocks ends about 230 million yr ago at the top of the Kaibab formation and at about page 13,857,143. Thereafter no record exists at Yaki Point. Hence the book has 3,285,714 empty pages except for the advent of humans.

Perhaps humans, immigrants from Asia by way of the land bridge of the Aleutian Islands during the last glacial period, first saw the Canyon about 25,000 yr ago. If such is the case, they would appear on page 17,142,500. There are only 357 pages left and when we leave Yaki Point we are recorded on the last page.

We leave Yaki Point in awe of a mighty chasm, The Grand Canyon of the Colorado River.

DEFINITION OF ROCK

In this text *rock* is defined as material that forms an essential part of the Earth's crust. It includes loose incoherent masses such as beds of sand, gravel, clay, and volcanic ash, as well as firm, hard, solid masses such as granite, sandstone, limestone, quartzite, and schist.

According to this definition and in keeping with the sense of rock excavation, rock also includes the metallic and nonmetallic ores and minerals which are usually contained within the rocks.

One example is the open-pit mining of chromite, in which the disseminated chromium ore is excavated like clay, as in California. Another example is strip mining of subbituminous coal, a sedimentary rock, by huge stripping and loading shovels in Wyoming. A third example is the quarrying of nickel in veins of gabbro at Sudbury, Ontario, by open-pit methods.

The rock excavator should know as much as possible about rocks, ores, and minerals. Success depends on the excavator's familiarity with the materials excavated.

CLASSIFICATION OF ROCKS

By definition, rocks vary from the hard Barre granite of Vermont to the loose granitic sands of California. There are three classes of rocks, containing or not containing ores and minerals, which are forming and have been forming almost from the beginning of the existence of planet Earth.

Igneous Rocks

These are the original rocks, formed initially at varying distances below the Earth's surface. There are two kinds, intrusive and extrusive.

Intrusive, Sometimes Called Plutonic These rocks originated as molten magma and they slowly cooled at great depth, with resultant large crystals. A well-known example is the majestic granite batholith of the Sierra Nevada of California and Oregon, as portrayed in Plate 1-1.

The rock to the left in Figure 2-2 is intrusive granite diorite formed 99 to 180 million

Figure 2-2 Igneous rocks: *(Left)* Intrusive, or plutonic, granite diorite from the Sierra Nevada of California. *(Right)* Extrusive, or volcanic, basalt from the Columbia River plateau of Oregon.

yr ago from the Sierra Nevada. The dark crystals are hornblende and the light crystals are quartz. Depending on the degree of weathering, the rock is soft to hard excavation.

Extrusive, Sometimes Called Volcanic These rocks burst forth at or near the Earth's surface and they cooled rapidly, with consequent minute crystals. An example is the gigantic lava flow of the Columbia River Plateau of Washington, Oregon, and Idaho.

The rock to the right in Figure 2-2 is extrusive basalt of the Columbia River Plateau, 2 to 13 million years old. The holes of the characteristic vesicular structure were caused by air entrapment during cooling, giving the rock a rough appearance. Again, depending on the degree of weathering, basalt may be medium to hard excavation.

Sedimentary Rocks

These rocks are the consolidated and cemented products of the disintegration and decomposition by weathering of the three groups of rocks, igneous, metamorphic, and even sedimentary rocks themselves. They also derive from limy constituents of lake and ocean waters, limestones and coquinas being examples.

The rocks are consolidated by mechanical cementation and by chemical precipitants, sometimes accompanied by pressure. They are found at or near the Earth's surface. If formed on land they are called continental or nonmarine. If formed in the sea they are called marine. Originally they were laid down in horizontal or near horizontal beds with thin to thick bedding. Subsequent crustal deformations may have tilted, folded, and faulted these original strata.

The rock to the left in Figure 2-3 is a soft, punky, well-weathered sandstone of the Del Mar formation of San Diego County, California. This 36- to 58-million-yr-old rock is part of an excavation for dwellings and it is soft excavation for medium-weight tractor-ripper fragmentation.

The rock to the right in Figure 2-3 is hard, unweathered dolomite from the Tule River Indian Reservation of southern California. This bluish-white rock is 230 to 310 million yr old and it requires blasting for fragmentation. It is a form of limestone and these chemically precipitated marine rocks make up hundreds of material pits and quarries of America.

Figure 2-3 Sedimentary rocks. *(Left)* Well-weathered soft sandstone of the Del Mar formation of San Diego County, California. *(Right)* Hard unweathered dolomite from the Tule River Indian Reservation of Tulare County, California.

Metamorphic Rocks

These rocks are formed from existing igneous, sedimentary, and metamorphic rocks. In the last case they have simply reworked themselves. All have been formed by heat, pressure, and attendant gases and liquids below the earth's surface. Two examples are shown in Figure 2-4.

To the left is slate, derived from shale. It is a marine metasedimentary rock of the Mariposa formation with an age of some 146 million yr. It is from a 30-year-old highway cut in Mariposa County, California. Although slightly weathered in the exposed slope, it represents extremely hard ripping for a heavyweight tractor-ripper or else requires blasting. Metamorphic rocks such as this slate are characteristic of the Mother Lode country, which is some 150 mi long and 50 mi wide.

To the right is a granite gneiss from the Marble Mountains of San Bernardino County, California. It is of the Archeozoic group of rocks, dated at about 1450 million yr of age. This rock in its unweathered state requires blasting.

Sometimes the demarcation between igneous and metamorphic rocks is ill-defined, and the process may be reversed. The metamorphic rocks may become molten and cool into igneous rocks. In turn, these may be weathered into sands and become sandstones. And sedimentary shales may likewise be partially metamorphosed to relatively soft shale-slates.

Obviously the igneous, sedimentary, and metamorphic rocks may pass through many cycles, frequently doubling back on themselves. Two examples of metamorphic rocks are shown in Figure 2-5.

To the left is a marble, derived from limestone. This hard rock is from the south portal of the excavation for the San Bernardino Tunnel, California. This Precambrian group of rocks is more than 570 million yr old. The marbles and associated gneisses and schists were blasted in the open-cut excavation.

To the right is finely laminated schist, and the schistose structure is visible above the scale. Schists are formed from igneous and sedimentary rocks or even from reworked metamorphic rocks, and they are generally extremely hard and of great age. This schist is in the foundation excavation for Auburn Dam on the American River, Placer County, California. This schist occurs in an ultrabasic formation made up largely of amphibolites and schists and it is about 160 million yr old. About 20 ft below natural ground or below the weathered rocks it was necessary to blast the rock formation.

The transition from an early stage of metamorphism to a late stage is shown in Figure 2-6. The rock to the left is greenstone, altered from intrusive igneous diabase. The three rocks to the right are serpentine, the one to the extreme right showing accessory asbestos fibers. The rocks are from the Franciscan formation of age about 135 million yr, located on the Redwood Highway, Mendocino County, California.

Figure 2-4 Metamorphic rocks. *(Left)* Extremely hard slate, a metasediment derived from shale. This durable rock is a part of the Mariposa formation of the California Mother Lode country, the formation being made up of slates, phyllites, and metasedimentary sandstones and conglomerates. Along with the Calaveras formation and several others, the Mariposa formation makes up an extremely complex assemblage of rocks covering some 7500 mi² of the gold country. *(Right)* Hard granite gneiss from the Marble Mountains of San Bernardino County, California. The rock is a well-weathered cobble, but in its natural state in the unweathered rock formation it would require blasting. The banded rock structure is a property of the gneisses.

Figure 2-5 Metamorphic rocks. *(Left)* Extremely hard marble, derived from limestone. Finely seamed structure. From the San Bernardino Mountains, San Bernardino County, California. This not too common rock in excavation is invariably hard and requires blasting in its natural unweathered state. *(Right)* Extremely hard schist, probably derived from metamorphic slates. Finely banded structure as contrasted with the gneisses. The rock is from Placer County in the Mother Lode country of California, and it came from a dam foundation where blasting was in progress.

Figure 2-6 Metamorphic rocks from early state to late stage of change. *(Left)* Greenstone, altered from igneous intrusive diabase. *(Right)* Three rocks of serpentine, altered from the greenstone. The serpentine rock to the right of the group shows accessory asbestos fibers. All rocks are from the Franciscan formation of Mendocino County, California.

The actual three-stage transition is from hard diabase, requiring blasting, to medium-hard greenstone, calling for hard ripping by a heavyweight crawler-tractor-ripper, to soft to medium-hard serpentine, generally fragmented by soft to medium ripping.

A common mineral for boron is ulexite, and Figure 2-7 shows two examples of this sedimentary rock from the open pit of the United States Borax and Chemical Corporation at Boron, California. The ulexite to the left shows the laminations typical of

Figure 2-7 Ulexite, a common mineral of boron. *(Left)* A typical laminated specimen of this sedimentary rock, showing a stratum or layer of fibrous structure. *(Right)* A common example of the white fibrous silken complexion of the ulexite. The minerals are from the open-pit mine of the United States Borax and Chemical Corporation, located in the midst of the Mojave Desert of California.

sedimentary rocks, and the ulexite to the right shows the common loosely intergrown fibrous crystals.

The overburden above the boron ore is several hundred feet in thickness and it consists largely of desert alluvia and clays. It is excavated and wasted several thousand feet beyond the rim of the huge 96-million-yd^3 excavation.

Mining practice is to blast this hard ulexite. It is then loaded out by electric shovels, hauled by diesel-electric trucks to the primary crusher, and elevated by conveyor from the floor of the open pit to the stockpiles and preparation plant near the rim of the pit.

ROCKS, ORES, AND MINERALS

The common rocks, ores, and minerals encountered in construction and in open-pit mining are summarized in Table 2-2.

TABLE 2-2 Rocks, Ores, and Minerals

Igneous Rocks	Formed by crystallization from an originally deep-seated magma.
Intrusive igneous	Formed below the surface by the slow cooling of magma, resulting in medium- to large-size crystals.
Diabase	Dark color. Small crystal matrix with large, elongated white crystals. Medium to hard excavation.
Diorite	Light to dark color. Medium, uniform-size crystals. Soft to hard excavation.
Gabbro	Dark color. Medium-size crystals. Medium to hard excavation.
Granite	Light to dark color. Medium to large crystals. Soft to hard excavation.
Porphyry	Dark color. Fine-grained crystal matrix with coarse light crystals. Medium to hard excavation.
Syenite	Light to medium color. Small crystals. Medium to hard excavation.
Trap rock	Quarry worker's term applied to such intrusive igneous rocks as diabase and gabbro. Dark color. Small- to medium-size crystals. Medium to hard excavation.
Extrusive igneous	Formed near or on surface from fissures or from volcanic eruption. Rapid cooling causes minute crystals and sometimes vesicles or small gas pockets.
Andesite	Medium to dark color. Fine-grained lava flow with thin to thick beddings. Sometimes called felsite. Medium to hard excavation.
Basalt	Dark color. Fine-grained lava flow with thin to thick beddings. Generally characterized by columnar structure of polygonal cooling joints and by vesicles. Medium to hard excavation.
Breccia, volcanic	Light- to dark-colored fine to coarse pyroclastic rocks with different degrees of consolidation due to welding. Originated by violent airborne eruption. Soft to medium excavation.
Obsidian	Dark color. Glassy with no visible grains because of instantaneous cooling. Thin to thick beddings. Medium to hard excavation.
Pumice	Light color. Porous because of release of gases during cooling. Soft to medium excavation.
Rhyolite	Light to medium color. Fine-grained lava flow. Thin to thick beddings. Soft to hard excavation.
Scoria	Light to dark color. Cindery and jagged because of release of gases during cooling. Rhyolite family. Thin to thick beddings. Soft to medium excavation.
Trap rock	Quarryworker's term for extrusive igneous rocks such as basalt and andesite. Medium to dark color. Small crystals. Medium to hard excavation.

TABLE 2-2 Rocks, Ores, and Minerals *(Continued)*

Sedimentary Rocks	Formed by degradation of existing rocks and organic remains and their subsequent deposition by natural forces.
Unconsolidated rocks	
Clays	Light to dark color. Extremely fine size, less than 0.005 mm. More or less plastic. Thin to thick beddings. Soft to medium excavation.
Silts	Light to dark color. Fine size, between 0.005 mm and 0.05 mm. Soft excavation.
Sands	Light to dark color. Fine to medium size, between 0.05 mm and 2.0 mm. Soft excavation.
Gravels	Light to dark color. Medium to large size, between 2.0 mm and 20 cm.
Boulders	Light to dark color. Large to extremely large size, 20 cm to larger than 10 m.
Consolidated rocks	Formed by cementation of deposited rocks by silica, calcium carbonate, or iron oxide, by chemical precipitation from solutions, and by the action of organic agents.
Agglomerate	Cemented angular, as contrasted with rounded, fine to coarse particles. Equivalent to breccia. Light to dark color. Thin to thick beddings. Soft to hard excavation.
Coal, bituminous	Dark color. Compressed organic material associated with limestones, sandstones, and shales. Medium to thick beddings. Medium to hard excavation.
Conglomerate	Cemented, rounded, fine to coarse particles. Light to dark color. Thin to thick beddings. Soft to hard excavation.
Dolomite	Carbonate rock both chemically and organically derived. Light to medium color. Thin to thick beddings. Hard excavation.
Lignite	Brown coal. Medium brown color. Mildly compressed organic matter associated with shales and sandstones. Thin to thick beddings. Soft to medium excavation.
Limestone	Carbonate rock both chemically and organically derived. Differs from dolomite by its lower magnesium carbonate/calcium carbonate ratio. Light to medium color. Thin to thick beddings. Hard excavation.
Limestone, bituminous	Limestone impregnated with bitumen. Medium to dark color. Up to 20% bitumen content. Thin to thick beddings. Medium to hard excavation.
Sandstone	Cemented sand. Light to dark color. Thin to thick beddings. Soft to hard excavation.
Sandstone, bituminous	Sandstone impregnated with bitumen. Medium to dark color. Up to 10% bitumen content. Thin to thick beddings. Soft to hard excavation.
Shale	Cemented clay. Light to dark color. Thin to thick beddings. Soft to hard excavation.
Shale, bituminous	Shale impregnated with bitumen. Medium to dark color. Up to 15% bitumen content. Thin to thick beddings. Soft to hard excavation.
Siltstone	Cemented silt. Light to dark color. Thin to thick beddings. Soft to hard excavation.
Metamorphic Rocks	Formed from existing rocks by heat, pressure, and mineral-bearing hydrothermal solutions, acting together or separately.
Coal, anthracite	Black color. Associated with altered limestones and shales. Hard excavation.

Metamorphic Rocks	
Gneiss	Formed from igneous, sedimentary, and older metamorphic rocks. Light to dark color. Coarse banded structure. Medium to hard excavation.
Greenstone	Formed from fine-grained igneous rocks. Light to dark greenish color. Medium to hard excavation.
Marble	Formed from limestone. Light to medium color. Hard excavation.
Quartzite	Formed from sandstone. Light to medium color. Hard excavation.
Schist	Formed from igneous, sedimentary, and older metamorphic rocks. Light to dark color. Fine banded structure, differing from gneiss in its tendency to split into layers because of its foliation. Medium to hard excavation.
Serpentine	Formed from igneous rocks. Light to dark greenish color. Lustrous and waxy. Soft to hard excavation.
Slate, or argillite	Formed from shale with original thin stratifications. Light to dark color. Medium to hard excavation.
Soapstone, or talc	Formed from igneous and metamorphic rocks. Light to dark color. Soft to medium excavation.

Ores and minerals	In this text an ore is a rock containing a mineral. It may be the entire ore body to be excavated or it may be a part of the total excavation. A mineral, metallic or nonmetallic, is any substance of definite chemical composition. It also is considered to be rock, part of the ore.
Aluminum ore	Bauxite, a hydrous aluminum oxide of 55% average aluminum content. Light to dark brown color. Decomposition product of igneous and metamorphic rocks. Occurs in lenses and pockets. Soft to medium excavation.
Asbestos ore	Chrysotile, serpentine, a hydrous magnesium silicate of 100% asbestos content. Light to dark color. Occurs in veins of serpentine. Soft to hard excavation.
Boron ore	Borax, a hydrous sodium borate of 11% average boron content. Light to medium color. Occurs as sediment in extrusive igneous and sedimentary rocks. Thin to thick beddings. Soft to hard excavation.
Chromium ore	Chromite, a ferrous chromic oxide of 47% chromium content. Dark color. Occurs in veins and disseminations in igneous and metamorphic rocks. Medium to hard excavation.
Copper ore	Chalcopyrite, a copper and iron sulfide of 34% copper content. Light to medium color. Occurs in veins and disseminations in igneous and metamorphic rocks. Medium to hard excavation.
Diatomaceous earth ore	Diatomite, a silicon dioxide of 47% silica content. Light color. Occurs in massive sedimentary beds of diatoms, microscopic forms of plant life. Medium to thick beddings. Soft to medium excavation.
Gold ore	Gold, an element, is yellow. Occurs in veins of igneous rocks, chiefly in quartzose rocks. Medium to hard excavation. Occurs also in placer deposits of alluvia as fine gold and nuggets. Soft to medium excavation.
Gypsum ore	Gypsum, a hydrous calcium sulfate of 32% lime content. Light color. Occurs in massive sedimentary beds associated with shales and limestones and in cavities of limestones. Medium to thick beddings. Medium to hard excavation.
Iron ores	Hematite, an iron oxide of 70% iron content. Medium to dark color. Occurs in thick beddings in sedimentary and metamorphic rocks. Soft to hard excavation.
	Limonite, a hydrous iron oxide of average 60% iron content. Medium to dark color. Formed as a residual by the decomposition

TABLE 2-2 Rocks, Ores, and Minerals ***(Continued)***

Ores and minerals	
Iron ores	of iron minerals. Thin to thick beddings. Soft to medium excavation.
	Magnetite, an iron oxide of 72% iron content. Dark color. Magnetic. Occurs in veins and in disseminations in igneous and metamorphic rocks. Medium to hard excavation.
	Taconite, an iron oxide composed of hematite and magnetite, of 71% average iron content. Occurs in ferruginous chert sedimentary rock associated with limestones and sandstones. Thick beddings. Hard excavation.
Kaolin ore	Kaolinite, a hydrous aluminum silicate of 100% kaolin content. Light color. Occurs as thick claylike beds. Soft to medium excavation.
Lead ore	Galena, a lead sulfide of 87% lead content. Dark color. Occurs in veins in igneous, sedimentary, and metamorphic rocks. Medium to hard excavation.
Magnesium ore	Magnesite, a magnesium carbonate of 29% magnesium content. Light color. Occurs in veins in serpentine, as dolomite, and in sedimentary deposits of medium to thick beddings. Soft to hard excavation.
Manganese ore	Pyrolusite, a manganese dioxide of 63% manganese content. Dark color. Associated with sedimentary and metamorphic rocks in thin to thick beddings. Medium to hard excavation.
Mercury ore	Cinnabar, a mercury sulfide of 86% mercury content. Medium red color. Occurs as disseminations in igneous, sedimentary, and metamorphic rocks. Soft to hard excavation.
Molybdenum ore	Molybdenite, a molybdenum disulfide of 60% molybdenum content. Medium to dark color. Occurs as small veins and disseminations in igneous, sedimentary, and metamorphic rocks. Medium to hard excavation.
Nickel ore	Niccolite, a nickel arsenide of 44% nickel content. Light to dark color. Occurs in veins in igneous and metamorphic rocks. Medium to hard excavation.
Nitrate ore	Soda niter, a sodium nitrate of 100% sodium niter content. Light to medium color. Associated with gypsum. Thin to thick beddings. Soft to hard excavation.
Phosphate ore	Apatite, a calcium fluorophosphate or calcium chlorophosphate, or both, of average 18% phosphorus content. Light to medium color. Occurs as veins in igneous rocks with hard excavation and in sedimentary deposits of thin to thick beddings with soft to medium excavation.
Potassium ore	Sylvite, a potassium chloride of 52% potassium content. Light to medium color. Associated with halite. Thin to thick beddings. Soft to medium excavation.
Salt ore	Halite, a sodium chloride of up to 100% purity. White to light color. Occurs in massive sedimentary beds associated with gypsum and potassium ores. Medium to thick beddings. Medium to hard excavation.
Silver ore	Argentite, a silver sulfide of 87% silver content. Dark color. Occurs in veins of igneous and metamorphic rocks, along with gold and nickel. Medium to hard excavation.
Sulfur ore	Occurs as a yellow element in medium to thick beddings in volcanic and sedimentary rocks. Up to 100% purity in native state. Soft to medium excavation.
Tin ore	Cassiterite, a tin oxide of 79% tin content. Light to dark color. Occurs in veins in igneous rocks, in which excavation is medium to

Ores and minerals	
	hard. Occurs also as rounded pebbles in stream alluvia, in which excavation is soft.
Uranium ore	Uraninite, a mixture of uranium oxides, averaging about 86% uranium content, and generally associated with other minerals. Light to dark color. Occurs as veins in igneous and metamorphic rocks and as disseminations in sedimentary rocks with medium to heavy beddings. Medium to hard excavation.
Zinc ore	Sphalerite, a zinc sulfide of 67% zinc content. Light to dark color. Associated with galena. Occurs in veins of igneous, sedimentary, and metamorphic rocks. Medium to hard excavation.

NOTE 1: Thicknesses of beddings for some extrusive igneous rocks and for sedimentary rocks, indicative of excavation methods, are given in this table.

Thin beddings	Up to 1-ft thickness.
Medium beddings	Between 1- and 3-ft thickness.
Thick beddings	Greater than 3-ft thickness.

NOTE 2: The excavation characteristics of rocks, given as soft, medium, and hard, are dependent largely on the degree of weathering, as discussed in Chapter 3. In terms of excavation methods, the characteristics are idealized in this table.

Soft excavation	Excavation without preliminary ripping or with soft ripping by heavyweight crawler-tractor-ripper.
Medium excavation	Excavation by preliminary medium to hard ripping.
Hard excavation	Excavation by preliminary extremely hard ripping or blasting.

Rocks are classified as igneous, sedimentary, and metamorphic. These three classifications are somewhat broad, as some igneous and sedimentary rocks may be in an initial stage of metamorphism, which is scarcely discernible.

Ores are classified according to the principal mineral for which they are mined, some ores providing several minerals and native elements. The most common mineral for the element desired is given. Table 2-2 includes metallic and nonmetallic ores most commonly mined in open-cut excavation.

The deposition of ores may be simple or complex. It may be contemporary, the ore being formed at the same time as the host rocks. It may be subsequent, the mineral compounds having been gathered from the host rocks and deposited therein. It may be replacement, the present ore having been deposited after removal of the original ore.

After the ore has been deposited, weathering and further enrichment of the ore body may create several zones within the deposit. An example is a top zone of complete oxidation, a second zone of leaching, a third zone of oxide enrichment, a fourth zone of sulfide enrichment, and a bottom zone of primary ore.

The forms of ore deposits are tabular deposits, veins, pockets, disseminations in the host rock, and residual bodies. All these forms may be excavated by open-cut methods provided the yield of the desired mineral is economical.

Ores usually, but not always, contain gangue. Gangue is any valueless material that is necessarily excavated along with the desired ore or mineral. Generally it is a part of the host rocks. Gangue is removed by initial concentration processes and by final smelting.

The descriptions of ores and minerals have been simplified for the excavator in terms of their occurrences and characteristics for excavation.

FORMATIONS

For the earthmover an understanding of the rock formation is fully as important as knowledge of the rocks that make up the formation. As defined and used by the U.S. Geological Survey, the *formation* is the ordinary unit of geologic mapping, consisting of a large and persistent stratum of some kind of rock.

Actually, in the language of the excavator and the construction and mining geologist, a formation is an individual rock or assemblage of rocks of igneous, sedimentary, or metamorphic origin that is sufficiently homogeneous or distinctive to be given a distinct name. The given name is typically the name of a locality where the rock is exposed and identified.

The Lineage of a Formation

Referring to Table 1-1, the formation is a subdivision of a series of rocks, which in turn is a part of the larger system of rocks. For example, the limestone and dolomite hard-rock excavation of the Osage River basin near Warsaw, Missouri, make up the Theodosia formation. The lineage of the Theodosia formation is as follows:

Theodosia	formation	*from*
Jefferson City	subseries	*from*
Canadian	series	*from*
Ordovician	system	*from*
Paleozoic	group	

The age of the fossil-bearing Theodosia formation is about 489 million yr.

Examples of Formations

The term formation is applied to igneous, sedimentary, and metamorphic rocks and their combinations. For example, the Topanga formation of southern California is an 18-million-yr-old Middle Miocene assemblage of sedimentary sandstones, conglomerates, and shales, with much interbedding of volcanic intrusive and extrusive rocks such as andesites and basalts. The well-known formation is of both continental and marine origin.

In such varied formations the excavator may find troublesome thin dikes and sills of hard igneous rocks separating masses of soft sedimentary rocks. As a result, systematic ripping and blasting of the formation are both sometimes made impossible by the attitude of the formation and the consequent excavation difficulties. Such a transition might be from a 3-ft thickness of shale to a 5-ft thickness of basalt to a 4-ft thickness of sandstone, all strata having extreme dip and the basalt sill requiring costly drilling and blasting.

A unique intrusive igneous rock is the granite formation known as tonalite. By definition it is simply quartz diorite of the granite family. However, when weathered it is characterized by up to room-size, rounded boulders, both on the surface and submerged beneath natural ground in a matrix of "DG," or decomposed granite. In this formation are easily rippable rocks containing "drill-and-shoot" boulders, all at the same plane of excavation. Figure 2-8 shows this approximately 99-million-yr-old troublesome formation in a 1,044,000-yd^3 cut for a freeway in Riverside County, California. Prior to excavation

Figure 2-8 **Massive boulder formation of granite.** Tonalite, a kind of granite distinguished by huge, weathered boulders, in Riverside County, California. In spite of the awesome appearance of this granite formation, this part of the 75-ft-deep cut was excavated by soft to hard ripping by heavyweight tractor-rippers. Grooves left by the ripper teeth are visible along the relatively smooth face of the cut, the flat face being in marked contrast to the rough natural ground. Outsize boulders were blasted before being loaded out to the fill.

the surface of this 75-ft-deep and 3700-ft-long cut resembled the area above the cut slope of the picture. In spite of the formidable appearance of the boulders and outcrops of granite, relatively little yardage of the million-cubic-yard cut required blasting. Below the grade of the cut, the weathered granite with its imbedded boulders is slowly succeeded by jointed semisolid and ultimately solid rock.

Basalt is an extrusive igneous rock, usually occurring in flows with characteristic columnar tabular formation. Figure 2-9 illustrates the outcroppings along the centerline of a 27-ft-deep cut for a proposed highway in Siskiyou County, California. The highway is in the shadow of young Mount Shasta, and one of Shasta's volcanic eruptions some 1 million yr ago produced the lava flow of this formation.

Another distinguishing feature of basalt, although not always present, is its vesicular surface, consisting of small holes left by the frothing cooling rock. Figure 2-10 shows this mottled surface. The area of the picture is about 2 ft^2, or about the size of the boulder in the foreground of Figure 2-9.

This basalt formation will be excavated by medium to hard ripping by a heavyweight tractor-ripper to about 4 or 5 ft above grade, and the remaining small prism of rock will be taken out by either extremely hard ripping or blasting. The choice of ripping or blasting will depend on the quantities involved and on the availability of drilling and blasting machinery and crews.

A not too common extrusive igneous rock of pyroclastic nature is tuff, which sometimes, since it is composed of small particles airborne from an erupting volcano, is mistakenly called a sedimentary rock. This is especially true because it resembles heavy-bedded buff and brown sandstones.

Figure 2-11 shows an existing freeway cut in the Bishop tuff formation of Mono County, California. This excavation involved four major cuts totaling 1.7 million yd^3 with a maximum 53-ft-deep cut in this approximately 2-million-yr-old formation.

The andesitic tuff rings with a pick blow and shows an ugly surface. And yet, all the roadway excavation was ripped out economically. Testimony to this less costly means of excavation is apparent in the smooth, clean-cut slope as contrasted with the jagged natural slope. On the left side of the road there appears to be a rough-cut slope. Actually, this is a natural slope on the other side of the gulch of Crooked Creek.

Figure 2-9 Well-weathered surficial basalt. Weathered basalt lava flow from Mount Shasta, Siskiyou County, California. The young formation, some 1 million yr old, will be ripped to near grade of the 27-ft-deep highway cut. A few feet above grade the blocky basalt will become semisolid jointed rock and it will call for either extremely hard ripping by a heavyweight tractor-ripper or expensive blasting because of the shallowness of the remaining prism of rock.

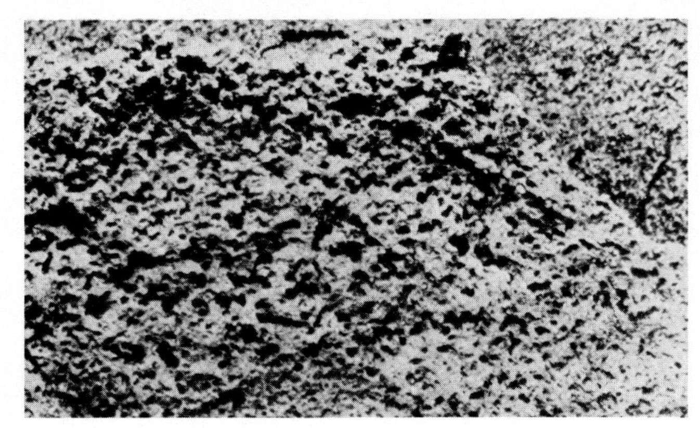

Figure 2-10 Basalt with air cells. Vesicular basalt. The area of the photo is a 2-ft square. The holelike vesicles have been formed by the rapid expansion of gases during the quick cooling of the lava flow. Sometimes after cooling the vesicles are filled by one or more minerals, and in that case the vesicular structure becomes an amygdaloidal rock. Basalt is not always vesicular but the exceptions with smooth surfaces comprise a small percentage of the total basalts.

Figure 2-11 Volcanic tuff. Bishop tuff, Mono County, California. This deceptive igneous pyroclastic formation of 1.7 million yd^3 in four cuts up to 53 ft deep was of soft to hard ripping and the yardage was ripped entirely by heavyweight tractor-rippers. The misleading nature of the formation is graphically portrayed in the contrast between the massive rocky natural ground and the ultimate smoothness of the cut slope when finished by the slope board of the crawler-tractor.

A troublesome formation for excavation is scoria, as shown in Figure 2-12. Scoria is a rough, rubblelike extrusive igneous rock, and it is usually formed by the extremely fast cooling of basaltic lava. This approximately 1-million-yr-old scoria formation is in the Owens Valley, Inyo County, California. Like many of the extrusive igneous rock formations of the Owens Valley fault area, it is of the Pleistocene series and among the youngest of California rocks. Further south in the Mojave Desert is the scene of equally spectacular Pleistocene volcanic activity.

A typically soft marine sedimentary formation is the Capistrano of California, consisting of thinly bedded friable sandstones, shales, and conglomerates. Figure 2-13 shows this formation in a highway cut in San Diego County. The first 10-ft depth of residuals was excavated without ripping and the rest of the excavation to grade was ripped easily by a medium-weight tractor-ripper.

Figure 2-12 **Cinderlike basaltic lava.** Scoria, an extrusive igneous rock formed from basaltic lava. This sharp, angular, abrasive rock is along the centerline of an 18-ft cut in a proposed freeway of Inyo County, California. Seismic studies and subsequent excavation revealed that the rubble merged into solid rock a few feet above finished grade. Accordingly the formation required expensive drilling and blasting commencing some 5 ft above grade. The overlying scoria is of medium weight and hard whereas the country rock, solid basalt, is both heavy and hard.

Figure 2-13 **Young sedimentary rock formation with little deformation.** The Capistrano formation of southern California is normally made up of sandstones, shales, and conglomerates. In this roadside cut of 20-ft depth are 10 ft of residuals overlying 10 ft of soft shales, sandstones, and conglomerates. Excavation was relatively soft in this San Diego County formation, requiring only mild ripping by a medium-weight tractor-ripper for full fragmentation.

In this picture are shown the medium-resistant shales of the upper portion; the middle, more resistant sandstones; and the lower, less resistant conglomerates. The hand pick shows the 8-in thickness of the hardest sandstone. The age of the Capistrano Upper Miocene formation is about 15 million yr.

Sometimes there are variations in the same formation with respect to the kinds of rocks and their respective consolidations, although generally a formation is fairly uniform. This variation in the Capistrano formation is illustrated in Figure 2-14. A 10-ft lens of San Onofre breccia forms an overhanging ledge above the softer sandstone. The breccia is made up of cemented particles of schist and interbedded grit and sandstone. Thus, the excavator would find hard breccia above the soft sandstone. Normally, in a downward sequence of strata the hardness of the formation increases with depth, but this is not necessarily so. The dimensions of this well-weathered river terrace may be visualized with respect to the man shown in the picture. Above the breccia are clay and sand residuals. The excavator would encounter:

1. Fifteen feet of easily excavated residuals.
2. Ten feet of breccia calling for hard ripping by a heavyweight tractor-ripper.
3. Twenty feet of siltstone requiring medium ripping by a medium-weight tractor-ripper.

This bank has been eroded and weathered by the San Juan River and by the winds and the rains.

A thick-bedded sedimentary limestone that forms the overburden of anthracite coal stripping in Pennsylvania is shown in Figure 2-15. This hard limestone of maximum 8-ft thickness requires blasting prior to casting by walking draglines. Only thinly bedded limestones of a relatively soft nature may be successfully ripped by heavyweight tractor-rippers. This limestone formation is of the Pennsylvanian series of the Carboniferous system. The Carboniferous is the anthracite-coal-bearing system. The Pennsylvanian series is about 295 million yr old and it has been contorted by the mountain building of the Appalachian Revolution sometimes into anticlines and rather deep synclines. This orogeny has complicated the art of coal strip mining in many areas of the Appalachian Mountains, resulting in high ratios of overburden

Figure 2-14 **Hard stratum interbedded in soft strata of sedimentary rocks.** The Capistrano formation with a variant—a 10-ft lens of San Onofre breccia. This breccia is hard and frequently it characterizes the formation in the coastal area. With reference to the man, it is 10 ft thick and it overhangs the less resistant light sandstone. Excavation would involve medium to hard ripping by a heavyweight tractor-ripper.

Figure 2-15 **Massive hard-rock overburden.** Overburden of limestone formation in eastern Pennsylvania anthracite coal field. About 80 ft of overburden is being cast by the walking dragline so as to uncover the coal vein. The hard rock is blasted prior to removal because of the heavy bedding of the limestone strata. Such massive limestone formations of the Carboniferous system of rocks are common in the eastern and central United States, occurring as outcrops in the mountains and as underlying beddings in the plains. *(E. I. du Pont de Nemours & Co.,* Blasters' Handbook, *1966, p. 309.)*

yardage to recoverable coal tonnage. In this case the limestone overburden is about 80 ft in thickness.

The excavation of sedimentary rocks such as shales, sandstones, and limestones depends on the bedding or lamination thickness and the vertical joints, even though the individual stratum of the rock may be hard and strong. These planes of weakness and their cubic dimensions determine the ease or difficulty of ripping and the necessity for alternate blasting. A generalization for these structural spacings is:

Spacings less than 12 in	Soft to medium excavation
Spacings 12 to 36 in	Medium to hard excavation
Spacings more than 36 in	Hard to extremely hard excavation

The metamorphic rocks are generally hard and durable. A common one is granite gneiss, or altered granite. Banded and wavy structures, due to heat and pressure, distinguish this fine-grained rock. It usually calls for extremely hard ripping or blasting for fragmentation, and its excavation characteristics are akin to those of granite except that it does not weather as rapidly as the coarser-grained granite.

A variable metamorphic rock formation, derived from marine sedimentary rocks, is the Bedford Canyon formation of California. Figure 2-16 shows schist outcroppings of this formation in a rather shallow highway cut adjacent to a proposed 880,000-yd^3 72-ft-deep cut for a freeway. These metamorphosed sandstones, siltstones, and conglomerates are fairly solid up to the grass roots, and the existing highway cut indicated what might be

Figure 2-16 **Minimum weathered outcrops of metamorphic rocks.** Bedford Canyon formation of metamorphosed sedimentary rocks. This California formation of phyllites, schists, argillites, and graywackes, located in San Diego County, is of the Upper Jurassic system of rocks, about 146 million yr old. It is an inlier of about 2 mi along the centerline of a freeway and it is surrounded by younger granites about 99 million yr old. The contacts between the igneous and the metamorphic formations are customarily in the swales or in the fill areas of the freeway. This picture shows a cut in the existing highway paralleling the proposed freeway cut. Semisolid rock is but a few feet below natural ground.

expected in the rock excavation for the proposed adjacent freeway. The correlation was good as the huge adjacent cut was ripped to an average 35-ft depth and blasted under this level to grade.

Another fairly common metamorphic rock is slate, derived from shale. In America the well-known slates used for commercial purposes are those of Vermont and New York State. Figure 2-17 shows the slate of the Slate River area of Virginia. In the midst of the massive slate there is a small fault, with its brecciated zone between the two hand picks of the picture. In the weakened zone rock has been crumbled and the adjacent faces have been weakened by bending. However, only a few feet from the fault the rock remains massive and hard. From a practical standpoint this 465-million-yr-old Ordovician formation will require blasting.

The beddings of slate vary from less than ⅛ in to a foot or more in thickness, and characteristically there is a minimum thickness of residuals and little weathered slate between the residuals and the solid slate. In the Mother Lode country of California there are many slate and other metamorphic rock formations, which trend generally north to south along the axis of the gold mining activities for a distance of some 160 mi. Such a slate formation is the Mariposa, which, in contrast to the Slate River, Virginia, formation, is generally without minor faults. Although the cleavage planes are apparent, they are not evidence of weakness, as the hard rock requires blasting. In a moderate-size highway cut of 40 ft maximum depth the "rock line" was but a few feet below the grass roots, and some 60,000 yd^3 of the total 75,000-yd^3 cut was blasted.

Attitude of Formations—Dip and Strike

Sedimentary and extrusive igneous rocks are laid down horizontally or nearly horizontally. Later the formation may be dipped or tipped by deformation of the Earth's crust. While in a horizontal or dipped attitude, the rocks may be converted to metamorphic rocks. It is important for the earthmover to recognize the *dip and strike* of the rock

Figure 2-17 **Brecciated fault zone of metamorphic rock.** The slate or argillite member of this Slate River, Virginia, rock formation is a hard, tough, metamorphic facies requiring blasting. Nature's weathering process on this hard, durable rock is manifestly slow. As shown between the two hand picks, there is a narrow brecciated zone of the small fault, but otherwise within a short (less than 2 ft) distance the rock is solid. Because of the contortions of metamorphism slates generally occur in steeply dipped attitudes, and with their associated schists, phyllites, and quartzites they present unknown and perplexing problems to the excavator. *(H. Ries and Thomas L. Watson,* Engineering Geology, *Wiley, New York, 1925.)*

formation because it is important in the fragmentation and the loading of the rock. It is also significant from a safety standpoint because of the possibility of dangerous slides even in moderately dipped rock formations.

Figure 2-18 shows the relationship of the dip and strike of a formation to a horizontal plane or datum plane of a body of water.

Three examples of the value of dip-strike information to the excavator are:

1. In the case of fragmentation by ripping, one should not rip at right angles to the strike and away from the direction of dip because the ripper points would slide along the inclined beddings and would not penetrate the stratum efficiently. It would be better to rip in the opposite direction so as to lift the stratum with the ripper points or to rip parallel to the strike. Naturally, the ideal approach for ripping must be modified sometimes to accommodate the physical aspects of the work area.

2. When drilling vertical blastholes, the drill steel is apt to be deflected and wedged in the partings of the strata if the dip is vertical or nearly vertical. It is better to drill off the vertical at an angle to the plane of the dip.

3. When loading soft sedimentary rock by scrapers without preliminary ripping, the problems with the cutting edge of the scraper bowl as it is forced into the stratum are the same as those with the points of the ripper. The solutions are the same, as in the case of the cutting edge the operator is trying to penetrate a stratum, just as the operator is trying to force the ripper points into the stratum.

Information concerning dip and strike is of great significance to the miner investigating the worth of an ore deposit and the cost of the proposed mining work. In open-cut excavation the length of the strike and the width of the deposit on the Earth's surface, together with the dip, determine the amount of ore that can be taken out economically by surface mining.

Correlation of Rock Formations for Excavation

Just as similar machines rip, blast, and load rock in like manner in the same rock formation, so the same types of formations, though many miles apart, behave in pretty much the same way when ripped, blasted, and loaded.

The Principle of Uniformitarianism of geologist James Hutton was enunciated by this Scottish man of vision in 1785 in his revolutionary book *Theory of the Earth.* Hutton stated that the present is the key to the past and that, given time, the processes now at work could have produced all the geologic features of the Earth.

In terms of the excavation of rock formations, this principle may be extended to the theory that rock formations of the same kind and of the same time on the geologic time scale tend to have the same characteristics for excavation. The theory is necessarily a generalization because the spatial aspects of the same rock formations vary according to locality.

This extension of the Principle of Uniformitarianism, supported by many correlations, is an example of the practical, useful application of an accepted theory. In a sense it is a basic assumption to which modifications may be added to complete the understanding of a new and questionable rock formation. The correlation of the unknown formation with a known formation in which the excavator has had experience is made by the use of correlation charts, as shown in Table 2-3. In this table the center section, embracing the Basin Ranges, Mojave Colorado Desert, Peninsular Ranges, and part of the Transverse

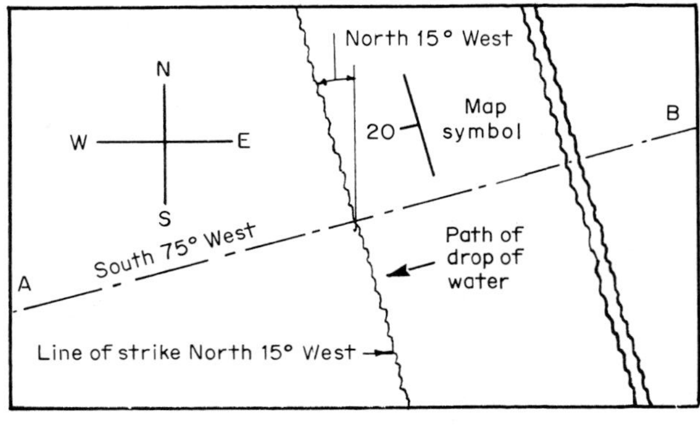

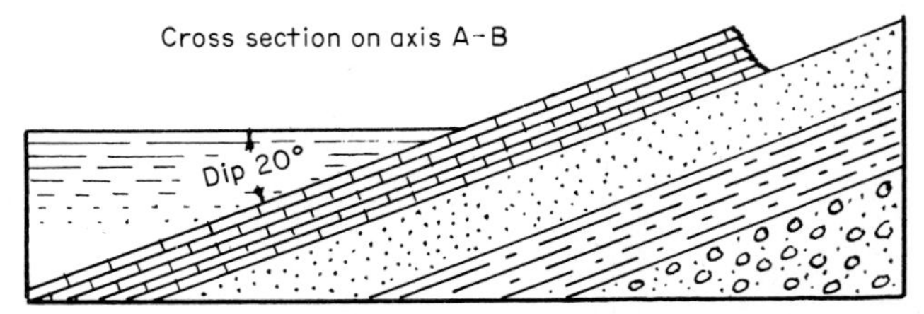

Figure 2-18 Strike and dip of rock formation. The illustration shows a dipped formation of sedimentary rocks along the edge of a body of water representing a horizontal plane.

Strike is the compass direction of the intersection of its bedding plane with a horizontal plane and it is measured in degrees.

Dip of a rock formation is a measure of its tilt or slope in the direction of its dip. The angle of dip is the acute angle which this direction makes with the horizontal plane and it is measured in degrees.

The strike of this formation is North 15° West or South 15° East and it is represented by the edge of the water. The dip is 20° below the horizontal plane of the water toward a compass bearing of South 75° West.

TABLE 2-3 Correlation Chart for Sedimentary Rock Formations of Southern California
(Interval of time is 135 million years)

NOTE: The authors' names that appear at the tops of the columns do not indicate responsibility for the contents of the columns (see text).

			SOUTHERN COAST RANGES				TRANSVERSE RANGES	
			Santa Maria Basin & Huasna Basin Woodring & Bramlette (50), Taliaferro (43)	Cuyama Valley Dibblee (52), Beck (52), Schwade (54)	Reef Ridge-Kettleman Hills Woodring, et al (40), Stewart (46)	S. ... Hod... Dib...	...s Basin-...es Hills ...oodford (51), ...oodring, (46)	Santa Monica Mountains Hoots (31), Durrell (54)
QUATERNARY	Pleistocene	Upper	Terrace deposits Orcutt sand	Fanglomerate	Older alluvium	Fan... ler...	non-...ce deps. sd.	Alluvial plain and marine terrace deposits
QUATERNARY	Pleistocene	Lower	Paso Robles formation	Paso Robles formation	Tulare formation	Tular...	sd. silt ...rl	Marine conglomerate
TERTIARY	Pliocene	Upper	Careaga fm Foxen fm		San Joaquin fm Etchegoin fm	forma...		Pico (San Diego) fm
TERTIARY	Pliocene	Lower	Sisquoc fm.	Morales fm	Jacalitos fm	Chano... forma...		
TERTIARY	Miocene	Upper	Sta. Margarita fm.	Santa Margarita fm Quatal fm.	Monterey fm. Reef Ridge sh.	Sta Mar... fm	...st. mbr. mbr.	Modelo fm.
TERTIARY	Miocene	Middle	Monterey fm. Point Sal fm	Monterey fm. Bitter Creek sd Caliente fm.	McLure sh.	Maricop... or Montere... fm.		Topanga fm.
TERTIARY	Miocene	Lower	Vaqueros fm	Painted Rock sd Soda Lake sh.	Temblor fm.	Temblor or Vaqueros		Vaqueros fm.
TERTIARY	Oligocene		Lospe fm Sespe fm.	Simmler fm.		Pleito fm		Sespe fm.
TERTIARY	Eocene	Upper			Kreyenhagen sh.	San Emigd... fm Reed sh		
TERTIARY	Eocene	Middle		Unnamed Eocene beds	Avenal ss.	Tejon fm Live... Uv...		
TERTIARY	Eocene	Lower						
TERTIARY	Paleocene		?	?				"Martinez fm."
CRETACEOUS	Upper		Asuncion (?) gp. Pacheco (?) gp.	Unnamed	Panoche fm.			"Chico fm." Trabuco fm.
CRETACEOUS	Lower		"Shasta" gp.	Cretaceous beds				

Portions of a correlation chart of sedimentary formations in Southern California are shown, with brackets drawn to indicate the Panoche and the Chico formations—See text for discussion.

SOURCE: Charles W. Jennings and Mort D. Turner, *Geology of Southern California*, California Division of Mines and Geology, 1954, chap. 3, plate 1.

Ranges, has been omitted for the sake of simplicity. Such charts are available from a number of private and public agencies, which include:

1. Geology departments of various colleges and universities.
2. American Association of Petroleum Geologists, Tulsa, Oklahoma.
3. State divisions of geology.
4. U.S. Geological Survey, Washington, D.C.
5. *History of the Earth,* 2d ed., by Bernard Kummel, W. H. Freeman and Company, San Francisco, Calif., 1970.

Some of these sources give valuable additional information in terms of rock types (igneous, sedimentary, or metamorphic), as well as data on stratification, dip and strike, folding and faulting, and even degree of weathering in some cases.

The use of rock formation correlations is typified in the following three illustrations.

1. An abundant sedimentary rock formation of northern California is the Chico, made up of thickly bedded sandstones and tightly bedded shales. This old formation, Upper Cretaceous dating back some 90 million yr, has the reputation of requiring much blasting in the heavy excavation of the northern California area over a long period of time.

In 1962 seismic studies were made for an 11-million-yd^3 freeway across Pacheco Pass of the California Coast Range. This formation is the Panoche and it also is of the Upper Cretaceous series. An early correlation was made by means of the chart of Table 2-3, linking the Chico and the Panoche formations. The later seismic studies and the excavation experience show that the contractor blasted some 65 percent of the roadway excavation.

In 1965 preliminary studies were made for an 8.5-million-yd^3 rock job for a freeway across Santa Susana Pass, some 260 mi south of Pacheco Pass and some 420 mi south of Chico, the city after which the Chico formation was named. The Santa Susana Pass formation is known as the Chico, and so there was no hesitation in correlating this rock work with that of Pacheco Pass. Seismic studies verified the assumption, as did the excavations for the job, during which 68 percent of total yardage was blasted.

Figure 2-19 shows dramatically the massive nature of the Upper Cretaceous formation at La Jolla, California. Both the miniature sea mount and the main rock structure show few planes of weakness.

2. After a lot of experience igneous rock formations may also be correlated, although there is more deviation in the attributes of two similar formations than in the case of the sedimentaries. In 1959 preliminary seismic studies were made of a proposed freeway excavation a few miles east of Barstow, California. The rock in one big cut was rhyolite,

Figure 2-19 Hard sedimentary rock characteristic of the Upper Cretaceous series. An unnamed Upper Cretaceous sedimentary rock formation at La Jolla, California. This formation, correlative to the Chico some 550 mi northward, is massive and it presents the same excavation problem of drilling and blasting. The Rosario formation of coastal Baja California, another 220 mi southward, correlates with the unnamed La Jolla formation. Straddling the Oregon–California boundary 160 mi north of Chico is the Hornbrook formation of Upper Cretaceous rocks in which blasting experience corresponds to that in other formations of the same geologic time. Between Hornbrook and Rosario there are 900 mi of near perfect correlations among six formations of the Upper Cretaceous series of sedimentary rocks, typical of the practical feasibility of rock and rock-formation correlations in excavation appraisals.

a 19-million-yr-old Miocene rock lava flow. The rock was moderately weathered and the rock line was judged to be about 9 ft below the ground surface. Practically all the 547,000-yd^3 cut was blasted.

In 1960 studies were made for a proposed freeway excavation across Golconda Summit, east of Winnemucca, Nevada. A study of the geology showed the excavation to be in Eocene rhyolite, which, although three times the age of the Miocene rhyolite of Barstow, was probably weathered under the same desert conditions. It was a foregone conclusion that ripping and blasting conditions would be the same. And they were in fact about equal, demonstrating that correlation was feasible over a distance of 400 mi and over a time span of 40 million yr.

Similar correlations prior to excavation have been made in the granite formations of the West. Agreements are closer in this intrusive igneous formation than in the extrusive igneous formations of basalt, andesite, and rhyolite.

3. As with igneous rock formations, geologists have cataloged well the metamorphic rock formations. The time determinations are based largely on associations with sedimentary rocks of like age and, of course, often on radioactive methods of dating. Within reason, correlations are entirely feasible.

In 1967 preliminary studies were made on 18 million yd^3 of freeway excavation in Cajon Pass, San Bernardino County, California. Of this, 6.9 million yd^3 was in Pelona schist, a Precambrian formation more than 500 million yr old. All schist excavation was in the rift zone of the active San Andreas fault and thus was subject to weakening deformations over tens of millions of years. The prediction of the studies and the experience of the contractor indicated that no drilling and blasting should be done, although the cuts ran up to 130 ft in depth.

In 1969 investigations were made on a 9.4-million-yd^3 reach of the California Aqueduct located west of Palmdale, California. About 2.5 million yd^3 of canal excavation was located in the same Pelona schist. And, by coincidence, the aqueduct section was also in the rift zone of the San Andreas fault, paralleling the fault some 8000 ft to the north. Here, too, the contractor's experience corresponded to that of the builder of the Cajon Pass, although there was a modicum of blasting at the bottom of one 60-ft-deep cut. As an observation on the influence of fault action, there have been two major earthquakes of magnitude greater than 6 on the Richter scale in the Cajon Pass during recorded history, whereas there have been none along this particular reach of the California Aqueduct.

Finally, in appraising the worth of correlation of far removed rock formations of identical ages, the equivalence of eastern Pennsylvania limestone at Allentown to western Texas limestone at Sierra Blanca, despite dissimilar physiographic and climatic conditions, should be cited.

In 1968 preliminary seismic studies and observations were made on 55,000 yd^3 of jointed and semisolid limestone for building-foundation excavation at Allentown. The formation was the upper Jacksonburg member of the Pennsylvanian series, some 295 million yr old. Almost all the rock required blasting, in keeping with the quarrying and road-building methods of the area.

Shortly thereafter studies were made for mountaintop microwave stations near Sierra Blanca, Texas. A mountain of white limestone was the site of one of these stations. This rock was also of the Pennsylvanian series. Again, both seismic studies and excavation methods were similar in these two formations, 1800 mi apart.

Thus it is evident that nationwide correlations of rock formations are feasible. Actually, worldwide correlations are entirely workable. Kummel, in his book *History of the Earth*, correlates the relatively soft Miocene marine sedimentary formations of America with the comparable Hofuf, Dam, and Hadrukh formations of Saudi Arabia. There is no reason to believe that the sandstones and shales of the Upper Miocene Dam formation would behave very differently under a two-shank heavy-duty tractor-ripper than the Upper Miocene Capistrano formation of southern California, even though Saudi Arabia and the western United States are some 10,000 mi apart. Of course, one can speculate beyond the Earth in making correlations. The exploration of the planet Mars disclosed the similarity between the basalt of the desert of Mars and the same rock in the Mojave Desert of California, even though an average of 135 million mi of space and untold aeons of time separate these comparable extrusive igneous rock formations.

SUMMARY

Just as the general picture of the geology of the excavation is important to the excavator, so also is the detailed picture of the rocks, ores, minerals, and rock formations equally essential to the understanding of the rock and earth excavation.

The earthmover should fix in mind the nature of the excavation, which is most important in the case of rock work, because in this basic knowledge is the key to the successful use of methods, machinery, workers, and money.

CHAPTER 3

Rock Weathering

PLATE 3-1

Uneven weathering of eroded plutonic rock. Irregular weathering of tonalite, a kind of granite, in an abandoned quarry of San Diego County, California. The blocky rock has many unpredictable highly weathered cooling joints, the largest being above the front of the car. This condition made it impossible for the quarry excavator to develop an economical drilling and blasting program without removing some 50 ft additional depth of rock below the present quarry floor. Such costly pioneering work was not feasible. The exposed face dramatizes the problems of the quarry excavator in this well-weathered granite formation.

Chapter 2 discussed pragmatically the weathered zone of rock in the thin regolith of the Earth's surface. It is this weathered zone that is of special interest to the earthmover because the degree of weathering to the final finished grade will determine the methods and the machinery for the excavation.

The contractor in public works such as highways and dams thinks in terms of the whole mass of excavation. The miners in open-pit mines and the stone excavators in the quarries consider overburden and stone, as well as ores, as materials for the entire excavation. Both these classifications of materials have weathering characteristics.

Figure 3-1 shows the stages of disintegration and decomposition in a steep granite cut beside a highway in Los Angeles County, California.

First is a thin 2-ft layer of residuals, consisting of clays and sands and including small cobbles. Second is a 10-ft layer of weathered granite, or decomposed granite (DG). Last is the semisolid and solid granite with its weakening cooling joints. As depth increases below the grade of the highway, the cooling joints will be fewer and thinner and the granite will become solid.

The geologist speaks of these three layers in terms of the degrees of weathering, as illustrated in Figure 3-2. While the geologist speaks of residuals, weathered rock, and semisolid and solid rock, the earthmover talks of "potato dirt," "hard pan," and "shooting rock."

The cross section of Figure 3-2 is idealized, being simplified into three zones which actually merge into each other imperceptibly. In many excavations the degree of consolidation or apparent specific gravity from loose humus to solid rock bears a nearly straight-line relationship to depth from ground surface to solid rock.

As must always be emphasized when appraising excavation from method and cost criteria, no factor is as important as the degree of weathering of the rock. Two extremes of granite weathering illustrate this truism.

Figure 3-1 Weathered boulders of tonalite formation embedded in decomposed granite. The degrees of weathering in a highway rock cut of granite in Los Angeles County, California. Dimensions may be visualized by the hand pick resting on the boulder in the middle of the picture. At the grass roots is a 2-ft top layer of clay and sand residuals containing some cobbles. Beneath the residuals is a 10-ft layer of weathered granite or DG in which is the embedded large boulder. This cubic-yard-size boulder is an example of spheroidal weathering, which commenced in the approximately cubical pattern of cooling joints. Beneath the weathered zone are the semisolid and the solid granite, which become increasingly more solid with depth and with the narrowing of the cooling joints.

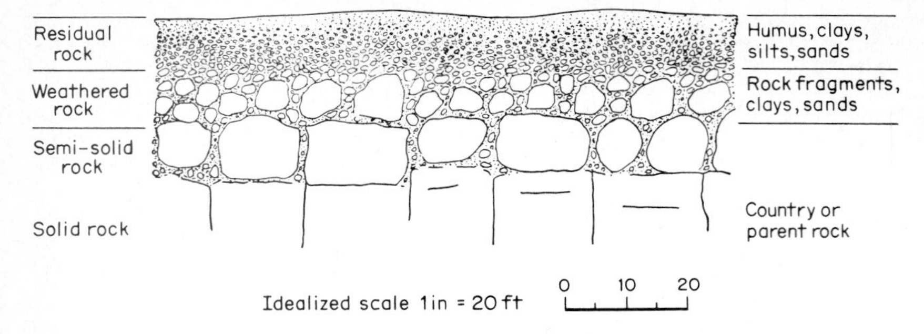

Figure 3-2 Rock excavator's conception of cross section of composite excavation.

Solid "country-rock" granite, relatively free of weakening joints, requires:

1. Fairly close pattern of drilling or spacing of holes. Rather expensive drilling.
2. Explosive yield of perhaps 1.5 yd³ per pound of explosive. Average cost for blasting.
3. Loading by shovel. Relatively expensive loading out of rock.
4. Hauling by rear-dump or side-dump trucks. Relatively expensive haulage.
5. Fill placement and compaction. Relatively expensive.

All five cost elements add up to high unit cost excavation.

In contrast, granitic sands, the ultimate stage in the weathering of granite, call for:

1. Loading and hauling by push-tractors and scrapers. Relatively inexpensive.
2. Relatively low cost for fill placement and compaction. These two cost elements add up to low unit cost for excavation.

Overall costs for these two kinds of excavation might be in the ratio of \$2.00/yd³ for the solid granite and \$0.75/yd³ for the derivative sand, the difference being solely due to the difference in degree of weathering.

Weathering is the work of Nature's continual efforts to lower the mountains and raise the valleys and thereby produce the ideal level land or peneplain. That all the land of the Earth is not completely leveled is due to the opposition between the raising of the mountains and the relative or actual lowering of the valleys. It is an ageless battle between creation and destruction.

FORCES OF WEATHERING

The weathering forces are chemical decomposition, the usually invisible destroyer, and mechanical disintegration, the normally obvious devastator.

Decomposition by Chemical Action

Decomposition results from hydration, hydrolysis, oxidation, carbonization, and dissolution, all mutually supporting the inexorable change from trap rock to clay.

Hydration example: Conversion of hard periodotite rock into soft serpentine by addition of water.

Hydrolysis example: Conversion of feldspar and mica from granite to clay by water containing mild acid.

Oxidation example: Breakdown of olivine basalt into soft iron compounds with characteristic brown color and stains.

Carbonization example: The breakdown of arkose or feldspathic sandstone into clays by the action of carbon dioxide and water.

Solution example: The dissolution of limestone by water, resulting in deep fissures and caverns.

When groundwater descends into and dissolves sedimentary rocks, it leaves a red clayey soil mantling the surface and filling the crevices and caverns. Figure 3-3 shows a cross section of this decomposition as exaggerated in a quarry face from which dimension building stone has been taken. Such karstlike topography is common in limestone areas of temperate and tropical regions. The manifest problems confronting the drilling and blasting crews are lost drill holes because of stuck drill rods and bits and

Figure 3-3 Chemical and mechanical weathering of sedimentary rock. Predominantly chemical decomposition of the Bedford limestone formation of southern Indiana. Surface and near-surface dissolution by descending groundwater has left a surface blanket and a subsurface seam- and fissure-filling deposit of red clayey residuals, as exposed in this face of a quarry for dimension building stone. The residual is called *terra rosea,* and it extends to considerable depth below the ground surface. *(William D. Thornbury,* Principles of Geomorphology, *Wiley, New York, 1961, p. 319.)*

lost explosive force because of soft pockets of residuals within the limestone formation.

Reiterating, the five agencies of chemical decomposition processes discussed above can be mutually and simultaneously active, all contributing to slow but insidious weathering.

Disintegration by Mechanical Action

In contrast to the decomposition of the Indiana limestone of Figure 3-3 is the predominantly mechanical disintegration of the Arizona limestone of Figure 3-4.

In this semiarid country with little rain and little ground cover, chemical forces are not significant when compared with mechanical energies. In this monolithic rock there are no topmost residuals and no intermediate weathered zones. Hard rock is at the surface. The limestone is of the Pennsylvanian series and about 295 million yr old, whereas the Indiana limestone is probably of the Mississippian series and 325 million yr old. They may be considered of about the same age, but they have been treated differently by the environments of their geographic locations.

Disintegration results from mechanical forces and these forces are at work at the same time as the energies of chemical decomposition. The mechanical breakup of rock is caused chiefly by the following agents.

Frost Wedging Water is generally present in the joints or interstices of semisolid rock. When water freezes into ice the volume increases by 9 percent. Terrific expansion forces are generated, serving to split the rock further. After the following thaw there is another freeze, and the breakup continues indefinitely. Radiocarbon dating indicates that such wedging of huge angular blocks of granite occurred thousands of years ago. Figure 3-12 shows the results of frost wedging, along with the weathering effects of rains, roots, and other agents.

Temperature Changes Daily and annual temperature changes disintegrate rock. In the United States these changes can vary through 75°F daily in the western deserts and thru 150°F annually in the northern states, causing compression and tension. These

Figure 3-4 **Predominantly mechanical weathering in arid climate.** Predominantly mechanical disintegration of Arizona limestone near Globe, Arizona. There is no cover of residuals and the hard rock is slightly weakened only by the irregular pattern of thin joints. The exposed face is about 30 ft wide and 40 ft deep. *(U.S. Geological Survey)*

changes act on rocks in both microscopic and megascopic ways. Minutely, a rock such as granite slowly disintegrates because of the differences in coefficients of expansion of the chief mineral constituents: quartz, feldspar, hornblende, and biotite. As these microscopic particles work against each other over the millennia, the rock slowly disintegrates and forms, ultimately, granitic sand.

The relationship of the different minerals with unlike thermal characteristics is illustrated in Figure 3-5. In this diorite with its little vein of aplite are chiefly light quartz, feldspar, and muscovite and dark biotite and hornblende.

Massively, temperature changes cause the spalling of rocks such as granite in more or less concentric layers. These layers are generally several inches thick. This spalling should not be confused with exfoliation, which resembles spalling but is caused by chemical decomposition.

Cooling In the process of cooling, igneous rocks such as intrusive granite and extrusive basalt contract and the masses pull apart from each other and form cooling joints. These joints are then invaded by chemical and mechanical agents, which further the weathering processes.

Figure 3-6 portrays stages in the weathering of granite, commencing with the formation of the jointing planes and ending with the final production of sand. The cooling joints of the distant rock formation were frost-wedged to produce the rubblelike mass. Accumulative disintegration and decomposition resulted in the boulders of the foreground and the sands of the desert floor.

This transition from solid granite to sand is taking place in the high desert country of Joshua Tree National Monument of southern California. Elevation is about 5000 ft and annual temperature range is about 75°F.

Figure 3-5 Disintegration due to different thermal characteristics of different minerals. Diorite from the Sierra Nevada of California. The rock is about 1 ft^3 in size. The vein of aplite, largely quartz, and the basic diorite minerals—quartz, feldspar, muscovite, biotite, and hornblende—have different expansion coefficients and these different rates of expansion and contraction slowly loosen the bonds between the different minerals. Eventually the diorite becomes a nearly incoherent mass, which can be reduced to sand by the blow of a sledgehammer.

Figure 3-6 Aeons of progressive weathering of igneous rock. Disintegration of granite in Joshua Tree National Monument of southern California. Commencing with the cooling joints in the distant ridge, a combination of the predominantly mechanical disintegration agents ice, rain, and wind has produced desert sand during the 81 million yr of age of this basement complex of the Sierra Nevada. *(U.S. National Park Service.)*

The excavation of the granite formation of Figure 3-6 for highway or railroad building would call for the following methods:

1. The cubic-yard-size weathered boulders of the foreground would be loaded out by rubber-tire tractor-shovel and hauled by rear-dump haulers. They would be incorporated into the fill provided the fill was sufficiently deep.

2. A few feet of the residuals of the desert floor would be loaded by push-tractors and hauled by rubber-tire scrapers.

3. The weathered zone of granite below the residuals and in the distant ridge would be fragmented by tractor-rippers and loaded out and hauled either by shovel and haulers or by push-tractor and scrapers. Choice of machinery would depend upon the degree of weathering.

4. The semisolid and the solid underlying granite would be fragmented by blasting, loaded by tractor-shovel, and hauled by rear-dump haulers.

A careful coordination of these different methods would be of prime importance to both efficiency and low cost.

Figure 3-7 displays vividly the characteristic vertical cooling joints of columnar basalt. This extrusive igneous formation of Le Puy, France, shows the following three zones of in-situ weathering:

1. Residual zone. The upper 10 ft shows the final effects of mechanical disintegration and chemical decomposition. The end result is the topmost foot of red clay.

2. Weathered rock zone. The middle 10 ft indicates the effects of all the factors working in the residual zone plus the breaking down of the hexagonal basaltic columns.

3. Semisolid and solid rock zone. The lower few tens of feet portray vividly this remarkable example of horizontal polygonal and vertical linear jointing of the cooling basalt. The rock is firm and hard.

At the bottom of the wall is a talus slope made up of mechanically weathered fragments of rock. The breakage of these cubic-yard rectilinear blocks has taken place along both vertical and horizontal cooling joints.

Figure 3-7 Unique weathering of columnar basalt. Weathering of columnar basalt near Le Puy, France. Three zones are present: The upper 10 ft are residuals; the middle 10 ft, weathered rock; and the lower few tens of feet, semisolid and solid rock. The individual columns are characteristically hexagonal and about 5 ft in diameter. *(H. Ries and Thomas L. Watson,* Engineering Geology, *Wiley, New York, 1925, p. 88.)*

The fragmentation procedure in such a basalt formation would be: first, no tractor ripping in the residual zone; next, soft to medium ripping in the weathered zone; and finally, extremely hard ripping and blasting in the semisolid and solid rock zone.

Breakage There are several visible and invisible mechanical movements causing rock breakage. One visible one is the fall of rock from a steep vertical face or escarpment or from a ridge. An example is shown in Figure 3-7, in which a talus slope is accumulating at the bottom of the face of columnar basalt.

When a turbulent stream undercuts the soft shale of a gorge, the overlying sandstones or limestones will break up in the gravity fall. This degradation is shown in Figure 3-8. Flexure along a jointing pattern also creates breakage, which accentuates the jointing pattern and sometimes results in definite cubic formations, as shown in Figure 3-8. This sandstone formation along the steep bank of the Green River in Canyonlands National Park in Utah has developed two-dimensional joints on the surface and invisible three-dimensional joints below and parallel to the rock surface. The percolation of river water and rainwater has widened the joints. The resulting cubelike blocks are about 20 ft on each side and weigh some 640 tons.

Near the surface and far below the surface these near cubes will be weathered along the jointing planes of weakness, as illustrated in Figure 3-9. Such spheroidal weathering takes place more commonly in igneous intrusive rocks such as granite. This action is illustrated by the huge embedded granite rock in the center of the picture of Figure 3-1. Spheroidal weathering is nearly nonexistent in igneous extrusive rocks such as basalts and andesites and in the metamorphic rocks.

In Figure 3-9 nearly final weathering has reduced the volume of the original cube

Figure 3-8 Massive cubical pattern of weathering in jointed sandstone. Breakage of falling blocks of sandstone along the gorge of the Green River in Canyon Lands National Park, Utah. Beneath this regional pattern of jointed sandstone, the underlying softer shales have been eroded by the swift water, causing the huge 640-ton blocks of sandstone to fall and break up. Stream degradation will reduce them ultimately to sand. *(U.S. National Park Service.)*

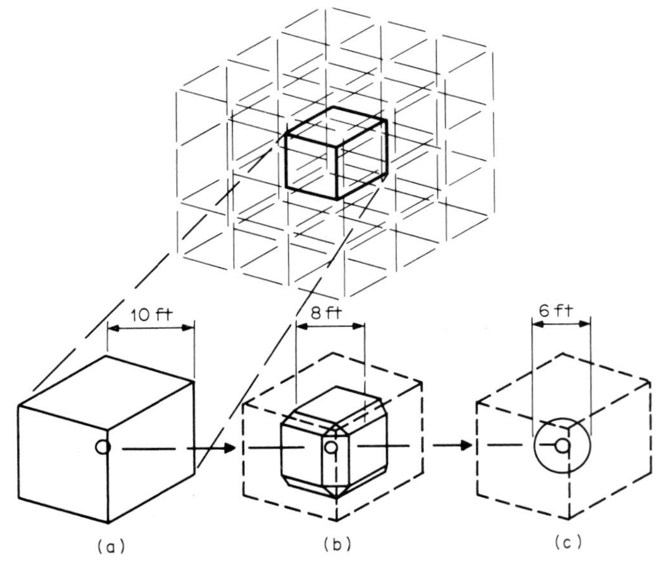

Figure 3-9 **The spheroidal weathering of rock.** Starting with initial weathering of huge prism of fresh cubically jointed rock made up of average 10-ft cubes; ending with near final weathering, with 6-ft spheres of solid rock surrounded by a well-weathered matrix of clays, sands, and small rock fragments. The matrix is 89 percent and the solid rock is only 11 percent of the original prism. The complete weathering will produce only clays and sands. *(a)* Cube—Initial weathering: beginning of chemical decomposition and mechanical disintegration; 37 yd^3 or 100 percent solid rock; 6 faces, 12 edges, and 8 corners for early weathering. *(b)* Cube-Sphere—Partial weathering: 16 yd^3 or 43 percent solid rock; 26 faces, 48 edges, and 24 corners for continued weathering. *(c)* Sphere—Nearly final weathering: 4 yd^3 or 11 percent solid rock; infinite number of minute area faces, no edges, and no corners; maximum area for weathering.

by 89 percent, although the overall reduction in the linear dimension is only 40 percent. When the 6-ft sphere becomes sand, the weathering will be complete.

An example of breakage by folding, showing dramatically the fracturing of the contorted strata, is apparent in Figure 3-10. An individual stratum of this Chickamauga limestone formation near Ben Hur, Virginia, has a maximum thickness of about 24 in. As the reverse fold becomes more pronounced, the strata will be more highly fractured and a minor brecciated fault will form. Such folds may include the entire volume of a huge cut and so their weakening effect may reduce a sizable yardage of apparently hard rock to rather easily ripped weathered rock.

Faulting fractures rock on both minor and major scales. A minor fault has been shown in Figure 2-17. In contrast to minor faulting there is the major weakening of an entire rock formation or several rock formations due to brecciation by a major fault over tens of millions of years.

Southern California is literally laced by major faults and this large-scale faulting is the explanation for several calamitous earthquakes during the past century. Figure 3-11 delineates this netlike fault system.

Most conspicuous is the famous and notorious San Andreas fault, stretching in its land appearance 660 mi from Point Arena, north of San Francisco, to the Mexican border. Geologists believe that the brecciated or gouge zone of the fault extends ½ to 2 mi on either side of the axis of the fault.

As discussed in Chapter 2, the appraisal of an 18-million-yd^3 freeway across the San Andreas fault in Cajon Pass, California, verified this contention. The excavation was investigated in 1967 by 63 seismic studies. Additionally, researches were made on the intensities and the magnitudes of four major earthquakes in Cajon Pass since 1857. From

Figure 3-10 **Weathering by severe minor faulting in rock structure.** Contorted strata in the Trenton limestone formation near Ben Hur, Virginia. The maximum thickness of an individual stratum is 24 in. Within the slightly reversed fold there is already a small brecciated zone of crumbled rock. As the reversed fold becomes fully developed, a minor fault will occur, producing a sizable brecciated and gouged zone. *(Virginia Division of Mineral Resources.)*

these analyses it was deduced that 4,300,000 yd^3 of granites, schists, and gneisses and 600,000 yd^3 of sandstones, all within the fault or rift zone, could be ripped economically. The contractor's excavation experience on the job corroborated the analyses. It was a striking example of the rock-weakening effects of major faults.

Plant Roots and Burrowing Animals Rock structures are weakened by roots and by burrowing. In faces of new cuts of decomposed granite, the tree and shrub roots extend down 30 ft or more in their strivings for moisture. Figure 3-12 shows the combined results of temperature changes, rains, frosts, and tree roots in the breakup or weathering of quartzite.

Combined Effects

In this metamorphic quartzite of Monroe, New York, the total combined effect of all destructive agents has been to reduce hard quartzite, normally requiring blasting, to rubble. Only ripping is required and not much of that if the cut is shallow. Cooperating Nature, by transforming this rock formation from the hardest to just about the softest, has materially reduced the work of the earthmover.

Climate has a profound influence on the weathering of rocks. The effects of different climates on the same basalt rock formation are dramatically illustrated by two pictures of the windward and the leeward sides of the western part of Maui Island, Hawaii. Figure 3-13 shows a vigorous early mature landscape on the windward side of the mountains. Both heavy vegetation and rapid erosion have resulted from the heavy annual rainfall of 200 in. Herein both mechanical disintegration and chemical decomposition are evident, but the chemical weathering has predominated.

Figure 3-14 displays the relatively arid climate of the leeward side of the mountains. Depth of weathering is less, streams are almost nonexistent, and vegetation is either scarce or absent. Mechanical disintegration is more in evidence than chemical decomposition.

Table 3-1 summarizes generally the degrees of weathering according to 11 contribu-

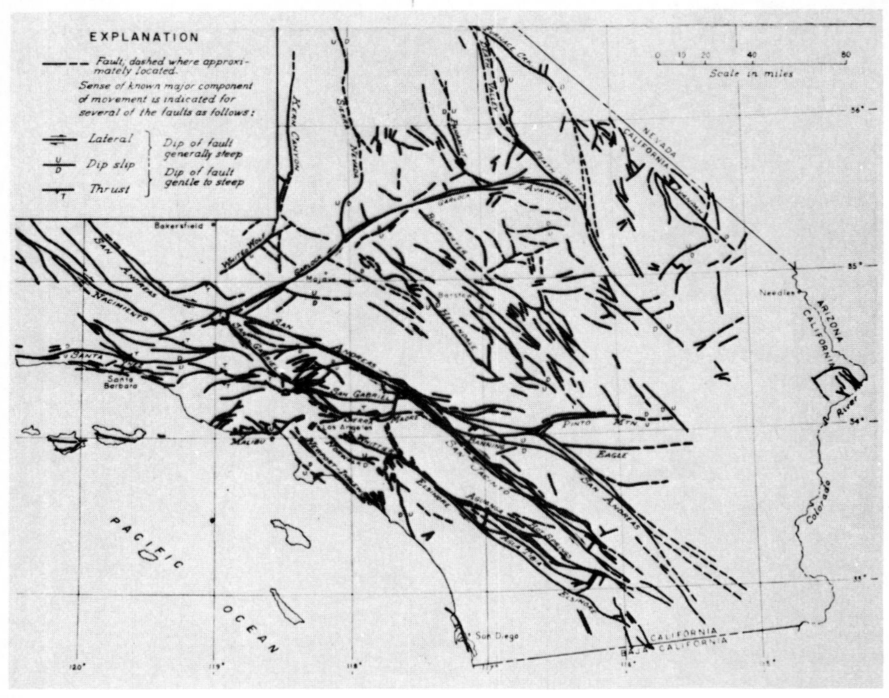

Figure 3-11 **Widespread system of faults in 13,200 mi² of California.** Major and minor faults of southern California. Notorious faults that have dealt death and destruction during the past century are Imperial, Newport-Inglewood, San Andreas, San Gabriel, San Jacinto, Santa Yñez, Sierra Nevada, and White Wolf. Others have been active before recorded history and presently are dormant. Throughout aeons all have been allies of the excavator by weakening the rock structures within their rift zones, which may extend up to 2 mi on either side of the fault axis. *(California Division of Mines and Geology.)*

tory factors. These influences are observable. Obviously, in using the table one must weigh the factors and use good judgment, as the table is idealized. And of course all factors are interrelated.

RATES OF WEATHERING

There are no real data concerning the linear rate of rock weathering in the natural state. The several thousands of years of recorded history are too brief for conclusions even if there had been observations.

Meager available information centers around a wide range of tombstones and buildings in England and in the New England states during the past 300 yr, around the 8000-yr-old ruins of Yucatan, and around surface rocks that show largely mechanical disintegration during the past 15,000 yr of the last interglacial period. These weatherings in igneous, sedimentary, and metamorphic rocks average about 2 in for the past 10,000 yr. These weatherings are all in exposed buildings and monuments and it is probable that they are the result of mechanical disintegration. Accordingly, the average rate does not reflect the more effective chemical decomposition of rock in the natural state below ground surface.

If chemical forces are twice as effective as mechanical agents, then a hypothetical rate of weathering might be 6 in, or 0.50 ft, in 10,000 yr for the combined weathering process for rocks in situ. If this rather vague rate is applied to the 10-ft cube of unweathered rock of Figure 3-9, the rock should become clay and sand in 100,000 yr.

Figure 3-12 **Cumulative effect of several mechanical agents on weathering.** Metamorphic quartzite of Monroe, New York, weathered and broken by temperature changes, frost, rain, and plant roots. Fresh quartzite requires blasting. This well-weathered quartzite requires little or no ripping if the cut is only a few feet deep. *(H. Ries and Thomas L. Watson,* Engineering Geology, *Wiley, New York, 1925, p. 225.)*

Figure 3-13 **Strong geologically early weathering in wet climate.** Vigorous early mature landscape on the windward side of the mountains of western Maui. In this subhumid climate rock weathering has been advanced by the annual 200 in of rainfall, resulting in chemical decomposition by the rich vegetation and in mechanical disintegration by the abundant streams. *(U.S. Air Force.)*

Figure 3-14 Weak, geologically late weathering in dry climate. Feeble, immature landscape on the leeward side of the mountains of western Maui. In this semiarid climate rock weathering has been reduced by limited rainfall, culminating in little chemical decomposition because of sparse vegetation and moderate mechanical disintegration by the scarce streams. *(U.S. Air Force.)*

Personal observations of surface and subsurface rocks over a long period of time in some 1 billion yd^3 of rock excavation indicate that the rocks most resistant to weathering are metamorphic, that those of average resistance are igneous, and that the least resistant rocks are sedimentary. This is a generalization, as may be inferred from the fact that sedimentary limestone in arid country is more resistant than igneous granite in semiarid, subhumid surroundings, and that granite is more unyielding than metamorphic serpentine in humid lands.

The depth of weathering varies greatly, being dependent upon the factors of Table 3-1. Based on observations of deep cuts in public and private works, the maximum depth appears to be about 400 ft. The rocks and ores of underground mines show weathering at considerably greater depths, but the chemical decomposition and replacement processes are the result of metamorphic actions rather than surface and subsurface weathering.

TABLE 3-1 Observable Factors Affecting Weathering

	Degree of weathering		
Factor	Minimum	Average	Maximum
Rainfall, average annual (in)	Up to 15	15 to 50	50 and up
Temperature, average annual (°F)	Up to 40	40 to 80	80 and up
Humidity, average annual (%)	Up to 40	40 to 80	80 and up
Topography	Level	Hilly	Mountainous
Rock exposure	South	East-west	North
Vegetation	Sparse	Average	Abundant
Residuals, depth	Thin	Average	Thick
Rock outcrops	Heavy	Light	None
Rock grain	Small	Medium	Large
Rock color	Dark	Medium	Light
Rock hardness or durability	Hard	Medium	Soft

SUMMARY

Rock weathering is an important ally of the earthmover in his efforts to excavate rock in the many stages of mechanical and chemical breakup. In all cases, to a greater or lesser degree, Nature has substituted generously for the tractor-ripper and for explosives in the fragmentation of rock.

Weathering studies are all significant to the excavator, as through the millennia Nature may very well have fragmented the rock down to the finished grade without the knowledge or understanding of the earthmover.

CHAPTER 4

Landforms and Geomorphology

PLATE 4-1

Obstacles to rock excavation along a bluff shoreline. Along the shores of submergence into an encroaching sea there are often accentuated promontories and canyons that present serious problems to highway and railroad builders. Plate 4-1 illustrates the highway construction enigma near Hecate Light of the rough Oregon coast. Characteristics are steep sidehill cuts and fills, bridges and tunnels, and ever-present confined working conditions. A compounding perplexity is the unknown stability of the weathered basalt rocks. These 49-million-yr-old lavas are sometimes of columnar jointing, making for dangerous uncertainties in the necessarily steeply sloped sidehill cuts. *(Brubaker Aerial Surveys.)*

Landscapes or geomorphology, physical geology that deals with the form of the Earth, are to the rock formations as the forest is to the trees. Landforms are the volumes of rock to be excavated and their understanding is important to the earthmover.

For example, is the gravelly river terrace of Figure 4-1 simply a comparatively level and deep alluvial fan washed down from the mountains? Or is it a thinly veneered horizontal sedimentary rock deposit? If the former, it is common excavation, moderately consolidated. If the latter, the rock may require blasting after the removal of the alluvium blanket. These questions are paramount for the construction of a highway or a railroad from the narrow valley to the top of the terraces.

The geomorphic processes that shape the earth and as a result control the excavator's methods are degradation, or the breaking down of the earth's crust, and aggradation, or the building up of the crust. Generally the same process is responsible both for degradation at higher elevations and resulting aggradation at lower levels. For the rock specialist, degradation uncovers the existing rock, making it visible, and aggradation obscures the present-age rock. Accordingly, the imponderable is generally the result of building up processes and the resulting landforms.

Degradation or subtraction and aggradation or addition of earth-rock are the result of six agents: ice, water, wind, volcanics, gravity, and crustal movements. All six forces may work individually or collectively and cooperatively.

ICE

Ice action on a large scale involves erosion, transportation, and deposition of earth-rock by glaciers. On a small scale the action involves frost or ice wedging, as discussed in Chapter 3. Although glacial actions are as old as Precambrian groups of rocks, the excavator is interested primarily in the landforms of those of the Pleistocene series, commencing about 2 million yr ago and ending about 18,000 yr ago. The last one, known as the Wisconsin in North America and the Wurm in Europe, influenced greatly the lives of our ancestors. It also released meltwater to form the Great Lakes and to excavate the rapids and basin of Niagara Falls.

Figure 4-2 illustrates the extent of these glaciers or ice sheets. They covered the

Figure 4-1 A possible misleading interpretation of surficial rock formation. Deceptive terraces of Cola Creek, Castle Hill basin, New Zealand. Unless there are exposed outcroppings of subsurface earth-rock structures along the banks of the creek, a cursory examination of the surface gravels of the terraces will not disclose the true nature of deep excavation for a highway or railroad ascending the escarpment. The simple question is whether the gravels extend to the grade of the excavation or are merely a veneer atop a rock formation. *(William D. Thornbury,* Principles of Geomorphology, *Wiley, New York, 1961, p. 159.)*

Figure 4-2 **Coverage of four glacial periods in North America.** Areal extent in North America of four glacial periods during the past 1 million yr. This vast and powerful action of moving ice resulted in transportation of incalculable amounts of valuable topsoils from Canada to the north central and northeastern United States, in excavation of the Great Lakes by the meltwaters, and in formation of the present-day glaciers of the Rocky Mountains and the Sierra Nevada. It is presumed that the Northern Hemisphere is presently in the midst of the fourth interglacial period, the Wisconsin in North America and the Wurm in Europe. *(U.S. Geological Survey.)*

northern half of North America, which included many of the northern states. In the 11 western states the glaciers covered much of the high elevations of the Sierra Nevada and the Rocky Mountains, and many mountain glaciers have survived into the present interglacial period.

There are four principal landforms resulting from the deposition of the earth-rock mixtures. All have one characteristic of materials composed of unsorted and unstratified till and drift materials made up of clays, sands, gravels, cobbles, and boulders. These particles vary in proportions and in shape from rounded to angular. The four landforms are *lateral moraines, terminal moraines, drumlins,* and *eskers.* They are pictured and discussed in Figures 4-3, 4-4, and 4-5. Since they are all formed by deposition from melting ice, their landforms sometimes overlap in configuration.

An analysis of these glacial landforms is important to the excavator both prior to construction and during the prosecution of the work. The following example points up the necessity for analysis.

In 1960 freeway building at Conway Summit on the east slope of the Sierra Nevada of California called for an 850,000-yd^3 cut of maximum depth 140 ft. Exposed granite surrounded the cut. Some bidding contractors believed that the gravelly surface material was an overburden atop country granite and they adjusted their bids accordingly. In one case the following analyses were made and conclusions drawn.

Figure 4-3 **The wide gradation of the conglomerate of a glacial moraine.** Glacial ground moraine near Lansing, Michigan, typifying a lateral or terminal moraine. The lateral moraine results from the deposition of materials when the ice melts along the edges of the glacier. The terminal one comes from placement at the end or tongue of the glacier. The eroded bank of till is about 30 ft high and it illustrates the heterogeneous nature of the moraine, ranging from silts to boulders of up to 2-ft^3 volume and laid down unsorted and without stratification. Over the aeons till is sometimes cemented by mineral agents and then it becomes tillite. Thus, a moraine may represent soft to hard excavation for the earthmover, but usually it is soft. *(U.S. Geological Survey.)*

1. A geology map showed the cut to be part of a terminal moraine dating back about 18,000 yr to the Wisconsin glacial stage. Such youthful age precluded any thought of even mild consolidation.
2. Limited exploratory borings showed some refusals to drilling, suggesting either bedrock or, more likely, interbedded boulders.
3. Seismic studies showed no shock-wave velocities greater than 2900 ft/s to finished grade.
4. The conclusions were:
   a. The gravelly overburden indicated a moraine rather than typical decomposed or weathered granite.
   b. The boring refusals were caused by large boulders and not by country rock or bedrock.
   c. Seismic studies, indicating uniform low velocities to excavation limits, forecast a gravel-boulder deposit with anticipated soft ripping by medium-weight tractor-ripper.
   d. Finally, it was wisely concluded that the huge cut could be taken out by methods for common excavation.

The beginning of a peneplain by the sculpturing of a glacier is shown in Figure 4-6. This aerial photograph shows a section of the glaciated Sierra Nevada of central California in the vicinity of Donner Pass. The eight-lane freeway section of the view is at about elevation 6000 ft and Donner Pass in the eastward distance is at elevation 7008 ft.

At this high elevation the surface of the bare granite is striated by the grooving action of the glaciers and it is strewn with rock particles torn from the mountainside by glacial gouging. The particles range in size from sands to room-size boulders. Along the distant horizon may be seen the remains of *cirques.* Cirques are generally shaped as steep amphitheaters and they are faceted where the tops of the slopes meet the natural slopes of the rugged mountains. Bear Valley floor to the left in the picture is made up of rather

Figure 4-4 Large rocks carried by glacier a short distance from the mountains. The terminal moraine of a cliff glacier on Mount Lyell of the Sierra Nevada, California. As compared with the two men of the picture, the largest of the massive glacier-borne rocks are room size. They were torn from the cirques and the gorges of the mountains and they were transported a relatively short distance, as attested by their angular form. Excavation of this rubble would call for shovel and trucks, with some blasting of the hugest fragments. *(U.S. National Park Service.)*

Figure 4-5 Wide gradation of well-traveled, well-rounded rocks of a glacial drumlin. Drumlins and eskers are deposits from melting ice within the body of the glacier, rather than at the extremities as in the case of the moraines. Clay, sand, pebbles, and boulders are mixed in this unconsolidated till of Oconomowac, Wisconsin. It is sometimes called boulder clay and it represents soft excavation. Drumlins are elongated hills or ridges up to 1 mi in length and 200 ft in depth, their axes being parallel to the movement of the glacier. Eskers are rather narrow, sinuous ridges, longer but lower than drumlins. The till or drift material of an esker resembles closely that of a drumlin, as they have a common origin, and they are oriented along the path of the glacier. *(U.S. Geological Survey.)*

Figure 4-6 **Glacially sculptured granite of Donner Pass of the Sierra Nevada of California.** Ageless leveling of granite and andesite of the great Sierra Nevada of California by aeons of glacial action. Rock particles are torn from the mountainsides by gouging action and are then transported by the ice down the mountainsides, grooving the rock surfaces and removing any loose rock fragments. An almost bald rock surface remains, with little or no surface residuals except in small vales. Such topography always creates abnormally high rock excavation costs. *(California Division of Highways.)*

coarse alluvium with little soil cover. In general the land is topped by a minimum thickness of residuals, which supports only sparse vegetation.

The rock excavation for this magnificent freeway called for much blasting in granite and andesite, as ripping of the rock by heavyweight tractor-rippers extended below natural ground by only a few inches or feet. Cuts ranged in depth from a few feet to several tens of feet, and the freeway alignment is mostly on the ridges. The preponderance of rock excavation, the short construction season, and the remoteness of the work area contributed to high cost to the state and low profit or loss to the excavators.

WATER

Running water has produced and is forming the greatest number of landforms, which make up excavations. It is the most powerful of the forces involved in Nature's eternal efforts to level the land. A trip by airplane across the United States from Los Angeles to New York unveils a myriad of landforms of interest to the earthmover. A few are:

1. The several-hundred-foot-deep waterborne fanglomerates of the San Gabriel Valley call for hundreds of millions of cubic yards of gravel excavation for the aggregates used in the Los Angeles area.
2. The abrupt water-carved topography of the San Bernardino granite mountains presents real rock excavation problems to the roadbuilders.
3. The desiccated floor and the mudflows of the Mojave Desert, the results of flash floods, require continuous cuts and fills in what appears to be a monotonously flat desert.

4. The nearly vertical rock walls of the Colorado River escarpment made the abutment excavation for Hoover Dam painstaking and costly.

5. The precipitous walls of the Black Canyon of the Gunnison River of Colorado called for tedious sliver-cut excavation for the Denver and Rio Grande Railroad.

6. The easily worked river-gravel terrace deposits of the central states afford building materials for the Mississippi Valley states.

7. The northeast to southwest trending ridges of the Appalachian Mountains call for hard-rock excavations for railroads and highways, unless location engineers can find water gaps and wind gaps.

8. The igneous rock escarpment of the Hudson River Palisades provides an abundance of trap rock for the needs of the trade area.

Running water degrades the landscape in the form of streams, ranging from little brook gullies to big river valleys, as in the combined watersheds of the Conewango Creek, Allegheny River, Ohio River, and Mississippi River, stretching 1100 mi from New York State to Louisiana. On the other hand, water aggrades the same land in this watershed all the way from the little gully outwash of Conewango Creek to the big Mississippi River Delta.

Valleys and Gorges

Figure 4-7 dramatizes the late stage in the development of a mature river gorge. Such a rock exposure in the walls, together with the rock cuts of the highway and railroad, indicates that the excavator will encounter soft, medium, and hard excavation and that there will be some blasting. If excavation is planned in nearby canyons of the tributaries

Figure 4-7 **Unmistakable visible harbingers of hard-rock excavation.** Mature gorge of the Cheakamus River, British Columbia. A powerful mountain stream is cutting this precipitous waterway, compounding the granite excavation for the sidehill cuts of the highway and the railroad. Exposed faces foretell problems of any future excavation in the Cheakamus gorge and the gorges of nearby tributary streams. *(Geological Survey of Canada.)*

of the Cheakamus River and if the rock formations in these resemble those of Cheakamus Gorge, then the proposed excavation will follow the same pattern for work.

Contrastingly, an earthmover who is considering heavy subsurface excavation for a reservoir in plains country and sees a fairly deep stream with slumping banks in the general area must assume that the reservoir will involve soft excavation. Quite naturally, the earthmover would wisely verify the supposition.

Floodplains

Just as the valley is the result of stream erosion, so the *floodplain* is the product of stream deposition or aggradation. Floodplains are generally fairly level and consist of clays, silts, sands, and gravels. To considerable depths they present no excavation problems.

Fanglomerates

A possibly deceptive landform is the alluvial fanglomerate, which is formed by material transported by water at the end of mountain valleys. The fanglomerate consists of fine to coarse aggregates, ranging from silts at the base to room-size boulders near the apex of the triangular fan.

In a fanglomerate the unknowns to the earthmover are these:

1. Does the fanglomerate exist down to and below the grade of the excavation, or is it a shallow deposit atop the pediment of the mountain rock?

2. If the fanglomerate does, indeed, exist to grade, does the invisible mass resemble the surface material as far as gradation is concerned?

Such questions were raised when the Owens Valley Aqueduct, supplying water to Los Angeles County from Bishop, California, was built. The 300-mi water carrier crosses many fanglomerates of the Owens and Mojave Valleys, the granitic alluvia being derived from the steep east escarpment of the Sierra Nevada. Complete investigations of these and other variables of the open-cut and tunnel excavations were made prior to construction.

Figure 4-8 outlines several fanglomerates from the Funeral Mountains near Furnace

Figure 4-8 Physical features of a typical geological basin-range province. Fanglomerates of Death Valley, southern California. Furnace Creek Inn is built upon a fanglomerate flanked by a fault escarpment. In back of the inn are several small alluvial fans debouching from the canyons of the foothills. Furnace Creek fault is at the base of the distant Funeral Mountains and many major fanglomerates issue from the large canyons. *(California Division of Mines and Geology.)*

Creek, Death Valley, California. These triangular aprons are being constantly built by the water transportation of well-weathered rock particles from the eroded mountains. Furnace Creek Inn is built upon a recent fault scarp and the major Furnace Creek fault is at the base of the distant mountains.

The unsorted materials of a fanglomerate are strikingly shown in Figure 4-9. The dissected wadi reveals up to cubic-yard-sized boulders. Such boulders would "refuse" an exploratory drill, perhaps erroneously indicating bedrock. Obviously this common excavation would present some problems, calling for shovel loading of trucks, but there would be no blasting in the loose alluvia.

Terraces

Excavation in abrupt, steep-sloped river terraces is generally obvious, as the formation shows up in the exposed bank. This is not the case with flat-sloped terraces, as shown in Figure 4-1. In these the true nature of the innocent-appearing slopes may be obscured by slope wash.

Mudflows

Mudflows are movements of water-saturated materials down streambeds and slopes. They may be mild or they may be of startling and even fatal proportions. A sudden heavy flash flood in the mountains, unknown to the careless camper in a sunny desert arroyo, may give the victim less than a 1-min warning roar before he is swept away.

Clays and silts of desert hillsides and volcanic ash of mountain slopes are common

Figure 4-9 **Massive, poorly graded conglomerate of an alluvial fan.** Stream channel crossing an alluvial fan or fanglomerate. Bank shows the characteristic texture of fine to coarse materials laid down by the waters. Some of the large boulders are in the middle of the arroyo and they have been moved only by powerful floodwaters. *(U.S. Forest Service.)*

materials of mudflows. Figure 4-10 outlines an ancient mudflow of volcanic breccia in the Mojave Desert near Victorville, California. The area is about 7 mi^2 and the volume of the mudflow is about 50 million yd^3. Such sudden mudflows have reached velocities of 60 mi/h, with devastating results.

Such mudflows in level or gently rolling country generally offer no excavation problems. However, this is not in the case in mountainous terrain, where the flow material may include larger than room-size angular and rounded boulders and where the underlying bedrock or pediment is of unknown and sometimes variable depth.

Mesas and Buttes

Mesas and buttes, typical of the 11 western United States, are formed chiefly by the combined agents of water and wind. Sometimes the ranges, or uplifted horsts of basin-range geologic provinces, appear to be mesas because of their steep escarpments. However, they are the direct result of vertical faulting, modified by water and wind. It is true that horsts, because of their configuration, present the same excavation problems as do the mesas.

Usually mesas and buttes are distinguished by a hard-rock cap that protects the lower, less resistant rocks and preserves the plateaulike surface of the land.

When a highway or railroad line must ascend the escarpment of a mesa from the valley below, the sidehill cuts may cross different rock strata and thereby may present excavation problems because of the varying hardnesses of rock and the tendency for slopes to slide.

Figure 4-11 shows a four-lane highway descending the side of a mesa to the lagoon of the Pacific Ocean near Del Mar, California. The escarpment of this mesa is not steep, having been eroded from the well-weathered sedimentary rock formation. This Del Mar formation, some 47 million yr old, is made up of multicolored mudstones, siltstones, and sandstones. There are oyster interbeds, testifying to its marine origin. It is of medium

Figure 4-10 Outsize but typical mudslide characteristic of desert country. Breccia mudflow in Mojave Desert, near Victorville, California. The magnitude of the mudflow of volcanic rock may be judged by the narrow ribbon of the road in the left foreground. The lobe is about 2½ mi across and about 7 mi^2 in area and totals some 50 million yd^3. Such flows, generally the result of prolonged mountain rains, are usually soft to medium excavation. They are an unsorted mass of many sizes of earth-rock particles, ranging from clays to room-size boulders. *(William D. Thornbury,* Principles of Geomorphology, *Wiley, New York, 1961, p. 93.)*

Figure 4-11 **Easy soft-rock excavation along gentle bluff of eroded mesa.** Four-lane highway descending the escarpment of a mesa along the Pacific Coast near Del Mar, California. The moderately sloped bluff, together with the medium-hardness Del Mar marine sedimentary rock formation, presented no excavation problems to the earthmover. The mixture of mudstones, siltstones, and sandstones is sufficiently indurated to allow for a designed backslope of ration about 1:4.

hardness, as indicated by the steep sideslope of the highway, and it calls for medium ripping by a tractor-ripper.

The oceans, with their shores, bays, inlets, coves, and estuaries, shape the land so as to cause problems for the building of excavations and embankments. Plate 4-1 outlines the complexity of highway building near Hecate Light on the Oregon coast. The rock is basalt and andesite and is subject to slides in the cut slopes and in the wave-carved shoreline. Here just about all excavation problems are present. After a railroad or a highway is built in such a shore of submergence, the external cause of failures, undercutting by the waves, will always exist.

WIND

Wind degrades and aggrades. Degradation or erosion of existing surfaces serves the earthmover by uncovering earth-rock formations that would be obscured otherwise by misleading residuals.

Deposition by winds results chiefly in loess and sand dune formation. Loess is a fine silt and dust deposit of the central United States. Deflation by the wind involves pickup from the prairie lands of fine particles, which are then windborne and deposited at considerable distances. Loess has accumulated to depths of 40 ft and it is characterized by near-vertical slopes when eroded by nature or excavated by humans. Being of recent making and having little cementation or pressure, it is not consolidated and it is soft excavation.

Sand dunes are common to shores of lakes and oceans and to desert lands. Figure 4-12 shows barchand dunes in the Imperial Valley of California. These ever-creeping landforms cover and uncover the old plank road, which was built early in this century.

Dunes are obviously soft excavation. They present only a moderate loading problem but generally they offer an aggravating hauling problem to the excavator. Figure 4-12 shows dramatically the haulage enigma because early automobiles could not pass over these sands without the plank or corduroy road, even though the natural road-building material was laid carefully. In this same country the new freeway called for moderate cuts and fills but the single-axle-drive rubber-tire scrapers could not tra-

Figure 4-12 **Eternally encroaching sand dunes.** Barchand sand dunes obliterating the old narrow plank road across the Imperial Valley of southern California. In the early part of this century, because of the ever-shifting sands this corduroy road was built between El Centro and Yuma, Arizona. Every 1/4 mi it was widened to allow passing. It served well until improved road-building methods made it possible to construct and maintain a first-class highway over the treacherous desert. Persistent migratory dunes no longer bedevil the building of freeways across the dunes of the seashores and the deserts. *(U.S. Forest Service.)*

verse the sandy haul roads. The problem of traction and flotation was solved by the use of two-axle- or all-axle-drive scrapers with twin engines, supplemented by liberal wetting of the haul roads.

A similar haulage problem confounded the builder of the Cape Cod Canal of Massachusetts some 40 yr ago. Rubber-tire bottom-dump haulers were used and only single-drive axles were available. Work was possible by liberally wetting the unstable sands of the cuts.

VOLCANICS

Volcanics interest the earthmover almost exclusively in terms of the unknown consolidation of cinder cones and breccia and the questionable hardness of lava flows. The certainty of hard excavation normally requiring blasting is evident in the lava flow of Figure 1-1. This "aa" or rough vesicular basalt flow may be excavated by hard to extremely hard preliminary ripping by a heavyweight tractor-ripper.

Tuffs and breccias differ only in the maximum size of their particles, both being pyroclastic or formed from airborne volcanic tephra. Tuff is of less than 1-in diameter and breccia contains fragments of more than 1-in diameter.

Moderately solid breccia rock is shown in Figure 4-13. The canyon of Rio de las Frijoles in northern New Mexico has been eroded in this thick deposit. Cliff dwellers quarried blocks for their cliff-hanging homes from this rock formation. Such tuff and breccia are hard, but they are usually rippable by heavyweight tractor-rippers. The formidable, massive nature of the formation is accentuated by reference to the man in the picture.

A typical Mojave Desert basalt cinder cone is shown in Figure 4-14. It is a lightweight-

Figure 4-13 **Canyon wall of finely graded medium-hard volcanic breccia.** Eroded pyroclastic breccia of the Canyon del Rio de las Frijoles in northern New Mexico. The massive vertical walls above the man at the left of the picture, suggesting extreme hardness, are to be contrasted with the weathered caverns at the right of the picture, suggesting softness. Excavation properties are in between, as the medium-hard igneous rock formation is usually rippable by heavyweight tractor-rippers. *(U.S. Forest Service.)*

aggregate source for an open-pit operation near Red Cinder Mountain. These cones are usually mildly consolidated and, in spite of a rather forbidding appearance, they are generally soft to medium-hard excavation.

GRAVITY

The land-forming effects of gravity are largely the results of the already discussed actions of ice, water, wind, and volcanics. However, visible examples of the work of gravity are provided by creep, slump, slide, and talus slope, for here are landforms only partially dependent on the aforementioned four agents.

Creep or slump in a hillside or river bank is a gradual process, usually but not always accompanied by a high moisture content and by horizontal slump ridges in the formation of the hillside. Figure 4-15 pictures the gentle movement of a slump that was preceded by creep. The vertical fall of the detached section of the bank is about 4 ft vertically and about 6 ft horizontally. The bank was weakened by the undercutting of the stream in flood stage.

Of significance to the earthmover in all creeps and slumps is an indication of an unstable earth-rock formation. Such indications also mean soft excavation to moderate depth. From a design standpoint they point up the necessity of flat back slopes of cuts and possibly a requirement for retaining walls in the cuts.

Of course slides, big or small, indicate highly unstable natural earth or rock. Sometimes slides occur in rock slopes which have been blasted because of weakening of the natural

Figure 4-14 Volcanic cinder cone of soft to medium-hard aggregate-size scoria. A huge cinder cone of the Mojave Desert in California is the source of lightweight aggregates for this commercial open-pit materials plant. The red volcanic cinders are ripped and bulldozed from the cinder pit *(CP)* to the screening plant *(P)* by a medium-weight tractor and then the sizes are stockpiled in the yard *(SY)*. Excavation is soft to medium hardness. Nearby is a sister cinder cone in which a 75-ft-deep sidehill cut was made for a freeway. The excavation experience of the earthmover was identical to that of the operator of the cinder pit. *(California Division of Mines and Geology.)*

Figure 4-15 Slump in residuals and well-weathered rock formation. Slumping of subsoil and soil in the bank of a stream near Spartanburg, South Carolina. The gentle slump was preceded by a gradual creep of the edge of the bank, which was undercut by the stream in flood stage. While more common in earthy materials, creep and slump are also common in weathered rock formations. *(U.S. Soil Conservation Service.)*

rock by the back-breaking or shattering effects of the blast beyond the designed prism of excavation. Such effects are common in weathered granite and serpentine.

In sedimentary rock formations the slide generally takes place along the moisture-lubricated stratum of clay or shale, the dip of the slide-prone strata sometimes being as low as 15° to the horizontal. Such a slide of dangerous proportions is dramatically shown in Figure 1-3. This huge mass movement was caused by moisture-laden earth-rock, weakened by removal of the natural buttress when the back slope of the highway was excavated. Like creeps and slumps, slides indicate a weak earth-rock formation and resultant soft excavation. In slide-prone country the hazards to construction are manifest. Not only are the hazards indigenous to the immediate work area or right-of-way, but they may endanger also the adjacent property above the cut, as shown in Figure 1-3, and below the fill because of surcharging of the natural ground.

Sometimes there is a monetary gain to the excavator in terms of a profitable unit price for excavation. Many slides in cut faces are caused by too steep back slopes, especially in the grading for building sites. If a slide occurs the earthmover excavates additional earth-rock, which is more easily handled than it would have been in its original state.

Talus slopes are typical of mountainous country. Sometimes they take the form of steep cones, and the instability of the fallen rock fragments is obvious. On the other hand, in the case of the moderate slope of a rock stream the talus slope would be relatively stable. Such a rock stream is shown in Figure 4-16.

The lower slope of the talus rock stream is not especially steep and yet the stream is continuing to flow down into the valley. Were a railroad or a highway to be built across this landform, the back slopes would be unstable and the roadbed would be displaced slowly downstream. Similarly, any fills would be in jeopardy. It is doubtful that any work done by humans could be located economically or safely on such ever-mobile land.

Excavation in this landform is soft and the poorly sorted but generally hard rock particles are a good source for aggregates.

Figure 4-16 Slowly advancing rock stream of graded conglomerate. Rock stream in Silver Basin, San Juan Mountains, Colorado. These streamlike talus slopes extend down into the valleys and they originated as rock falls from glacially eroded steep cliffs of the mountains in the distance. Slowly moving rock streams would make for instability in either excavation or embankment and in any building raised thereon. The inherent instability is synonymous with soft to medium excavation characteristics. Like alluvia and fanglomerates, rock streams are rarely consolidated by any cementing agency. *(U.S. Geological Survey.)*

CRUSTAL MOVEMENTS

These kinetic phenomena in rock formations are the result of internal forces within the Earth's crust or regolith. They are usually the results of ages of work by the pressure within but sometimes they result from swift deformations by earthquakes. They are subsurface activities and the surface manifestations have generally been obliterated by years of weathering. Now and then one sees a recent scarp or ridge resulting from an earthquake fault.

Faults

Figure 4-17 shows transverse cross sections of common vertical faults. As indicated by the legend of Figure 4-17, these sometimes concealed faults are misleading even after careful field investigations.

In the case of the normal fault, a ditching contractor who based the estimate for bidding on borings to the west of the fault would possibly have quadrupled the estimated cost when working to the east of the fault.

On the other hand, by the same prebidding estimate the contractor in the case of the reverse fault would probably find the costs quartered when excavating to the east of the concealed fault.

Of course there are more variations and complexities in the vertical faults, but these normal and reverse faults typify the excavation problems.

Another type of fault is the lateral one, in which the sides of the fault move more or less horizontally with respect to each other. If they are truly lateral and not complicated by vertical movement, they are generally not misleading. Perhaps the greatest number of vertical and lateral faults, some with displacements of 20 ft, are in California. Their extremely extensive distribution is depicted in Figure 3-11. These have created both engineering and excavating problems. Sometimes the age-old movement creates brecciation in rift zones thousands of feet wide. These zones are made up of fractured rocks and clays rather than of the adjacent solid rocks from which they were derived. The salutary effect of of the rift zone on the cost of excavation was discussed in Chapter 3.

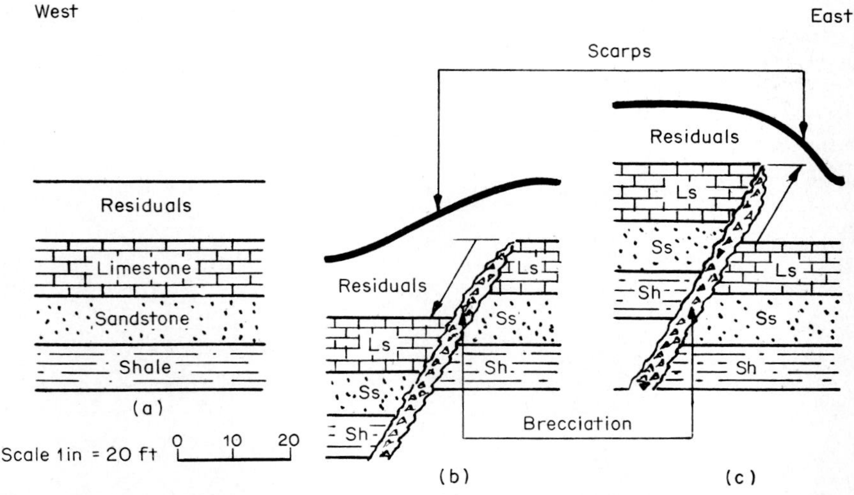

Figure 4-17 **Types of vertical faults.** *(a)* Unfractured beds. Undisturbed strata prior to faulting. *(b)* Normal fault. Tens of years after the faulting. Rather gentle slope after years of weathering. Fault is concealed by deceptive north-south trending ridge. An excavator digging a 20-ft ditch from west to east would run into about 14 ft of additional rock at the fault line. *(c)* Reverse fault. Tens of years after the faulting. Rather steep slope after years of weathering. Fault is concealed by deceptive north-south trending ridge. An excavator digging a 20-ft ditch from west to east would run into all earth at the fault line. Brecciation is made up of angular rocks torn from the sides of the plane of the fault.

A deformation equally as important as faulting of the rock formation is *folding*. Figure 4-18 shows the progressive deformation of sedimentary rock strata.

In a general way folds may have nearly vertical, inclined, or nearly horizontal axes. They may be of a minor nature, several inches to several tens of feet, or they may be of major proportions, up to tens of miles in length.

In Figure 4-18 an anticline has been slowly warped into a horizontal or recumbent fold, the upper half being shown in the lower cross section. At this juncture of development the strata could not resist the lateral forces, and the recumbent fold was faulted horizontally and the zone of brecciation created.

Such deformation presents to the earthmover obvious problems in estimating and subsequent excavation. Synclines, anticlines, folds, thrusts, and faults are common in the anthracite coal fields of eastern Pennsylvania. Here the coal, usually interbedded with limestones and shales, has been contorted in such manner as to complicate greatly the open-pit mining of the coal. Figure 2-15 shows such a coal stripping mine in which the rock formation has been only mildly deformed, affording the miner unusually good

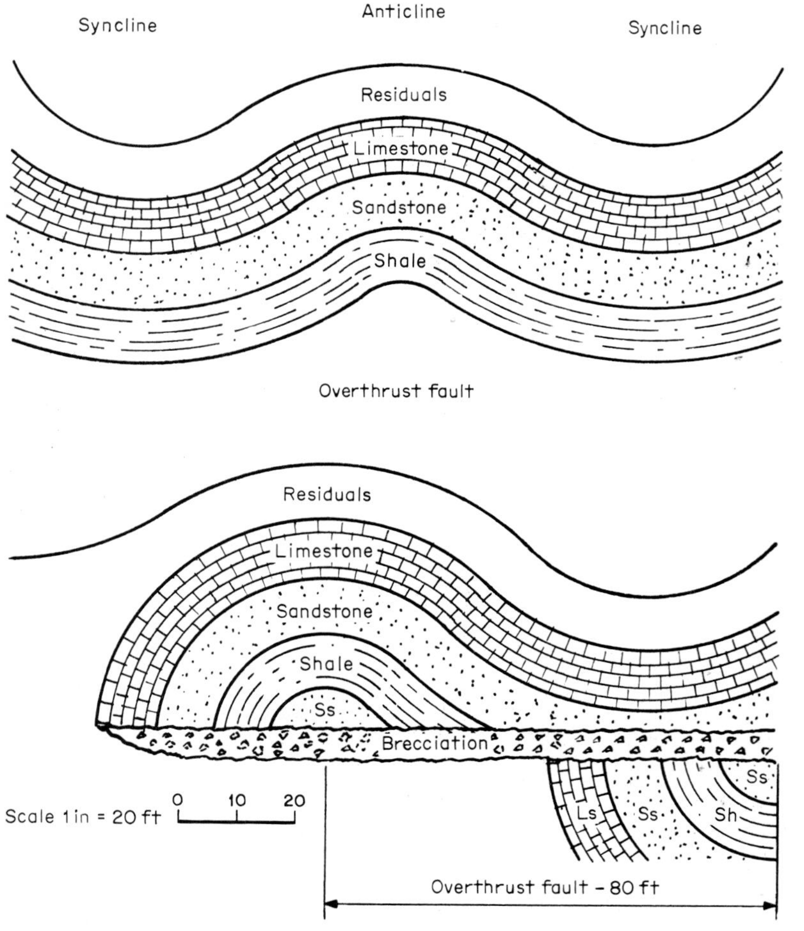

Figure 4-18 **The progressive deformation of sedimentary rock strata.** The original anticline became a recumbent fold and the recumbent fold is now an overthrust fault as the result of intense deformation. Brecciation is made up of angular rocks torn from sides of the plane of the fault.

Figure 4-19 Extreme deformation and weakening of a sedimentary rock formation. Folding and faulting of a sedimentary rock formation within the rift zone of the San Andreas fault near Palmdale, California. In this 1,146,000-yd^3 cut the sandstones and the shales were so weakened by ages of continuous flexuring that the excavator was able to fragment the rock by only soft ripping by a tractor-ripper. Nature, with infinitely patient work, had substituted for the machines of humans.

working conditions. There is a fairly fixed ratio of near horizontally bedded overburden to coal in terms of cubic yards per ton.

There is one benefit from these deformations. The rocks are necessarily severely fractured in the flexuring. The limestones, which normally might require blasting, may be ripped for scraper operation and may be sufficiently fragmented by Nature for shovel and dragline work.

A classic example of folding and faulting in the rift zone of a major fault is portrayed in Figure 4-19. The rift zone is perhaps 1 mi in width on either side of the San Andreas fault.

This 1,146,000-yd^3 76-ft-deep cut is within the rift zone near Palmdale, Claifornia. The Anaverde formation of sedimentary rocks includes sandstones and shales which were laid down some 8 million yr ago. During this epoch the originally horizontal strata were contorted into these picturesque faults and reverse folds. Because of rather thin bedding and the weakening effects of the continuous flexuring, the rocks were fragmented by soft ripping.

In contrast to the soft ripping in this freeway cut, the same Anaverde formation in a railroad cut a few miles to the east and outside of the rift zone required medium to hard ripping by the same heavyweight tractor-rippers.

This chapter ends the discussion of geology, rocks, ores, minerals, rock formations, rock weathering, and landforms. It would be interesting and intellectually and financially rewarding for the rock excavator to pursue these fascinating subjects far beyond the elementary treatment in this book.

SUMMARY

The body of excavation is always a landform, and the study of landforms, or geomorphology, will assist the earthmover in his appraisal of the excavation. Appearances of landforms are deceptive, especially when they are covered by a mantle of weathered material which may not indicate the nature of the excavation below the mantle. In many cases an understanding of the simple landforms and recognition of them will help in the initial examination of the excavation.

CHAPTER 5

Open-Cut Excavations

PLATE 5-1

Largest open-pit mine of the world. The earth's largest excavation by humans is the Bingham Canyon, Utah, open-pit copper mine of the Kennecott Copper Corporation. This huge bowl averages 2 mi in width and 0.5 mi in depth. The excavation was commenced in 1904, and the historical and logistical statistics are awesomely impressive.

1. A total of 3.6 billion tons of porphyry copper ore and overburden or waste rock has been excavated and hauled to the 20-mi distant smelter and to the nearby waste-pile areas. Of these 3.6 billion tons, ore made up 1.31 billion tons and overburden accounted for 2.29 billion tons as of 1976. The stripping ratio, or ratio of overburden to ore, is 1.75:1.

2. The pit is 2.25 mi west to east, 1.50 mi north to south, and 0.50 mi deep. The average train of locomotive and 16 ore cars carries 1390 tons and is about 565 ft long. To haul 3.6 billion tons the continuous string of trains would stretch 277,000 mi, or once around the Earth, then to the Moon, and finally around the Moon.

3. A close scrutiny of Plate 5-1, showing the "thumbprint" of humans on the earth's surface, reveals 55 benches of average 45-ft height and average 80-ft width. Rail haulage takes place below the 6340-ft level and truck haulage is above this level.

4. Fragmentation of ore and overburden is by blasting, and for every pound of total explosives used 4.30 tons, or about 1.89 yd^3, of material is removed. Annually 15,300 tons of explosives and blasting agents is used, filling a string of box cars about 3 mi long.

5. Machinery inventory for the average daily production of 433,000 tons during 1975 in a normal triple-shift day was:

a. 100 mi of standard gauge railroad track.
b. 54 miles of track in 3 mine-tunnel entries.
c. 18 blast-hole drills of 8- to 12-in drill-hole diameters.
d. 37 electric power shovels of 6- to 15-yd^3 dipper capacity.
e. 60 electric and 2 diesel locomotives of 70- to 125-ton classification.
f. Approximately 1000 ore and overburden hopper cars of average 87-ton capacity.
g. 97 rear-dump haulage trucks ranging from 65- to 150-ton capacity.

6. On October 13, 1974, a world's record for excavation was set. In 24 h 504,167 tons of ore and overburden was loaded and hauled away. This tonnage is equivalent to 222,000 yd^3 of solid granite. It may be compared with the hard-rock excavation for a good-sized freeway cut or dam spillway.

7. The excavation for the Panama Canal was 268,000,000 yd^3 of rock-earth, or about 502 million tons. Accordingly, it would require approximately the material from seven Panama Canals to fill the colossal open pit of Bingham Canyon.

(*Kennecott Copper Corporation* and *Utah Power and Light Company.*)

There are nine principal kinds of open-cut excavations, those for: airports, building sites, canals, dams and levees, highways, pits and open mines, quarries, railroads, and trenches.

All these excavations generally combine both common and rock materials, including residuals, weathered rock, and semisolid and solid rock. The rock work is generally of a lesser yardage when compared with the common excavation, but it is of greater unit cost and it presents more problems.

The frontispiece and the accompanying description of the huge open pit of the United States Borax and Chemical Corporation vividly portray open-cut excavation.

AIRPORTS

Ideally, airports are located in flat or gently rolling country for obvious reasons of safety and construction costs. Actually, airports involving sizable rock excavation are a small minority of all airports.

Examples of those having 100 percent common excavation are numerous. Three such examples are cited below.

1. The Baltimore airport in Maryland was built years ago in the swamps of Chesapeake Bay. The borrow pit for the fill material, some 2 mi away, was common excavation of silts and clays.

2. The Kansas City airport, Missouri, is built on a flat alluvial plain at the confluence of the Missouri and Kansas Rivers. Grading consisted of a mild leveling process in the sand and gravel river terraces.

3. The Seattle-Tacoma airport of Washington is built on a plain. Considering its 11,600-ft runways, proportionately little excavation, all common, was involved.

One must turn to hilly and mountainous country for examples of rock excavation, where sometimes the only place for an airport is atop a rocky mountain. Such an airport is the Wheeling airport, West Virginia, built during World War II under emergency winter conditions.

1. Figure 5-1 shows the necessary bad-weather beginning of this 3-million-yd^3 rock job. Because of snowy and icy conditions and consequent excavation and haulage problems, the residuals and well-weathered shales and limestones were moved by crawler-tractor-drawn scrapers rather than by rubber-tire or wheel-type scrapers. Hauls averaged 2000 ft one way downhill, calling for maximum traction on the uphill empty return.

Figure 5-1 Airport building in icy, snowy wintry weather. World War II winter emergency construction of an airport in the mountains near Wheeling, West Virginia. The 2.5-yd^3 shovel is loading 15-ton rear-dump trucks with frozen weathered limestones and shales. This is the opening cut of the 3-million-yd^3 job. The work is icy and costly and it will not improve until the end of the spring thaws. During the thaws the steepening haul roads will be of mud rather than ice.

The weathered rock was loaded without blasting by the 2½-yd³ shovel into 15-ton rear-dump trucks. Solid rock was blasted before loading. The winter picture shows three of the four trucks shuttling without turning on the initial 300-ft one-way level haul. An average of 4.5 buckets was loaded in 2.0 min for a 7.5-yd³ payload. By elimination of two turnings cycle time was reduced to 4.7 min, and the three-unit fleet production was about 200 yd³h.

After the bad weather the excavator brought in wheel-scrapers for the summer work and the overall production for the job improved greatly.

2. The San Clemente Island airport off the coast of southern California was hewn from rhyolite-dacite volcanic solid rock. The maximum depth of cut was about 65 ft. The weathered rock was ripped to about 45-ft depth and the solid lava flow was blasted to grade.

The runways, stretching completely across the north end of the island, are 11,000 ft long and 500 ft wide and the excavation was 10 million yd³ of moderately weathered and solid rock. Two 2.5-yd³ shovels and eight 22-ton rear-dump rock haulers were used for the loading and hauling of the huge yardage.

BUILDING SITES AND LARGE FOUNDATION EXCAVATIONS

These sites for homes, huge buildings, and plants may run to millions of cubic yards of common and rock excavation. As in most excavation, the common is generally less than the rock work.

One exception to this generalization was the 1967 rock excavation for an additional plant of the Mack Truck Company at Allentown, Pennsylvania. Blasting was used for 140,000 yd³ of the Pennsylvanian series limestone, totaling just about all the excavation. This rock formation typified excavation with minimum residuals and weathered rock.

On the other hand, Figure 5-2 shows a granite rock excavation in which blasting of solid rock was at a minimum when contrasted with the total excavation. The maximum depth of cut in this notoriously well-weathered granite of southern California was 60 ft. Of a total of 1,943,000 yd³, only 198,000 or 10 percent was blasted.

The track drill of Figure 5-2 drills 4½-in-diameter holes at the rate of 610 ft of hole per 8-hour shift. Hole spacing pattern is 7 ft by 7 ft, the average production of the drill rig being 138 yd³/h. Ammonium nitrate explosives were used, the combined powder factor of the explosives and blasting agents being about 0.75 lb/yd³ of granite.

Figure 5-2 **Hard-rock excavation for mountain home sites.** A 1,943,000-yd³ excavation in granite for a tract of home sites in the San Raphael Hills of Los Angeles County, California. In spite of negligible residuals, the volume of weathered granite was extremely high with respect to the solid rock. Only 198,000 yd³, or 10 percent, of the total yardage was blasted.

After blasting, this granite was loaded out and hauled by wheel scrapers, a common practice in such formidable-appearing solid granite. In five representative housing tracts in the same granite formation, only 269,000 yd^3, or 8 percent, of the total 3,429,000 yd^3 was blasted. This low ratio of "shooting rock" was obtained in spite of the fact that the maximum depth of cut was 125 ft and the average depth was 83 ft.

CANALS

Classic canals excavated out of rock are numerous. Three are described below.

Panama Canal Of some 175 million yd^3 of common and rock excavation, about 130 million yd^3 was in dry work and the remaining 45 million yd^3 was dredged hydraulically. The canal was completed in 1914 and of course heavyweight tractor-rippers and wheel or rubber-tire scrapers would today handle much of the so-called rock which was blasted and loaded by railroad-type shovels of the early twentieth century.

The total bid price for the prism excavation of the canal was $138 million, resulting in a unit price for unclassified excavation of about 79 cents. It is probable that today's gigantic earthmovers could excavate such a canal for a bid price in the range of $0.75 to $1.00 per cubic yard.

St. Lawrence Seaway This mammoth rock excavation with channels, canals, and locks stretching some 100 mi from Ogdensburg, New York, to Montreal, Quebec, involved several million cubic yards of solid beddings of limestone. In addition to fragmentation of the rock by conventional blasting, it was necessary to give the walls of the rock excavation a relatively smooth surface. Preshearing or presplitting was developed to a remarkable degree during the period of the 1950s and the 1960s. In this operation, depending on the kind of rock and the smoothness desired, drill holes of 1½ to 4-inch diameter are spaced on 1-ft to 4-ft centers along the line of the face desired and the explosives are detonated along with the main charges.

Excavation for the canal was in a glaciated area in which residuals and weathered rock were at a minimum. Consequently solid rock excavation predominated, far outweighing the common excavation.

California Aqueduct The main channel of this lengthy canal stretches some 600 mi from Oroville Dam in Butte County to Auld Valley Dam in Riverside County. The open-cut excavation ranged from the residuals and alluvium of the valleys to the solid rocks of the mountains. The rocks varied from soft sedimentary siltstones through medium igneous granites to hard metamorphic marbles. However, of the tens of millions of cubic yards, less than a million cubic yards of igneous and metamorphic rocks was blasted.

Figure 5-3 shows one of these rare exceptions to the general practice of rock ripping as the means of fragmentation. Rock is in the intake channel of the Tehachapi pumping plant in Kern County, California. This 1966 excavation included alluvium overlaying sandstone of the Tejon formation, which in turn led into the country rock, the granite.

The alluvium required no ripping and the sandstone was ripped out without difficulty. However, the granite ranged from medium ripping through hard ripping to blasting. The picture shows the phase of hard ripping, as a heavyweight tractor-ripper and a tractor-dozer are delivering the rock from the upper level of the ridge down to the loading shovel. The 6-yd^3 shovel is loading three 35-ton rear-dump rock haulers. Shortly thereafter, as the ridge was brought down, it became necessary to blast the granite rock. Of the total rock yardage of the intake channel, some 15 percent was blasted. In spite of the small percentage of blasted rock, this case is atypical of the usual case of rock excavation in the aqueduct. This significant distribution between rippable and "drill-and-shoot" rock for all the excavation for the California Aqueduct is typified in the following summary of excavation for the Fairmont to Leona reach, located in the foothills of the San Gabriel Mountains.

1. Length of reach, mi		18
2. Excavation, yd^3		9,400,000
3. Number of cuts through the ridges		114
4. Average yardage of cuts, yd^3		82,000
5. Maximum depth of cut, ft		76
6. Distribution of excavation, west to east, according to rock formations, yd^3.		
a. Granite. All was ripped.	36%	3,400,000

b. Anaverde formation of sandstones and conglomerates. All was ripped.	4%	380,000
c. Alluvium of clays, silts, sands, and gravels. No ripping.	32%	3,100,000
d. Dacite. All was ripped.	1%	30,000
e. Pelona schist. About 50,000 yd³, or 2% of total 2,490,000 yd³ was blasted, 98% being ripped.	27%	2,490,000
Total yd³ of reach		9,400,000

Compared with the 600 mi of other reaches, this one in the pediment formations of the mountain range was considered to be rocky. Yet only 50,000 of the total 6,300,000 yd³ of rock was blasted, and this is but 0.8 percent of the total rock.

DAMS AND LEVEES

The open-cut excavations for dams may be subdivided in a general way into eight classes.

1. *Foundation excavation* or the removal of residuals and weathered rock so as to form a semipervious base for an earth-fill dam and an impervious setting for a concrete dam. Some of this excavation, especially in the case of a concrete dam, may require blasting or "dental" excavation if fault zones of breccia and gouge are present.

2. *Cutoff wall excavation.* This trench work is carried well below foundation excavation and is taken down to bedrock if bedrock is present. If bedrock is not present, as in the case of many earth-fill dams, it is carried down into firm material.

3. *Inlet and outlet portals* for the diversion tunnel. This excavation is characteristic of both types of dams, but more especially in the relatively rough topography of the setting for a concrete dam.

4. *Diversion channel excavations,* in lieu of diversion tunnels, are usually associated with earth-fill dams as a means of dewatering the work area of the dam.

5. *Spillway channel excavation,* as distinguished from the integral spillway of the dam. Since concrete dams are associated with mountainous country, rock excavation is usually present, whereas in the case of earth-fill dams the excavation tends to be common rather than rock.

6. *Power-plant and tailrace excavations,* down in the bed of the stream, may be either rock or alluvia or a combination of both materials.

7. *Borrow pits* located in residuals, alluvia, or weathered rock provide fill materials for earth-fill dams and concrete aggregates for the concrete appurtenances. Usually they

Figure 5-3 Excavation for canal to pumping plant of aqueduct. Rock excavation for the intake channel of the Tehachapi pumping plant, a part of the California Aqueduct. Although the yardage of all the rock blasted in connection with the total tens of millions of cubic yards of rock involved in the building of the canal was insignificant, this rock excavation in the granite pediment of the Tehachapi Mountains accounted for about 15 percent of the rock of the inlet channel. Almost without exception heavyweight tractor-rippers substituted for explosives in the rock excavation for the 600-mi canal.

are common excavation, although the cases of weathered rocks and cemented rocks such as caliche sometimes call for fragmentation by ripping. For concrete dams borrow pits may provide concrete aggregates and sundry materials.

8. *Quarry excavations* provide aggregates and haul-road metal for both earth-fill and concrete dams. They produce rock for semipervious, pervious, and blanket-rock zones for earth-fill dams.

There are many minor examples of rock excavation in the building of dams such as highway and railroad relocations, access roads, haul roads, building sites, and staging area excavations.

The enormous rock excavation quantities for two big dams under construction in 1975 are set forth in the following tabulations.

1. *New Melones Dam, Stanislaus River, California*

Foundation	1,380,000 yd³
Intake structure	50,000
Slide areas	1,910,000
Spillway	15,000,000
Total rock excavation	18,340,000 yd³

This dam is the rock-fill type and the total rock excavation will go into rock fill. These quantities do not include a sizable amount of rock removed during the first stage of construction for the diversion tunnel and other appurtenances. The rocks are metamorphic or altered basalts, sandstones, and shales of the Mother Lode country.

2 *Auburn Dam, American River, California*

Foundation, for both main dam and cofferdam	4,620,000 yd³
Power plant	220,000
Service spillway	1,100,000
Auxiliary spillway	800,000
Tailrace and channel improvement	650,000
Quarry riprap	30,000
Total rock excavation	7,420,000 yd³

This dam will be a double-arch concrete dam, and the first contract is for the excavation preliminary to the actual building of the dam. The dam is located in a slightly faulted area of the Mother Lode country, and the foundation excavation includes 82,000 yd³ for fault zones that involve closely controlled lines and grades or "dental" work. The quantities are generally a mixture of residuals and weathered and solid amphibolites and schists. The rock for the power plant and for tailrace and channel improvements and the quarry riprap are solid. All rocks are typically metamorphic in the work area.

The Balsas River Dam on the Río de las Balsas of Mexico is a part of the Infiernillo electric project and it was built during the 1960 to 1970 period. The 900,000-kW generating dam is of the rock-fill type with a crest length of 175 m (574 ft) and a height of 152 m (498 ft). Five million cubic meters (6.5 million yd³) of quarry rock excavation in a basalt-andesite formation was required.

Four 4-yd³ electric shovels and two 4-yd³ diesel shovels loaded out a fleet of 22-ton rear-dump rock haulers with an average production of 15,080 m³ day of double shifting. The average rate is 157 m³ (205 yd³) hourly for each shovel.

In this mixture of weathered and solid igneous rock such production is excellent, indicating good management and good machinery availability. The average hourly productions suggest a shovel performance in keeping with the following deduced tabulation of factors.

Shovel cycle	25 s
Shovel dipper factor, ratio of payload to capacity	0.60
Overall job efficiency	60%

Such quarry excavation on a hillside, as shown in Figure 5-4 is typical of foundation excavations for the abutments of concrete and rock-fill dams, although in foundation excavations generally only weathered rock is removed. During this quarry excavation, the undesirable residuals and well-weathered rock were removed initially and wasted,

Figure 5-4 **Dam building in precipitous mountain gorge.** Quarry excavation for the dam on the Río de las Balsas of Mexico. Two 4-yd^3 shovels are opening up the excavations by means of a "sliver cut" and later an additional four shovels will be used for full-scale quarry production of 15,000 m^3 (19,600 yd^3) each double-shift working day. In this pioneering work of residual and weathered rock removal, shovel efficiency is necessarily low because of the confined working conditions and the instability of the slide-prone rock formation. *(Bucyrus Erie Company.)*

then the acceptable slightly weathered rock was removed, and finally the desirable solid rock, requiring blasting, was taken out and the operation proceeded efficiently.

HIGHWAYS

Everybody is most familiar with the everyday sight of the work for roads, highways, expressways, and freeways. Excavation ranges from a few thousand cubic yards of common classification per mile for a "turnpike" road across a flat desert playa all the way up to several million cubic yards of rock per mile for an eight-lane freeway through rugged mountains.

In 1 mi of freeway building through the San Raphael Hills of Los Angeles County, 8,140,000 yd^3 was removed with a maximum 210-ft depth of cut in this all-granite job. In spite of the great depth in this hard rock only about 1.5 million, or 9 percent, of the total 11.6 million yd^3 was blasted.

The magnitude of highway excavation may be inferred from a summary of 506 public and private earth-rock and rock jobs analyzed during the past 20 yr. Of these varied jobs, 301, or 60 percent, were highway building. Since some 1 billion yd^3 made up these 506 projects, there was some 600 million yd^3 of highway excavation, and this represents only the experience of one engineer.

Figure 5-5 illustrates a typical rock job for a freeway of 3,510,000 yd^3, made up of about 1,350,000 yd^3 of granite, 1,424,000 yd^3 of schist, and about 736,000 yd^3 of residuals of granitic and schistose origins. The freeway is located in the mountains of San Diego County, California.

Figure 5-5 Huge saddle cut of freeway construction. Granite rock excavation in San Diego County, California, in 1975. This huge, 1,006,000-yd³ cut is about half excavated. It is 930 ft from top of slope to top of slope and 110 ft in maximum depth. About one-third of the total excavation was ripped to a depth of about 25 ft and hauled by wheel-type scrapers and by rear-dump hauler loaded by wheel-type shovels. The remaining two-thirds of the yardage is being systematically blasted prior to shovel loading. Four track drills are working at the 600-ft-wide elevation. Two 8-yd³ bucket loaders are loading out a fleet of 40-ton rear-dump rock haulers. A rock cut of such size will require about 8 months from the initial pioneering work to the final finishing to rough grade.

Analyses for bid preparations for the contractors disclosed that the rock distribution was in accordance with the following tabulation.

Kind of rock by cuts	Total rock, yd³	Maximum depth of cut, ft	Depth to rock, ft	Cubic yards of rock to be blasted, yd³
Granite:				
Cut 1	344,000	67	25	150,000
Cut 5	1,006,000	110	25	626,000
Schist:				
Cut 2	359,000	56	38	29,000
Cut 3	880,000	72	32	246,000
Cut 4	185,000	39	35	18,000
Residuals (Cuts 1, 2, 3, 4, 5, and 6)	736,000	110	Variable	
Totals	3,510,000			1,069,000

This formidable-appearing excavation with heavy outcrops of granite and schist includes only about 1,069,000 yd³, of "drill-and-shoot" rock, or 30 percent, of the total 3,510,000 yd³.

Among the 301 highway rock excavation jobs studied in the United States during the past 20 yr have been the following two outstanding hard-rock projects involving high percentages of necessary blasting.

1. Santa Susana Pass Freeway in the Santa Monica Mountains of southern California, totaling 8,533,000 yd³. About 5,800,000 yd³ of thickly bedded sandstone of the Chico formation, or about 68 percent of the total yardage, was blasted.

2. Pacheco Pass Freeway in the Coast Range of southern California, totaling 11,190,000 yd³. About 5,100,000 yd³ of heavily bedded sandstones and schists, or about 46 percent of the total yardage, was blasted. By coincidence the bulk of the "drill-and-shoot" rock was in the Panoche formation, also an Upper Cretaceous marine formation of sedimentary rocks correlating with the Chico formation of the

Santa Susana Pass Freeway. The correlation of rock formations was discussed in Chapter 2.

It is revealing that the rock excavations with highest percentages of yardage requiring blasting are in the heavily bedded sedimentary rocks rather than in the igneous or metamorphic rocks. Preexcavation analyses and subsequent excavation experiences show this to be true in all parts of the United States. Of some 1 billion yd³ studied, the three kinds of rock make up these percentages, as detailed in Table 5-1 and summarized below.

Igneous rocks	39%
Sedimentary rocks, thickly bedded	20
Metamorphic rocks	11

The average percentage of "drill-and-shoot" rock in the so-called rock highway job is about 10 percent of the total roadway excavation. This minimum of blasting, as contrasted with the much higher figure of a generation ago, is due to rock fragmentation by modern heavyweight class I tractor-rippers. These ripper- and bulldozer-equipped machines have up to 700-hp, 190,000-lb working weight and 76,000-lb usable drawbar or ripping-shank pull.

PITS AND OPEN MINES

The description of the Bingham Canyon copper mine of Kennecott Copper Corporation, Plate 5-1, dramatizes the biggest rock excavation made by humans. However, on a mundane, less grandiose scale many hundreds of pits and open mines produce a huge variety of building and construction materials and metallic and nonmetallic ores for human needs. These include sands and gravels, bauxite, phosphorite, iron, copper, coal, and a host of other materials and commodities.

Excavation varies from loose sands to ultrahard taconite iron ore. Most of the coals and the ores are beneath overburdens so that there is a dual excavation of waste material and desired product.

The bituminous-coal strip mining of Figure 5-6 illustrates the sequence of overburden removal by casting of the 50-yd³ stripping shovel, coal loading from the vein by the 8-yd³ coal shovel, and coal hauling from the mine to the preparation plant by the 25-ton bottom-dump haulers.

In this central Illinois mine the coal vein is 5 ft thick and the highwall to the right of

TABLE 5-1 Percentages of Kinds of Rocks and Depths to "Drill-and-Shoot" Rock as Encountered in About 1 Billion yd³ of Excavation in 508 Public and Private Works of Excavation

Kinds of rock	Percentage of rock	Depth to rock line, ft
Igneous:	39	
Intrusive, as granite	25	49
Extrusive:		
Thick-bedded, as rhyolite	7	43
Amorphous, as breccia	7	49
Average, weighted		46
Sedimentary:	50	
Nonbedded	9	
Noncemented, as gravel (*)	(5)	(120)
Cemented, as conglomerate	4	52
Bedded:	41	
Up to 1-ft bedding, as shale	6	90
1- to 3-ft bedding, as sandstone	15	71
3 ft and up bedding, as limestone	20	53
Average, except (*), weighted		59
Metamorphic:	11	
Massive, as gneiss	6	41
Laminated, as slate	5	43
Average, weighted		42
Grand total	100	
Grand average, except (*), weighted		55

Figure 5-6 Open-pit areal coal mine. Bituminous-coal stripping mine of central Illinois. The sequence of operations for this typical open-cut mine is as follows.

1. Sixty feet of limestone and shale overburden, previously blasted, is cast by a 50-yd³ stripping shovel from the highwall to the right to the waste area to the left from which the coal vein has been removed.
2. An 8-yd³ shovel loads out the 5-ft coal seam, which may or may not have been blasted, into a fleet of six 25-ton bottom-dump coal haulers. The single-shifting coal-loading shovel follows the triple-shifting stripping shovel.
3. The coal haulers haul on the mucky shale floor of the pit for 0.2 mi and then ramp up to the 0.9-mi main haul road to the preparation plant.

Every day the stripping shovel casts about 38,000 yd³ of overburden for the 3500-ton coal production of the plant.

the stripping shovel is 60 ft high. The overburden mixture of limestone and shale requires blasting and it bears a volume ratio of 12:1 to the coal or a ratio of 10.5 yd³ of overburden per ton of coal.

The stripping shovel cycles in 55 s and its consistent round-the-clock production is about 1800 yd³/h. It is on triple-shift operation so as to cast about 5400 yd³ for every 500 tons/h of single-shift coal production. The coal haulers are equipped with large single tires so as to have flotation and traction in the muck of the pit floor and on the sometimes soft, spongy haul roads. Haul is 1.1 mi, one way, and net cycle time is 11 min. Travel speed during the wet-season operation averages only 13 mi/h loaded and 22 mi/h empty. Sustained production for the fleet of six coal haulers is 3500 tons of coal per 7-h shift. During the dry season the rolling resistance of the coal haulers will be reduced greatly and cycle time will drop down to 9 min.

QUARRIES

The quarries of the United States produce hundreds of millions of tons of rock products annually. The more important nonmetallics from quarry-type operations are asbestos, bituminous rock, feldspar, fluorspar, gypsum, limestone and dolomite, manganese, perlite, rocks for crushed stone and riprap, shale, sulfur, talc and soapstone, and uranium ores.

Quarries vary greatly in size from the little mica quarry producing 50 tons/day to the big limestone fluxing-stone quarries producing 50,000 tons/day on round-the-clock triple shift operation.

A typical medium-size commercial quarry is shown in Figure 5-7. This granite quarry of San Diego County, California produces some 5000 tons of decomposed granite and

Figure 5-7 **Hard-rock quarry.** Quarry operations of a medium-size commercial rock producer. In the foreground are two track drills putting down blast holes in the solid granite. In the far background is an upper face of decomposed and well-weathered granite, which does not require blasting but which can be used for plant products. An 8-yd³ tractor-loader-type shovel is loading out the two 35-ton rear-dump rock haulers. A downgrade haul road leads to the primary crusher. Quarry output averages about 5000 tons of decomposed, semisolid, and solid granite from the quarry every 8-h day. *(South Coast Asphalt Products Company.)*

crushed granite every 8-h day. A unique feature of the operation is that the overburden is largely salable. It is made up of decomposed granite, called "DG", and well-weathered granite. These materials, requiring little blasting, are used for road base and simply as excellent random fill for works requiring more embankment.

The major use of the crushed rock is for two concrete plants and for select base materials for road building. One plant is the asphaltic cement concrete plant owned by the company, and the other plant is the Portland cement concrete plant owned by an associated company.

Crushed rock is sold to concrete plants without owned sources for aggregates and to a variety of industries requiring decomposed and crushed rock for road building and for the concrete of home and industrial buildings.

The quarry is located in the same tonalite granite formation as the abandoned quarry of Plate 3-1. Because of more favorable economic conditions and because of perhaps better planning, the present quarry has reached its present stage and it will continue into greater development and rock production.

Figure 5-8 shows the complex of material-handling machinery and plants that process the rock from the primary crusher.

RAILROADS

Because of the generally maximum 1 percent grades of railroads, which call for relatively deep cuts as contrasted with maximum highway grades of about 6 percent, the ratio of rock excavation to common excavation is high in hilly and mountainous country.

Such a new railroad is the Palmdale to Colton cutoff of the Southern Pacific Company across the Mojave Desert and through Cajon Pass of southern California. Its length is 78 mi, of which 24 mi are in the formidable-appearing rocks of Cajon Pass. From north to south through the pass these rocks are sedimentary shales, conglomerates, and sandstones, igneous granites, and metamorphic schists.

The total excavation was 6 million yd³ of which 1.6 million was in the Cajon Pass rock area. In spite of the array of imposing formations, the rocks were well-weathered and the rock requiring blasting did not exceed 50,000 yd³ of the most question-

Figure 5-8 Sequence of quarry work. Sequence of quarry-to-plant rock excavation and distant plant. *Center:* One of two drills putting down holes for blasting. *Foreground:* Blasted granite from the last shot of 32,000 tons. *(Right)* Loading out one of two 35-ton rear-dump rock haulers. *(Left)* Section of the 1600-ft well-maintained haul road with 11 percent downgrade to the primary crusher. *Far background:* Complex of primary crusher, secondary crushers, screening and conveyor equipment, asphaltic cement concrete plant, and Portland cement concrete plant of the efficient multiproduct industry of approximately 5000-ton/day capacity. *(South Coast Asphalt Products Company.)*

able 567,000 yd³ of the Cajon Pass excavation. This 567,000 yd³ was within the rift zone of the San Andreas Fault, and unquestionably the age-old weakening effect of the fault action resulted in minimum "drill-and-shoot" rock. This subject is discussed in Chapters 3, 4, and 5.

Figure 5-9 pictures the rugged nature of the east escarpment of Cajon Pass. The new line leaves Cajon Creek and ascends a steep 5-mi grade to the summit, whence begins the descent to the Mojave Desert.

In 1965 the Santa Fe Railway Company's relocation of its main line around Galisteo Creek Dam, New Mexico, required 900,000 yd³ of rock excavation in 4 mi. Cuts were up to 60 ft in depth. The excavation was in sedimentary rocks made up of heavily bedded Navajo sandstones, medium-bedded Dakota sandstones, and Pierre shales.

About 330,000 yd³ of sandstone, 37 percent of total yardage, was blasted. The rock was loaded by two 6-yd³ wheel-type tractor-shovels and hauled by six 18- to 32-ton rear-dump rock haulers. Residuals and weathered rock were ripped by two medium-weight tractor-rippers, push-loaded by two heavyweight push-tractors, and hauled by six 40-yd³ wheel-type scrapers.

Both percentage of "drill-and-shoot" rock and machinery types were in contrast on these two railroad rock jobs in formidable mountainous areas.

The Cajon Pass job of the Southern Pacific Company required but 9 percent of the questionable yardage to be blasted and all excavation was handled by conventional tractor-rippers and wheel-type scrapers. On the other hand, the excavation of the Galisteo Creek area required considerable blast-hole drilling machinery and both push-tractor–scraper spreads and shovel–rear-dump hauler spreads of machinery.

All the rock of the Santa Fe Railway Company job was questionable and, proportionately, four times as much rock was blasted.

SANITARY FILLS

Hundreds of millions of cubic yards of both common and rock excavation are moved annually in the building of sanitary fills, the euphemistic term for the old trash dumps. These "cut-and-cover" fills are of two types, those in level country and those in hilly or mountainous country. They differ, not in the manner of building the fills, but rather in the methods used for excavation.

Figure 5-9 **Railroad building.** Parallel relationship of the new Southern Pacific Company line to the existing Santa Fe Railway Company line through Cajon Pass of the San Bernardino Mountains of southern California. The new line is to the left or north and follows through the area which has been pioneered with access roads and cleared and grubbed of chaparral. Cajon Creek is in the middle of the picture at elevation 3680 ft and the summit in the far distance is at elevation 4260 ft. In this approximately 5-mi section of line the ruling grade is 2.2 percent. The rock east of Cajon Creek is the Crowder formation of soft friable fanglomerates, conglomerates, and sandstones, requiring only soft ripping by heavyweight tractor-rippers for all excavation. The rock west of Cajon Creek, in the foreground, is the Punchbowl formation of massive shales, conglomerates, and sandstones, requiring medium to heavy ripping. Southward are several sidehill cuts in granite and schist with a maximum centerline cut of 102 ft, in which only about 50,000 yd³ of hard rock was blasted. *(G. Sheley (ed.), "Building 78.4 Miles of Railroad,"* Western Construction, *April 1967, p. 71.)*

Figure 5-10 shows a dump in level country near Long Beach, California. Fill material, consisting of clays, silts, and sands, is taken from a wide, deep trench by push-tractor-loaded wheel-type scrapers and hauled upgrade to the fill area, where it is bulldozed in a 1.5-ft layer atop the 4-ft compacted layer of trash. Crawler-tractors serve in both bulldozing and compaction operations.

The two 21-yd³ wheel scrapers in Figure 5-10 are hauling about 300 yd³/h of cover material, or enough for about 3600 tons of trash daily. Accordingly, this dump will require some 600,000 yd³ of common excavation annually. Level-country dumps seldom involve rock excavation because they are usually in residuals and alluvia.

After this long trench is excavated, the accumulated trash atop the high wall will be bulldozed into the trench and then another sequence of alternate layers of cover and trash will be added until natural ground level and perhaps a higher elevation are reached. Lines and grades and construction methods for the fill are controlled closely.

As contrasted with the Long Beach sanitary fill in level country, the dump of the Santa Susana Mountains, some 35 mi to the north, calls for the leveling off of siltstone-sandstone ridges for the cover material. The sedimentary rock formation is well weathered and requires only soft to medium ripping by medium-weight tractor-rippers.

Figure 5-11 shows a medium-weight bulldozer cascading cover material to the fill some 80 ft below. This dump, much smaller than the Long Beach fill, requires only three medium-weight combination bulldozers and tractor-rippers for all the work, as the dozing distances on the fill are generally not great. When distances are too long to be economical, a crawler-tractor-drawn scraper is put into service.

About 1000 yd³ of cover material per 8-h day is sufficient for the some 1500 tons of trash handled daily. In this dump the three crawler-tractors handle both the excavation of the cover material and the trash compaction work in the fill.

Figure 5-10 **Excavating cover material for a sanitary landfill.** Sanitary fill construction in level country. Silts and sands are being excavated from this wide, deep trench for cover for the "cut-and-cover" dump. At the top of the high wall is a pile of accumulated trash. Beyond this pile the alternate layers of 4 ft of compacted trash and 1.5 ft of cover are being placed. After the trench is excavated to grade, it will be filled with the alternate layers of trash and cover and the cover will come from another trench parallel and adjacent to it. In large cities there is about 1 ton of trash annually for every person. A ton of trash compacts to about 2.5 yd³ and requires about 0.6 yd³ of excavation. Accordingly, this Los Angeles County sanitary fill can accommodate about 1 million population.

Figure 5-11 **Delivery of cover material for a sanitary landfill by gravity.** Bulldozing cover material of siltstone-sandstone from a ridge down to a sanitary fill. This dump is located in the Santa Susana Mountains of Los Angeles County, California. Construction methods for the alternate layers of trash and cover are the same as those for an operation in level country, as only the sources of the excavations differ. After a fill is completed and landscaped, the area is generally used for recreational purposes, for example as a park or small golf course. No buildings of more than one story are built because of the low density of the fill. However, a generation of building controlled sanitary fills indicates that the fills are remarkably stable.

Both sanitary fills are privately owned and both are controlled by the governmental body with local jurisdiction over methods. Such strict controls almost always prevent three objections to dumps—odor, rats, and spontaneous combustion. All are eliminated by compacting the alternate layers of trash and earth or rock to as high a density as possible. On compaction 1 yd^3 of loose trash becomes ½ yd^3, and 1½ ft of loose cover becomes a 1-ft thickness.

TRENCHES AND SMALL-FOUNDATION EXCAVATIONS

Trench excavations vary greatly in both quality and quantity. A trench for a telephone conduit in loam may be only 18 in deep and 12 in wide, resulting in soft excavation and only 6 yd^3 per station of 100 ft. Such excavation is an easy job for a small sidecasting vertical-boom trencher.

On the other hand a trench for a 6-ft-diameter concrete storm drainpipe in rock country may be 18 ft deep and 12 ft wide, calling for blasting and a minimum of 800 yd^3 per station. And, if the worker handling the powder is not diligent and allows too much backbreak from the charges, the yardage could well run to 1200 yd^3 of rock per 100 ft of length. Such a trench is a hard excavation job for a large, full, revolving 2½-yd^3 backhoe even after good fragmentation.Common trench excavations include those made for the following applications:

1. Government, commercial, and home utilities such as water, gas, electricity, telephone, sewers, and storm drains.
2. Large commercial pipelines such as those used for water, gas, oil, and coal transportation.
3. Farm applications, such as surface-ditch irrigation and drainage and subsurface pipe drainage.
4. Highway construction, such as side and lateral ditches, culverts, and diversion ditches.

Many other kinds of trenches are excavated as a part of the many types of open cuts described in this chapter.

Figure 5-12 illustrates vividly a particularly tough trench excavation. It is part of the San Bernardino Valley Municipal Water District's master distribution system for this California county.

The heavy fanglomerate material is described as "85 percent over 8 inch mechanically locked material." This notation is equivalent to saying that the alluvium is a high-density, tightly cemented conglomerate with up to boulder-size particles. The fanglomerates from the San Bernardino Mountains cover great areas and they extend to considerable depth.

The trench through this troublesome formation was 22 ft deep and 13 to 25 ft wide. The length was 6 mi and so some 500,000 yd^3 was excavated at the rate of about 80,000 yd^3 mi.

The excavation methods were changed occasionally during the work, but they were essentially as follows:

1. The first few feet of depth being characteristically gravels and cobbles, this zone was taken out by a 2½-yd^3 dragline assisted by two medium-weight bulldozers to a depth of 6 ft.
2. A 2-yd^3 backhoe then excavated to an average depth of 20 ft. In this zone oversize boulders were drilled by jackhammers and blasted.
3. To provide for bedding material beneath the 78-in-diameter steel pipe, the backhoe excavated an additional 8 to 12 in below grade.

This trench is an extremely rough example of excavation in cemented conglomerates made up of outsize fanglomerate boulders.

Analyses of many open-cut rock excavations warrant valuable conclusions as to kinds of rock and depths to "drill and shoot" rock in the several classifications. These data are as given in Table 5-1.

The table is based on 506 projects of public and private excavations for which complete seismic studies were made before excavation and in which there were many followups of the experiences of the earthmovers and seismic studies of the work in progress. These correlations during the actual work indicate that the excavator in rock country com-

Figure 5-12 **Trench excavation for main pipeline of water system.** Tough trench excavation in tightly cemented boulder-size fanglomerate of the San Bernardino Mountains, California. After blasting of the largest of the embedded boulders, the remaining 16 ft of the 22-ft-deep trench is being excavated by the 2-yd^3 backhoe. The bulldozer in the trench is leveling bedding material for the 6½-ft-diameter steel pipe. The top 6-ft-deep zone of up to cobble-size alluvium was excavated by a 2½-yd^3 dragline assisted by two medium-weight bulldozers. About 0.5 million yd^3 of this "mechanically locked" fanglomerate was excavated in 1975. *(San Bernardino Valley Municipal Water District.)*

mences to blast the rock when the shock-wave velocity gets up into the range of 6000 to 7000 ft/s.

Inferences from the table that are most significant to the rock excavator are:

1. Minimum depth to "rock line" or depth to blasting below original ground is in the metamorphic rocks, averaging 42 ft.
2. Intermediate depth is in the igneous rocks, averaging is 46 ft.
3. Maximum depth is in the sedimentary rocks, averaging 59 ft.
4. Depths in the sedimentary rocks vary inversely according to thicknesses of bedding.

Up to 1-ft thickness of stratum	90-ft depth
1- to 3-ft thickness	71-ft depth
3 ft and greater thickness	53-ft depth

5. The average depth to "drill-and-shoot" rock in all kinds of rock is 55 ft.

Of course, the tabular data are averages and as averages they are to be properly interpreted. One knows that the "rock line" is of variable depth even in the same kind of rock and in the same rock formation of the same job, let alone in like rocks and formations of different jobs. Nevertheless, the data are valuable generalizations.

The master curves from which the table derives show that the average 55-ft-deep "rock line" varies from 27- to 111-ft depths, or from −51 percent to +102 percent of the average value. This fact lends emphasis to the inconsistency of the weathered zone of the Earth's regolith and the urgency for the earthmover to learn all about the rocks and the rock formations before commencing rock excavation in open cuts.

SUMMARY

In this chapter a few examples of the nine principal kinds of excavation have been given and these may be called medium and large projects. Of course there are many kinds of small projects involving excavations, such as those for a swimming pool or for a small irrigation ditch on a 160-acre farm. Unfortunately there is no summary of total annual excavation in the United States, but it is probable that it is some 2 billion yd^3 bank measurement. Consequently, excavation is a big business.

CHAPTER 6

Examination of Excavation

PLATE 6-1

Rugged terrain for the examination of rock excavation. Topography of a four-lane freeway through the Antelope Valley of southern California. The rock excavation includes 7.7 million yd³ in 10.8 mi of mountains and valleys. Nine principal cuts are typified by the cut of Plate 6-1, in which a cross section of the cut and the centerline of the freeway are delineated. The cut is 907,000 yd³ with a maximum depth of 130 ft, and it is a symmetrical through cut with two 1100-ft benches on each side. Rocks include massive sandstones of the Vasquez formation and medium-bedded sandstones and conglomerates of the Mint Canyon formation. There are minor interbedded basalts in the sedimentary rocks of the Vasquez continental formation. In this cut 565,000 yd³, or 62 percent of the total roadway excavation, was blasted, the rock line being about 40 ft below natural ground. For the entire job 5.8 million yd³, or 75 percent, was ripped and 1.9 million, or 25 percent, was blasted.

The shaggy terrain accentuates the problems of walking the centerline of mountain jobs, but it also emphasizes the desirability of field reconnaissance. Such were the experiences and the policy of the efficient excavator of the large rock job of 1960 to 1962. *(California Division of Highways.)*

For both practicability and efficiency the examination of excavation in the office and in the field should precede the exploratory work in the field. Exploration of the excavation is the subject of Chapter 7.

There are two phases of examination, to be conducted in the following order:

1. Office work in terms of securing and analyzing topographical and geological data and historical excavation experiences in the area of excavation.

2. Field work in terms of "walking centerline" with literal down-to-earth examination of the rock job.

The office work should precede the field work because, metaphorically, it is best to view the forest before seeing the trees. A typical office examination of a 17.9 million-yd^3, 12-mi freeway excavation located in the mountains is used as an example.

OFFICE WORK

1. Analyses of plans and specifications provide a general outline of the examination. Cuts and fills are then outlined on detailed plan sheets, together with notations of yardages, depths of cuts, and height of fills.

Cuts and fills are then transferred to a master plan sheet or title sheet showing the entire job. Usually this master plan sheet is a reproduction of a United States Geological Survey (USGS) map of either the 7½-min series with a scale of 2000 feet to the inch or the 15-min series with a scale of 1 mile to the inch. The large-scale, 7½-min-series map is preferable because of greater detail and ease of drafting work on the map. However, since master plan sheets are generally reduced in size, the scale of the sheet may very well be an odd one and it may not be scaled precisely with the regular engineer's scale. In such a case the use of proportional dividers will solve the problem of transfer of distances to the odd-scale map.

2. The master plan sheet is now transferred to the topographical USGS map. If available, the 7½-min-series map should be used, and this series of maps is generally available except for remote areas of the United States.

Now it is possible to visualize the job by means of this topographical map. One sees clearly the relationships of centerline and cuts and fills to the mountains, valleys, and streams. Relationships are established also to the works of humans, such as roads, railroads, towns, dams, canals, mines, quarries, pipelines, electrical transmission towers, and the like.

3. The final step on the working plan sheet is to transfer the local geologic map to the sheet so as to divide up the map according to kinds of rock and rock formations. In keeping with the purpose of examination of rock excavation, step 3 is the most important.

Generally one or more geologic maps are available for the area from the following sources:

1. County: Engineering and environmental departments in the county seat.
2. State: Divisions of geology, public works, and/or environmental engineering in state capitals.
3. Federal: USGS in Washington, D.C., as well as other offices in principal cities.
4. State tourist information bureau in the state capital.
5. National Park Service, Washington, D.C.
6. Geological Society of America, Boulder, Colorado.
7. American Association of Petroleum Geologists, Tulsa, Oklahoma.
8. Libraries of the larger U.S. cities.

Scales of geological maps range from 50 miles to the inch to 2000 feet to the inch. Again, one should use the largest-scale map available so as to ensure precision. When the scale is small, as in the maps of the American Association of Petroleum Geologists, the small area can be enlarged for greater accommodation.

Precision is especially important when there are several rock types and different formations in the excavation area. The maps usually are divided according to the geologic series of rocks but sometimes they are divided into formations, corresponding to the fourth division of geologic time.

Generally the maps are accompanied by a description of the kinds of rocks in the series or formation. This information enables the engineer to anticipate what rocks will be met with in the field reconnaissance and in the rock work.

As has been mentioned, when transferring from the plan sheet to the USGS topographical map and from the local geologic map to the same topographical map, it is efficient and precise to use proportional dividers rather than the engineer's scale unless the scales happen to be the same. Of course all completed originals should be copied for use in the field.

Figure 6-1 shows the transfer of the centerline of the freeway to the geologic map of the Cajon Pass, the location of the rock excavation. From this map, which shows the distribution of the rock formations, one can now transfer the contacts between the rock formations to the map for field reconnaissance shown in Figure 6-2.

Figure 6-2 shows the working map or plan for the examination of the freeway. Only the middle section, about 4 mi long, is portrayed. The centerline on either side of the San Andreas Fault includes the rift zone of the fault. It includes 9,850,000 yd^3 of the roughest excavation in the 17,900,000-yd^3 job.

The rough topography of Cajon Pass is accentuated by the 3100-ft high elevation of the 3,471,000-yd^3 cut between stations 186 and 214 and the 2700-ft low elevation of the fill between stations 214 and 224. From top of cut to bottom of fill is a 400-ft vertical drop and 1400 ft laterally, and this will make for difficult pioneering of the excavation. Scrutiny of the master plan sheet, Figure 6-2, discloses other aspects of the field reconnaissance and subsequent field exploratory work.

1. The centerline of the rift zone of the San Andreas fault crosses the freeway centerline at station 235 in the center of a 1,761,000-yd^3 cut of conglomerate-breccia. This longtime faulting accounts for the breccia. The conglomerate is really a fanglomerate from the San Bernardino Mountains to the east.

2. If, as supposed, the rift zone extends 1 mi or so on both sides of the axis of the fault, then one may expect weathered rock formations from at least station 182 to station 288. An investigation of earthquake activity in Cajon Pass during the recorded history of the last century disclosed the following disturbances:

a. Old Cajon Pass Road, which followed approximately the present alignment of U.S. Highway 66, was covered by slides and debris by a violent earthquake in 1899. Its intensity on the Mercalli scale was 8 to 9, equivalent to 7+ on the presently used Richter scale.

b. In 1907 there was another powerful quake in the pass. Its magnitude was 6 and it was accompanied by heavy slides in the area of the freeway.

While the recorded seismic activity is less than a century old, one must keep in mind that the San Andreas fault has been active for over 150 million yr. Accordingly, the rift zone has been weakened continually for aeons and it is still being weakened.

These studies made prior to field reconnaissance suggested exactly the findings of the seismic studies for the excavation and the experience of the earthmover when the rock excavation was taken out. There was medium and hard ripping in the rift zone but there was no blasting of roadway excavation.

It is probable that more seismic studies of excavation were made on this freeway than on any other similar rock work in the United States. The California Division of Highways made available to the bidding contractors 39 studies and the writer made 28 studies. Total studies were 67, or one study for every 147,000 of the 9,850,000 yd^3 of questionable rock work.

3. Judging by the paucity of trails and roads through the work area, field reconnaissance must involve jeep travel and considerable hiking over the rough country.

4. The high-walled sidehill cut along the existing Highway 66 at Blue Cut should give indications of the nature of the 3,471,000-yd^3 schist cut to the southeast and on a bearing paralleling the fault. Overbend excavations for the pipeline near the freeway will be indicative of the excavation to the same depths. Likewise, data on the tunnel work to the east of station 237 will be helpful.

5. An important facet of the office work is the correlation of the rock formations of the freeway with the same formations with which the earthmover is familiar. This association of formations has been discussed in Chapter 2. Proceeding upstation, the cuts may be correlated in this manner.

Examination and exploration of the rock excavation for the Cajon Pass Freeway took place prior to and during 1966. The work was done during the 1967–1969 period.

The schist cuts are between stations 224 and 263. This same Pelona schist in sizable cuts

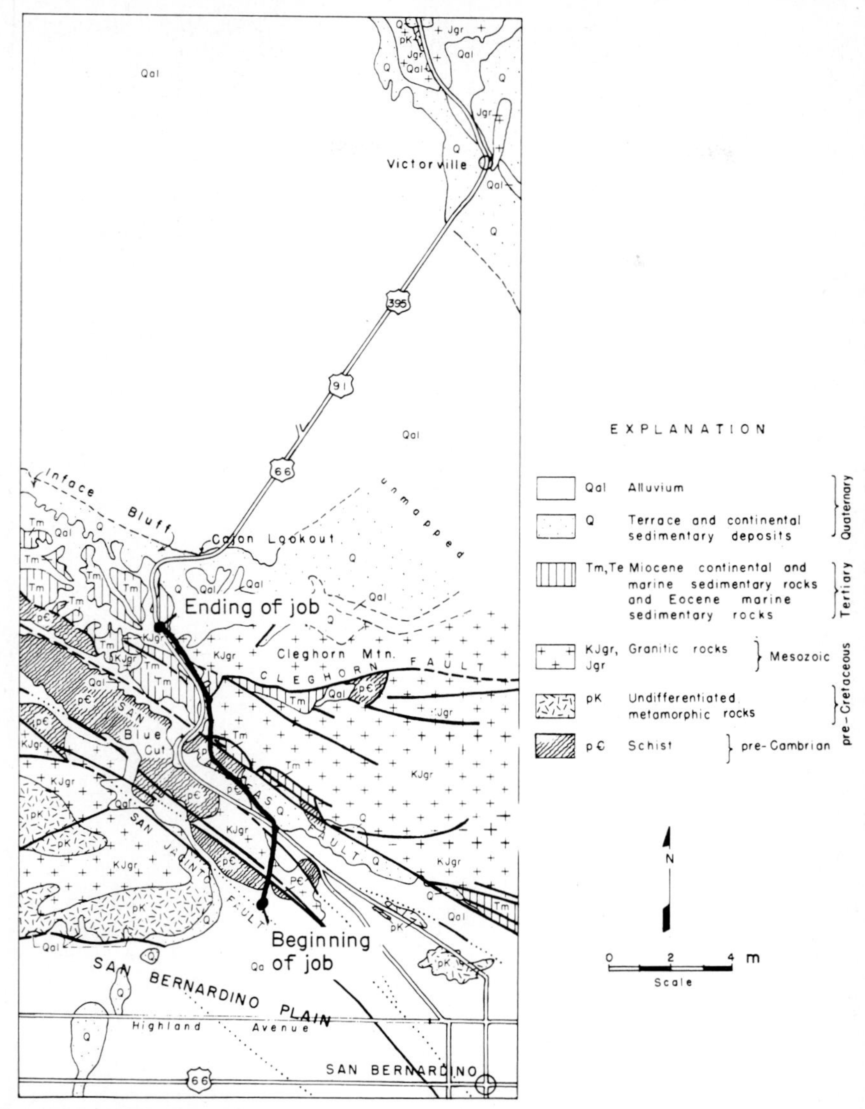

Figure 6-1 Geologic map of the Cajon Pass area, San Bernardino County, California. Centerline of entire Cajon Pass Freeway is superimposed on the geologic map of the area. Areal map shows the distribution of the rock formations. *(Olaf P. Jenkins,* Geology of Southern California, *California Division of Mines and Geology, Geologic Guide no. 1, 1954, p. 48.)*

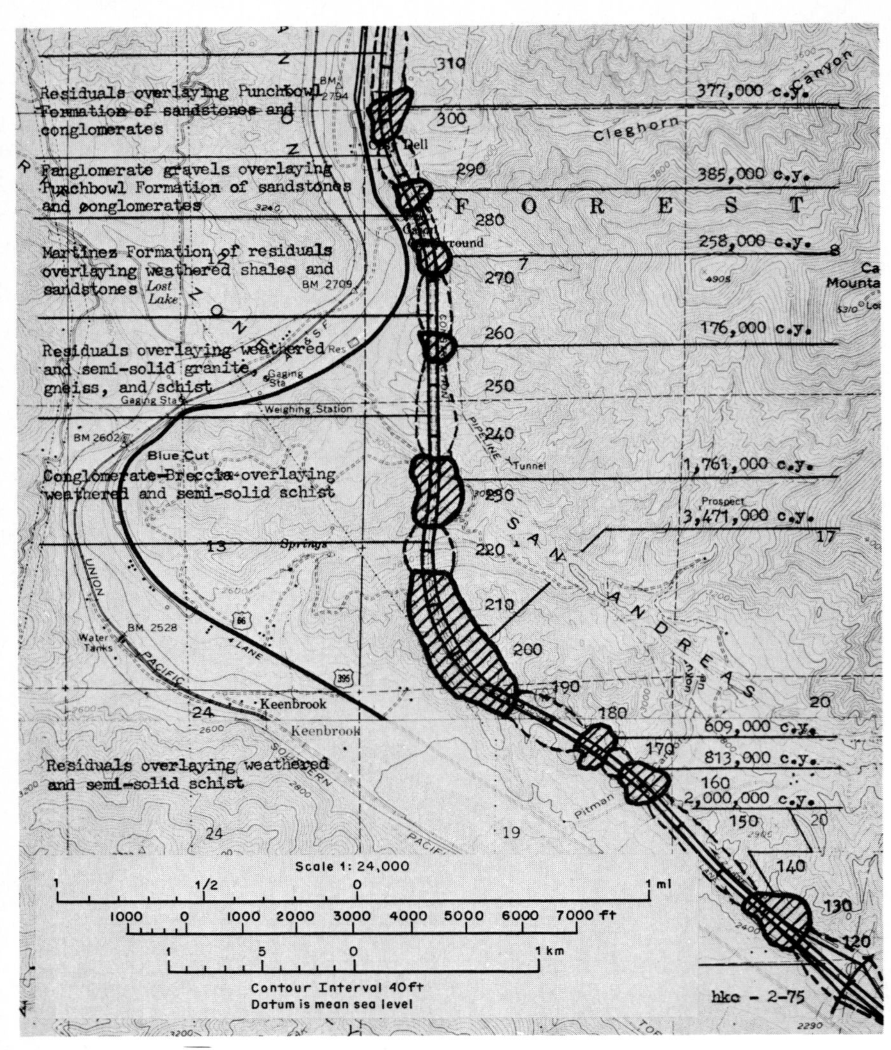

Figure 6-2 **Map for field reconnaissance of Cajon Pass Freeway of southern California, middle section.** Station 115 to station 318. 9,850,000 yd³ of roadway excavation. *(U.S. Geological Survey, Cajon 7½-min Topographic Quadrangle Map, 1968.)*

was excavated just a few years previously in the Bouquet Canyon Road across the San Andreas fault rift zone just west of Palmdale, California. There was no blasting, as the weathered rock was ripped by tractor-rippers of less weight and horsepower than those of 1966.

The 258,000-yd³ cut is in the Martinez formation of sedimentary rocks. About 1961 up to 145-ft-deep cuts were taken out by hard ripping on Highway 150 through San Marcos Pass, Santa Barbara County, California, in the same Martinez formation.

The two northern cuts, totaling 762,000 yd³, are in the Punchbowl formation of the Cajon beds. Unfortunately there is no correlation for these undivided or unclassified Miocene continental sedimentary rocks in terms of excavation experience. It is known, however, that the soft though massive Miocene rocks are rippable to a depth of 100 ft.

The lower and upper sections of the Cajon Pass Freeway, omitted in this discussion, are similarly correlated with recent work by machinery of the past 5 yr.

Generally an office appraisal of the job would include the history of the excavation for the existing U.S. Highway 66 and for the Southern Pacific Railroad to the east across Cajon Creek. However, the highway and the railroad were built several decades ago when rock ripping was not in vogue and fragmentation was by explosives. Consequently, an investigation of the excavation methods was considered to be not helpful.

In large public works the contracting officer or agency usually supplies the bidders with rather complete materials and soils data as part of the specifications, special provisions, and plans. This valuable information may include logs of test hole borings, as well as cores if obtained, exploration trench or open-pit and quarry data, and surficial geologic rock-type and formation data.

Sometimes in large private works the builder may supply similar information, although usually not as complete as that of the agencies. This collective material on rock excavation results in a better understanding of the job and better balanced bidding. Four examples are explained below.

1. Figure 6-3 is a portion of a standard soil survey sheet for a mountain highway in the Sierra Nevada of California.

Test holes 4, 5, and 6 were drilled to below grade in the 160,000-yd^3 cut between stations 567 and 585. All three logs record easy drilling for the auger drill in the well-weathered volcanic rock. This information was augmented by private seismic studies of the job. Studies 1 and 2 at stations 572 and 576 showed up to 5300-ft/s shock-wave velocities to 40-ft depth and 9200 ft/s to greater depths. These studies supported the test hole data.

Since the maximum centerline cut was 34 ft, the excavation was fragmented by methods ranging up to hard ripping for a heavyweight tractor-ripper.

2. Figure 6-4 is part of a geologic plan and profile for a reach of the California Aqueduct through the northern foothills of the San Gabriel Mountains of southern California.

The plan of areal geology shows the 238,000-yd^3 cut between stations 1852 and 1858 to be in Pelona schist, an old metamorphic rock the excavation history of which is well known. Rotary-core-drill hole 33 was put down to 25 ft below the canal invert elevation in the 95-ft-deep cut. Cores from the cut were available to the bidders. A private seismic study, 23, shows velocities and velocity interfaces and it suggests rippability to grade. Bucket-drill hole 34 to a depth of 43 ft was put down through weathered schist to 10 ft below grade, where refusal was met. The California Department of Water Resources' seismic profiles SP-68 and SP-67 are plotted in the smaller cut between stations 1863 and 1869.

There is good conformity between the agency's data and private seismic work. The rock was fragmented by methods ranging up to hard ripping for a heavy-weight tractor-ripper.

3. Figure 6-5 is part of the geologic map for the spillway of New Melones Dam on the Stanislaus River of Central California.

These some 15 million yd^3 of complex metamorphic rocks were investigated thoroughly by agency and private means. The agency means, as delineated on the complete map for the 6000-ft-long, 400-ft-wide, and maximum 237-ft-deep spillway included core drill holes, exploration drifts, and a sizable, 32-ft-deep test quarry. As is customary, core samples were available to the bidders.

Thirteen private seismic studies of the spillway rock excavation were made for the several bidders. Averages indicated a rock line at 6-ft depth, where a 6000-ft/s shock-wave velocity changed to 15,100 ft/s. The contractor's ripping and blasting experience corroborated the inferences of the agency and private investigators.

4. Several years ago a huge tract for homes in the San Raphael Hills of Los Angeles County, California was appraised for the granite rock excavation of 1,943,000 yd^3. There were four principal steps in the investigation.

a. Test holes of 4-in diameter were drilled by track drills to a maximum 70-ft depth of cut. The test holes evaluated the hardness of the rock, as well as determining the important drilling speeds according to the depth of drilling in the event that blast-hole drilling should be necessary.

b. Uphole seismic studies were made in the eight drill holes, thus establishing shock-wave velocities in the several zones of depth.

c. Twelve refraction seismic studies were run so as to cover the entire excavation area down to the grade of the finished cuts.

d. The foregoing explorations provided an estimate that 198,000 yd³, or 10.6 percent of total yardage, would require blasting. All data were made available to the bidders and consistent, reasonable bids were obtained.

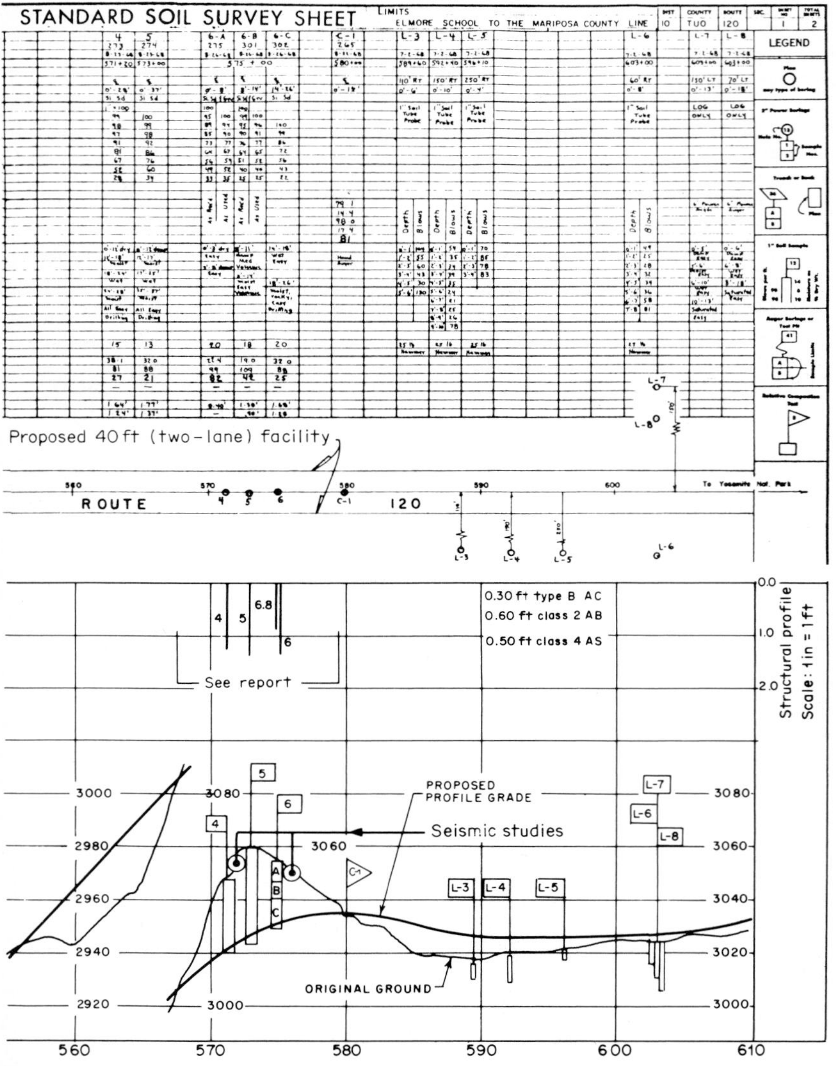

Figure 6-3 Route 120 Highway—FH-39-151 Mariposa and Tuolomne Counties, California. *(California Division of Highways.)*

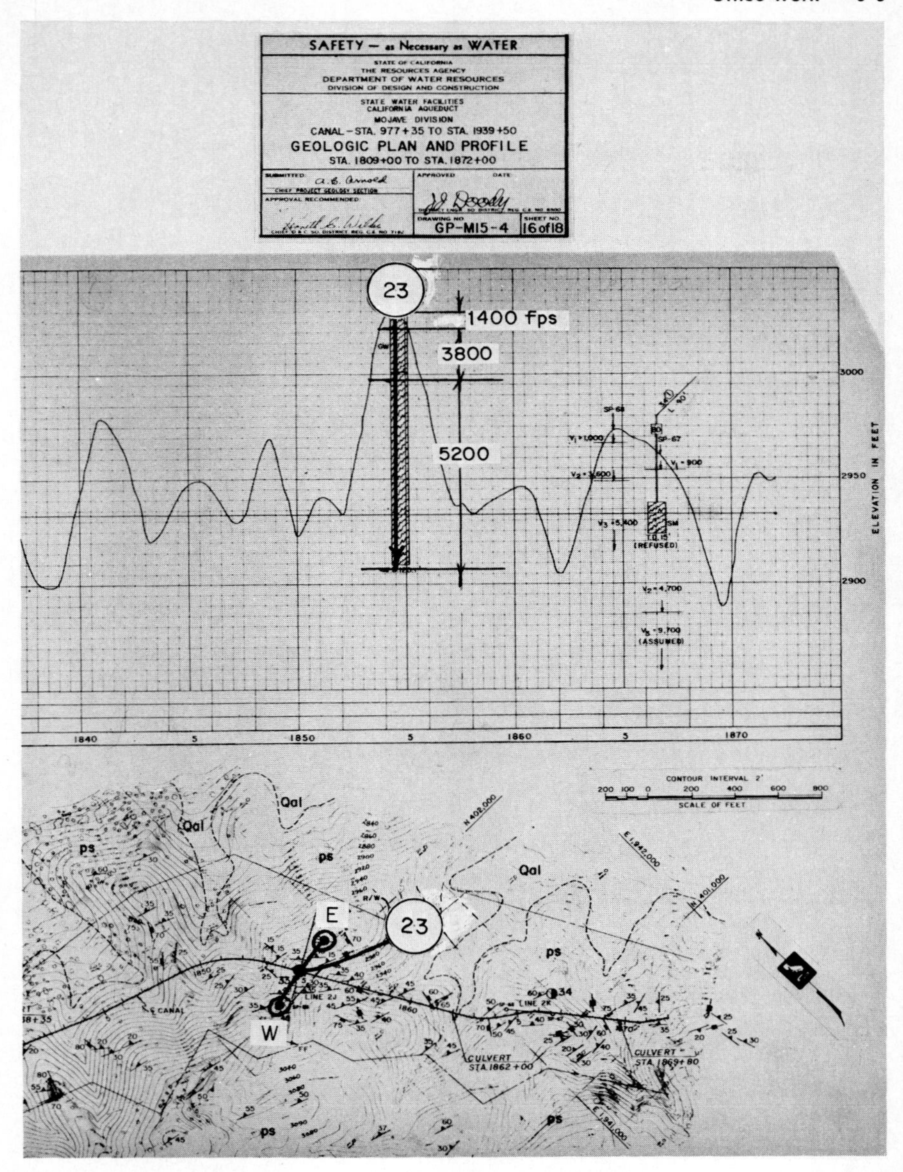

Figure 6-4 **Section of Fairmont–Leona Siphon reach, California Aqueduct.** Significant symbols of area of illustration: (Qal) alluvium of Quartenary system; (ps) Pelona schist of Precambrian group; (RD) test hole by rotary core drill; (BDP) test hole by bucket drill; (23) seismic study or profile by the writer; (SP-67, SP-68) seismic profiles by Dept. of Water Resources. *(California Dept. of Water Resources.)*

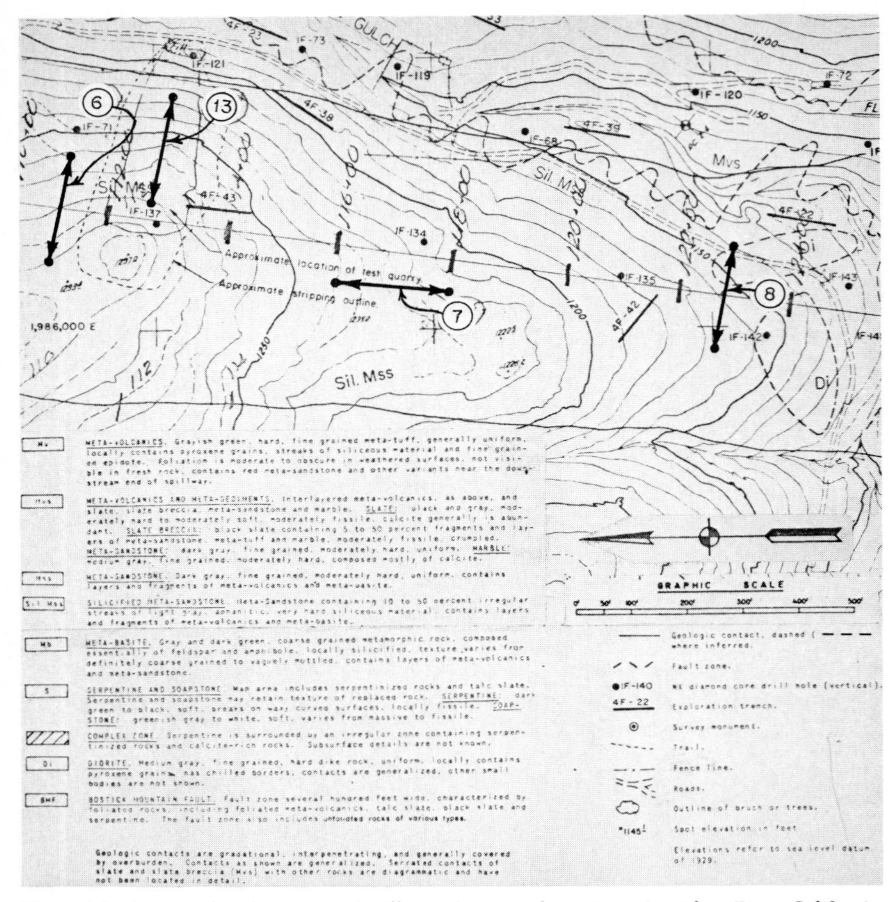

Figure 6-5 Section of geologic map of spillway of New Melones Dam, Stanislaus River, California. Four of the thirteen private seismic studies of the 15-million-yd³ spillway rock excavation are shown. Symbols of the complex rock formations include all within the work for the spillway. Not all are represented in this spillway section. *(U.S. Bureau of Reclamation.)*

This description of office work highlights the necessary prelude to field reconnaissance. A story of field work for another job emphasizes the importance of this correlated office and field effort.

FIELD WORK

Equipment

As in the case of office work, a worker is as good as the available tools, and for "walking centerline" the following items of equipment are suggested.

1. Brush-turning, hard-texture, loose-fitting clothing suitable for the prevailing weather conditions. In rainy weather "breathing" rain gear is important rather than the sweat-producing and ventilation-lacking clothing made from synthetic fabrics. If brush is thick and if rocks are sharp, let alone if barbed-wire fences are encountered, horsehide gloves should be worn both summer and winter. Boots should be waterproof high cuts, preferably engineer's type boots with over-ankle strap and relatively smooth composition

soles. Lace boots are time-consuming to put on, the hooks get caught in brush and barbed wire, and the boots are relatively heavy.

2. A light pack, of the heavy-duty Boy Scout type, for carrying plans and equipment. Shoulder straps should be broad and there should be a belly band. One's hands should be free when hiking over rough country and climbing sharp and slippery rocks.

3. Lunch pail. The regular plastic pail with 1-pt vacuum bottle is excellent.

4. Two canteens are advisable. In desert hiking with greater than 100°F temperatures, one can consume the water of two canteens in 4 h. The general-issue or "GI" English quart size is good. One of these should be worn on the belt for convenience.

5. Plans, specifications, and master plan sheet should be placed if possible in a three-ring binder so as not to be defaced.

6. Field notebook, slide rule or pocket computer, and engineer's scale.

7. Special equipment. These items are well nigh indispensable.

a. Camera, instant type.
b. Binoculars. Wide-angle and seven-power magnification.
c. Engineer's tape. Nonmetallic of 100-ft length.
d. Clinometer. Military pocket-type, reading in percentage of grade or inclination.
e. Pocket altimeter. Precision to ± 10 ft of elevation.
f. Compass. Boy Scout pocket-type.
g. Magnifying glass. Pocket-type.
h. Geologist's hand pick. Belt-type.
i. Small first-aid packet and snakebite kit.
j. Revolver, if in order, as protection against "man and beast."

These items will give a pack weight of about 15 lb, and the hands are free for the roughest of hiking.

In addition to the "tools of the trade" the engineer must consider his safety, especially in remote mountain and desert areas. Under extreme conditions it is best to have a fellow hiker in the event of such potentially fatal mishaps as broken limbs and heat prostration. If it is necessary to hike or even travel by jeep alone, always tell associates and the hotel manager what plans are afoot for the day.

Purpose of Field Reconnaissance

The purpose of field reconnaissance is to find out everything possible that relates to the rock excavation. When the entire construction or mining job is considered, quite naturally other matters pertaining to the entire work must be investigated. In this discussion only relevant phases of the excavation are considered. These items include the following major operations:

1. Pioneering and opening up of the job.
2. Clearing shrubs, chaparral, and trees to the lines of plans.
3. Excavation.
4. Haulage of excavation.
5. Fill operations, including placement and compaction.

Related to these five items are such operations as these:

1. Borrow pit and waste area locations.
2. Development of water supply.
3. Building of access roads, sometimes over privately owned lands.
4. Establishment of offices and machinery yard.
5. Relationships with property owners who may or may not have concluded negotiations with the contracting officer or quarry or open-pit area owner.

Thus the field operator must look to many job factors when making a survey. These factors are discussed below in relation to a mountain highway job.

Figure 6-6 is part of a plan–profile sheet for the Cleghorn Road to Crestline Highway on the north slope of the San Bernardino Mountains of southern California. Field investigation of this job, combined with seismic studies of the roadway excavation, was made between February 4 and 11, 1969. It was a rainy, snowy winter period and no field work could be done except during three days of the period, as the average elevation was 3600 ft at the rain-snow line.

Ten seismic lines were run, requiring 10 h, so that the hiking and reconnaissance work

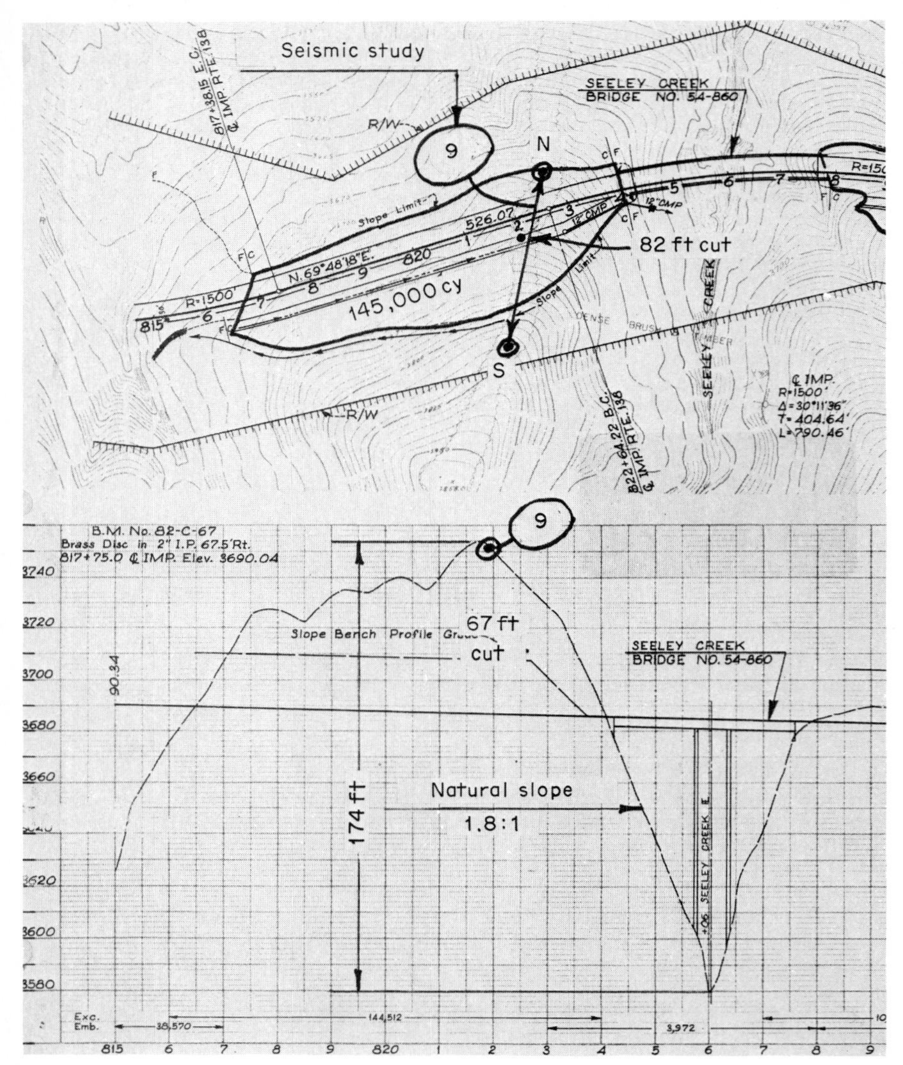

Figure 6-6 **Plan and profile, station 815 to station 829, Cleghorn Road to Crestline Freeway,** San Bernardino County, California. *(California Division of Highways.)*

required about 20 h for the 5.1-mi job. As shown by Figures 6-6 and 6-7, topography was rough and weather was bad. There was no access by jeep, so that total hiking was about 15 mi, including twice the length of centerline plus lateral distances from the existing highway. This highway was 0.3 mi to the north along the East Fork of the Mojave River and about 300 ft below the elevation of the freeway. Hiking for the work not pertaining to seismic studies was at the rate of 0.8 mi/h, a general average for the unfavorable conditions.

General Observations

The following observations relate specifically to this job, and in principle to all excavation jobs of the same general nature.

Pioneering or Opening Up of the Job Two factors make for expensive work. First, the job is in the San Bernardino National Forest and work is subject to both state and federal government controls. Federal control means minimum environmental damage to existing trails and minimum building of new access roads. Second, the rough topography and the granite rock require ripping out of rock instead of normal bulldozing for the pioneering and access roads.

Clearing Brush and Trees The job is in average mountain growth of flora in granitic soil. However, no burning is permitted, and so a disposal area must be found within reasonable distance or else the chaparral and small trees must be shredded and mulched. Again, the rough country makes clearing quite difficult. There may be merchantable timber so that free-of-charge logging may be a bonus in the form of lowered clearing costs.

Excavation Of paramount importance is the rock excavation. Office work with geology maps showed the earth-rock structure to be thin to thick residuals overlying average weathered granite, so-called "DG," which in turn overlies jointed and semisolid granite.

The granite is of the tonalite variety, characterized by up to room-size weathered boulders on the surface and embedded in a matrix of decomposed granite beneath the surface.

While geologic maps are of primary value in determining the nature of the rock excavation, a good secondary source is an aerial picture, such as that shown in Figure 6-8. This vertical view in San Saba County, Texas, shows definite relationships between vegetation and surface residuals of different rock types. Low mesquite grows on Mississippian Barnet shale (MB). Cedars, live oaks, and other trees grow on Pennsylvanian Marble Falls limestone (PMF), and upon Ordovician Honeycut dolomite and limestone (OH).

Sometimes different shrub and tree types grow on opposite sides of a fault separating two different rock types. For example, north of Stony Gorge Reservoir in Glenn County, California, the Stony Creek fault separates the western unnamed formation of metamorphic serpentines from the eastern Knoxville formation of sedimentary shales, sandstones, and conglomerates. Tall pines and chaparral favor the serpentine soil, while

Figure 6-7 Walking the centerline of proposed mountain freeway. Field reconnaissance hiking in the San Bernardino Mountains of southern California in the wintertime. Elevation is 3600 ft, marking the transition from rain to snow at the snow line. All possible jeep trails have been obliterated by washouts of the mountain ravines and slides of the water-saturated slopes. "Walking the centerline" for about 15 mi was at the rate of ¾ mi/h. By an interesting coincidence, this is just about the same rate which obtains for hiking in the desert in summer heat of 120°F. Physical discomfiture is about equal.

medium-size manzanita and tall grass prefer the shale, sandstone, and conglomerate residuals. The fault line literally separates the the western dark foliage from the eastern light leafage.

The U.S. Production and Marketing Administration, which furnished the picture of Figure 6-8, has done remarkable work in correlating vegetation with types of soil. It is merely an extension of these principles to correlate vegetation with types of parent or country rock beneath the soil cover or residuals.

Long-Range Surveillance

Job surveillance from the existing road to the north along the East Fork of the Mojave River indicates the following job facts.

1. Canyons are steep, particularly that of Seeley Creek, probably calling for the use of crawler-tractor-drawn scrapers rather than wheel-tractor-drawn units.

2. Canyon streams, tributary to the Mojave River, are running full over boulders of up to cubic-yard size and exposed granite. Initial construction of fills and building of bridge abutments and piers will be difficult and expensive.

3. Brush is from medium to thick and good-size trees are fairly abundant, varying from 6 in to 3 ft in diameter.

4. Cuts on the north slope of the mountains are and will be wet, regardless of season, and the slopes are slide-prone because of saturation of weathered granite. Drill holes for blasting will be wet, probably necessitating use of more costly dynamite instead of ammonium nitrate explosives. Likewise, the wet conditions will cause less job efficiency than that obtainable when working on a sunny slope.

5. About 1.5 mi west of the little hamlet of Cedar Springs is an alluvial flat area along the West Fork that is suitable for offices and yard. However, it is subject to spring flooding and a diversion channel should be excavated.

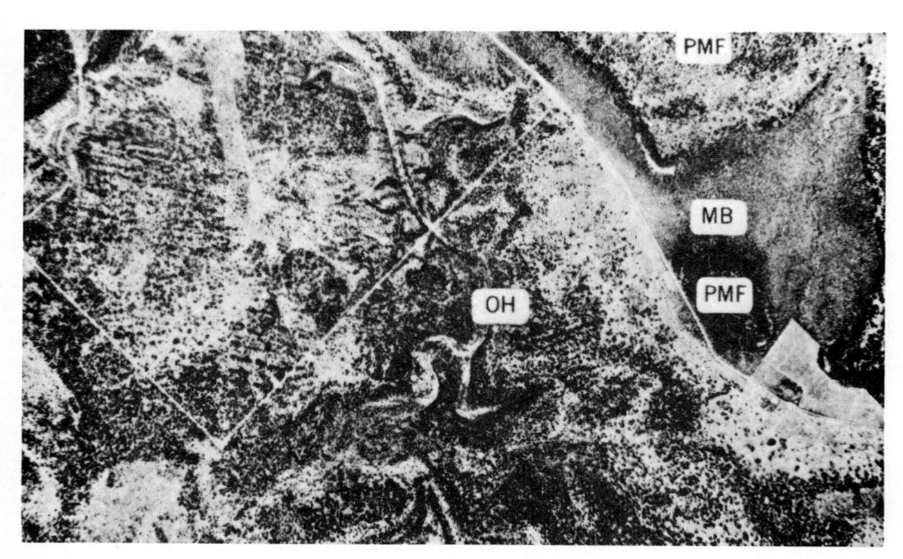

Figure 6-8 Correlations of kinds of rocks with kinds of vegetation. Correlation of vegetation with types of soil in San Saba County, Texas. Mesquite favors the Barnet shale (MB). Cedars, live oaks, and other trees are partial to Marble Falls limestone (PMF) and to Honeycut dolomite and limestone (OH). The residuals supporting the vegetation are the extremely weathered results of the immediately underlying rocks and thus the presence of these rocks may be inferred from the vegetation. However, in a rock excavation there may be only a surface zone of shale supporting the mesquite. The mass of harder rock beneath the shale might be limestone, in which case the deduction that the excavation is softer shale would be erroneous and costly. Such field reconnaissance observations must be a part of the mosaic of examination and exploration of rock work. *(U.S. Production and Marketing Administration.)*

Walking Centerline—Detailed Observations

Walking centerline results in the following notes concerning the cuts and fills according to station limits. All these notes will be analyzed by the office engineers, the estimators, selected job superintendents, and other affected personnel.

Stations 657 to 687

Three minor through cuts, totaling 90,000 yd^3. Maximum centerline cut 24 ft.
Scattered to dense brush. Some medium-size timber. Streams running full.
Well-weathered rock. Soft to medium ripping for heavy-duty tractor-rippers.
Use of wheel scrapers on maximum 3:1 natural slopes.

Stations 687 to 693

One through cut of 32,000 yd^3 with maximum 36-ft cut.

Average brush and small trees. Streams running full. Well-weathered rock. Soft to medium ripping. Possibly crawler-tractor-scrapers on account of 1.6:1 slopes.

Stations 693 to 701

One through cut of 109,000 yd^3 with 80-ft cut.

Medium brush and few medium trees. Full streams. Top of cut and slopes show moderately weathered rock. Blasting is estimated below 40-ft depth, requiring about 15 percent or 16,000 yd^3 of blasting. Soft to hard ripping to 40-ft depth.

Slope is 1.7:1 from top of cut to bottom of fill, 110 ft below. Use tractor-scrapers if rock breaks up satisfactorily or wheel loaders and rear-dump rock haulers if rock fragments are too big for scraper work.

Stations 708 to 714

One through cut of 70,000 yd^3 with 40-ft maximum cut.

Moderate brush and some small trees. Full streams. Cut is moderately weathered. Soft to hard ripping. A 4:1 maximum slope calls for use of wheel-type scrapers.

Stations 720 to 725

One 93,000-yd^3 through cut with maximum 57-ft cut. Sawpit Canyon bridge adjoins station 725 to the east. Rock will move to west down 2.2:1 slope.

Moderate brush and medium-size trees. Sawpit Canyon stream running full over outsize boulders and bedrock. Abutment and two-pier construction will be difficult and costly.

Partially weathered rock outcrops. Soft to medium ripping to 50-ft depth, suggesting minimal blasting. Crawler-tractor-scraper work.

Blasting estimated to be less than 5 percent, or 5000 yd^3. Probably to be hauled by tractor-scrapers.

Stations 730 to 738

One through cut of 184,000 yd^3 and maximum 81-ft cut. Moderate brush and fair amount of medium-size trees. Full streams.

Rock appears to be well weathered with maximum 3:1 slope on centerline. Soft to hard ripping.

Use of wheel scrapers with little likelihood of having to substitute crawler-tractor-scrapers.

Stations 747 to 754

One through cut of 99,000 yd^3 and maximum 67-ft cut.

Dense brush and moderate number of medium-size trees. Full streams of usual 2-ft depth and average 15-ft width.

Average weathered rock and maximum 3.5:1 natural slope. Soft to hard ripping and haulage by wheel-type scrapers.

Stations 760 to 766

One through cut of 134,000 yd^3 and maximum 90-ft cut.

Moderate brush and fair number of medium-size trees. Full streams.

Rock outcrops on top and slope of cut. Steep centerline slope of 1.7:1 down 205 ft vertically. Soft to hard ripping to 60-ft depth. Blasting below 60-ft depth, estimated at 5 percent, or 7000 yd^3. Use crawler-tractor-drawn scrapers.

Stations 768 to 776

Three through cuts, totaling 192,000 yd^3, with maximum 75-ft cut.

Medium brush and average number of medium-size trees. Three feet of water in turbulent boulder-strewn canyons.

Rock is moderately weathered. Soft to hard ripping to 25-ft depth or 75 percent of

192,000 yd³. Haulage of 145,000 yd³ by crawler-tractor-drawn scrapers with maximum 1.5:1 slope. Remaining 47,000 yd³ to be blasted, loaded by wheel-type shovels, and hauled by rear-dump rock haulers.

Stations 778 to 784

One through cut of 86,000 yd³ with maximum 72-ft cut. Heavy brush and moderate-size trees. Canyons carrying 2 ft of water over boulder-strewn bedrock.

Rock is average weathered, and westerly slope is 1.6:1 down 147 ft to bottom of fill.

Soft to hard ripping. Use crawler-tractor-drawn scrapers.

Stations 789 to 794

One through cut of 93,000 yd³ with maximum 65-ft cut at inside ditch line.

Dense brush and average number of medium-size trees. Easterly Burnt Mill Canyon running 2-ft of rough water. Building of 320-ft bridge over canyon will require costly work for abutments and two piers.

Rock will move to west down maximum 3:1 slope. Soft to hard ripping to 40-ft depth with use of wheel-type scrapers. Blasting estimated at 8 percent or 8000 yd³. Use wheel shovel and rear-dump haulers.

Stations 799 to 809

Two through cuts of total 99,000 yd³ with maximum 44-ft centerline cut.

Moderate brush and medium-growth small trees. Streams running to 2 ft of water.

Rock is well weathered. Soft to medium ripping. Haulage by wheel scrapers, as steepest natural slope is 3:1.

Stations 816 to 824

One through cut of 145,000 yd³ with 82-ft maximum cut.

This cut and adjoining westerly Seeley Creek bridge are shown in Figure 6-6. Centerline slope to east is 1.8:1, forecasting costly excavation and bridge building in the steep, rocky gulch.

Heavy brush and average-height trees. Creek running continuous 4-ft depth and 10-ft width. Cut appears rippable to half depth, resulting in 15 percent or 22,000 yd³ of blasting. Soft to hard ripping to 41-ft depth. Maximum centerline slope is 3:1 for westward moving excavation.

Wheel scrapers for ripped rock and wheel loader and rock trucks for blasted rock.

Stations 820 to 854

Five through cuts totaling 83,000 yd³ with maximum 30-ft centerline cut.

Medium brush and small trees. Brooks running full with about 1 ft of water.

Rock is well weathered and slope is maximum 3:1. Soft to medium ripping. Haulage by wheel-type scrapers.

Stations 855 to 865

Two through cuts of total 114,000 yd³ with maximum 40-ft centerline cut.

Heavy brush and timber of average height. Brooks running full with about 1 ft of water.

Centerline slopes average 3:1, and rock is well weathered. Soft to medium ripping. Haulage by wheel-type scrapers.

866 to 873, End of Job

One through cut of 29,000 yd³ with deepest cut of 28 ft.

Dense brush and trees of average height. Small canyons are running about 2 ft of rough water. Well-weathered rock. Easy slopes of maximum 4:1 slope. Soft to medium ripping for heavyweight tractor-rippers. Haulage by wheel-type scrapers.

Conclusions from Field Work and Subsequent Excavation History

The following general conclusions, after hiking centerline through this series of 16 cuts, are of great value to the office engineers and estimators.

1. Clearing should be gone over carefully as there may be some merchantable timber. Tree count is desirable, including tree diameters.

2. Abundant water is available for fill compaction and haul road maintenance either from wells in the river alluvium or from dams across the tributaries.

3. All ripping should be done by heavyweight tractor-rippers so as to take out marginal rock without blasting.

4. On the steepest slopes much rock can be bulldozed to bottoms of fills.

5. Consideration should be given to crawler-tractor-drawn scrapers for short, steep hauls, especially in view of wet, muddy haul roads in early and late summer.

6. Most of the rock will be hauled by wheel-type scrapers. An analysis of the mass diagram for the excavation, showing haul distances, will establish zones of work for the two types of scrapers.

7. Preliminary field estimate for blasting is 105,000 yd^3, or 6.2 percent, of total 1,700,000 yd^3 of roadway excavation. This rough estimate is to be checked by exploratory drilling and by seismic studies, accompanied by final quantity takeoffs from the plans.

8. Building of three bridges, each with two piers, with the longest to be 360 ft and the highest to be 105 ft, will be unusually costly in the rough, wet canyons.

Excavation jobs in the area and in the same granite rock formation were made objects of close study and these examinations resulted in these notes.

1. North portal excavation for the San Bernardino Tunnel, located northerly from station 745. Maximum centerline cut was about 90 ft. Cut was in granite gneisses and was taken out without blasting, although ripping ranged up to hard and extremely hard for heavyweight tractor-rippers. Excavation in 1967 and 1968.

2. First section of the relocated Cleghorn Road, located to the west of job. Work is in progress. It consists of five principal cuts with maximum centerline depth of 110 ft, as compared with 90 ft of this job. Topography is not as rough.

In the cuts where blasting is or has been taking place, the rock line appears to average about 40 ft below natural ground. Cuts show a fair amount of embedded boulders, typical of tonalite granite, which are difficult to handle without blasting and costly to blast by blockholing.

3. Cedar Springs Dam, 2 mi to the north, is being built. Spillway in the same granite rock was excavated to 90-ft depth by soft to extremely hard ripping in the last 30 ft of depth. Presently at the 90-ft depth the use of heavyweight tractor-rippers equipped with single shanks is about ended. The embedded boulders are up to room size, and they are surrounded by a little matrix of decomposed granite. The rock is becoming semisolid. Systematic drilling and blasting has commenced for this 210-ft-deep spillway. The diameter of the blast holes is 6 in and the hole spacing or pattern is 10 ft by 10 ft. Thus far ammonium nitrate is being used instead of dynamite, the holes being fairly dry.

With these field notes covering the reconnaissance of the job, together with rock exploratory work if deemed necessary and with further field investigations of clearing and water supply, the estimators will be ready for bid preparation on those items pertaining to the rock excavation.

Figure 6-9 shows two of the cuts being taken out in the Cleghorn Road to Crestline Freeway. The cut in the foreground is between stations 789 and 794. This 93,000-yd^3 cut with a maximum 65-ft depth has been brought down to about 30-ft depth by tractor-rippers and wheel-type scrapers. Although there was estimated to be 8000 yd^3 of blasting in this cut, all excavation was taken out by the tractor-rippers.

The distant cut is between stations 778 and 784 and is 86,000 yd^3 with the maximum cut of 72-ft depth. It is almost down to grade and it was taken out with wheel-type scrapers instead of crawler-tractor-drawn scrapers as suggested in the notes of the field reconnaissance. Actually, all excavation except blasted rock was hauled by wheel-type scrapers. After the cuts were opened up by bulldozers, it was found possible to lessen the grades of the haul roads by switchbacks, and thus the more economical wheel-type scrapers could be used for the job.

Figure 6-10 illustrates the finished freeway together with the topography of the mountainside job. The cut in the foreground and the three cuts in the distance are between stations 789 and 824. These cuts appear to be sidehill-type rather than through-type. However, all four cuts have small slopes on the downside or outside ditch line and so they are technically through cuts.

The picture is taken toward the west and the Mojave River is to the north. Tributaries run northward and their erosion has caused the ravines and canyons and the many northward trending ridges through which the freeway passes. The growth represents average density of brush and average size of trees.

Reiterating, it is well to investigate every excavation job on foot, whether it be the centerline of a highway or canal or the many lines and grades of a dam or open-pit mine. What may appear formidable from a distance may be easy rock work on closer examina-

Figure 6-9 **Progress of construction of mountain freeway.** Excavation for the Cleghorn Road to Crestline Freeway in 1969. These two cuts total 179,000 yd³. All granite rock is being ripped by heavyweight tractor-rippers, in spite of a maximum depth of cut of 72 ft. Wheel-type scrapers are negotiating up to 25 percent grades on switchback haul roads. Location is on the north slope of the San Bernardino Mountains of southern California. Elevation is 3750 ft and weather is bad during the springtime. The Mojave River lies to the north.

tion. For example, in desert areas what appears to be a dark lava flow from a half-mile view may very well be a harmless, gently sloping fanglomerate with a superficial coating of dark desert varnish. Conversely, the apparently easygoing alluvium may be a thin veneer of gravels on a gently sloping granite pediment.

Perhaps the greatest advantage in hiking is the unhurried opportunity to size up a job without interruptions, as well as the therapeutic value of a nice walk. In this hiking it is advisable to walk the job in both directions since hindsight is sometimes better than foresight.

In bidding on a highway job it is said that, because the costs of materials, machinery, and workers are fairly fixed and because methods are fairly uniform, the low bidder must have an "angle." The "angle" generally involves the excavation and especially rock work, be it for roadway, subbase, base, or quarry stone.

Perhaps, as shown in Figure 6-11, midway of the job and 100 yd upstream from centerline an eroded gully barely conceals an ample source of good rip-rap for the 100,000-yd³ revetment item of the bid schedule. The specified acceptable source may call for a 10-mi-longer haul. Development of the midjob quarry could mean a savings of 1 million yd·mi of haul or possibly $200,000 in haulage costs, giving the perceptive bidder a significant advantage and a greater profit margin.

Such an "angle" could have been discovered only by the hiker on centerline with wide-angle vision.

SUMMARY

Two correlative investigations form the basis of the analysis of excavation. These are examination of excavation in the office and exploration of excavation in the field, the subject of Chapter 7. The one is not truly valuable without the other and they go hand-in-hand, as should always be the case with interrelated efforts.

Figure 6-10 Completion of construction of mountain freeway. Completed Cleghorn Road to Crestline Freeway, summer of 1972. Total excavation, 1.7 million yd^3 with 22 major cuts of maximum depth 90 ft in the 5.1-mi mountainside job. The distant high-sloped cuts in ridges trending northward to the Mojave River illustrate the topography of the north slope of the San Bernardino Mountains. The freeway traverses a series of many ridges and ravines, resulting in short, steep hauls for the granite rock formation. The northward exposure and the rather luxuriant growth of brush and trees, coupled with the high elevation, resulted in moisture-laden rock formation and a short construction season in spite of southern California's semiarid climate.

Figure 6-11 Discovery of source of materials by walking centerline. A youthful gully being deepened into a deep ravine by weathering of the bedrock. Spalls of metamorphic schist are being removed by hydraulic action. This exposure of solid rock may suggest a source of quarry stone for use in the nearby projected highway. The exposure also indicates the depth of the shallow residuals and weathered schist. Such a discovery only comes from the inquisitive mind of the field reconnaissance man while walking centerline.

It is preferable for both examination and exploration to be performed by the same individual or, at least on a large project, for all the persons involved to be associated closely. There are unfortunate examples of lack of cooperation between office personnel working on examination and field personnel carrying on exploration. Under such conditions the result is not the best even though the individuals are the finest.

CHAPTER 7

Exploration of Excavation

PLATE 7-1

Part of working areas of foundation explorations for dam. Switchback trails built for the foundation explorations for the south abutment of Auburn Dam, American River, Placer County, California. The outline of the projected dam is 700 ft above the river. The correlated investigations included trenches, tunnels, drill holes, and seismic studies of the complex metamorphic rock formations and required about 10 yr. The unusually complete studies were necessary because of minor faults and shear zones within these typical Mother Lode country rock formations. The concrete arch dam requires solid foundations in this earthquake-prone country.

The excavation quantities for the preliminary foundation excavation for Auburn Dam are given in Chapter 5. About 1.7 million yd^3 is being excavated for the left or south abutment shown in picture.

The rocks are metamorphic schists and amphibolites and metasedimentary siltstones and sandstones. Although the exploration of excavation for the abutment by the contracting officer was largely for the purpose of design of the dam, the investigations fixed the hard-rock line at about 25 ft below original ground. The 25-ft-thick upper zone was made up of residuals and rippable weathered rock.

The outline of the concrete arch dam as shown in Plate 7-1 shows only about one-tenth of the 4000-ft-long crest of the 6.5-million-yd^3 structure. The combination irrigation and hydroelectric dam combines 2.3 million acre·ft of water capacity with 750 MW of ultimate electric power.

Exploration of rock excavation in the field is the sequel to examination in the office and field. These prior activities determine the scope and the methods for the subsequent exploration.

The methods used are manual, mechanical, and instrumental.

1. Manual methods may be as simple as the use of pick and shovel or a sounding rod. More refined methods include hand auger, wash boring, and soil sampler.

2. Mechanical methods include the use of bulldozers and angledozers, backhoes and other excavators, and many kinds of drills.

3. In present-day practice instrumental methods are confined to geophysical testing by seismic timers which includes ground-surface refraction studies and *uphole* studies of shock-wave velocities and interfaces between the different shock-wave velocities.

All these methods have their proper spheres and they may be used individually or collectively. All are forms of sampling and the results may be expressed axiomatically: The greater the sampling extent, the less the error in exploration of rock excavation.

MANUAL MEANS

The use of pick and shovel is almost as old as civilization and it needs no explanation. One simply picks and digs until one finds rock. Early mineral prospectors used this method to most profitable advantage.

The sounding rod is a series of 4-ft-long and 3/4-in-diameter pipes with couplings. The first section is fitted with a removable drive head and a drive point. A heavy sledge is used and increased depth is attained by adding lengths of the pipes. No samples of materials can be taken, and so the nature and the degree of consolidation of the residuals and the weathered rock depend on individual judgment. Obviously, the depth of investigation is limited to a few feet, generally less than 10 ft.

Figure 7-1 illustrates the three common kinds of hand tools.

Drilling a Hole with a Simple Open-Spiral Auger without Casing A degree of sampling is possible by examining the materials brought up by the auger and estimating of the depth from which the material was obtained. The degree of consolidation of the earth-rock structure may be judged by the ease or difficulty of augering. Augers range from 2- to 6-in diameter. In sands depths to 100 ft are possible. The maximum depth depends upon the leverage and down pressure applied to the auger. When small rocks are encountered, as in the case of weathered rock, a chopping bit may be substituted for the auger. If the material caves badly, as in the case of alluvium, a casing may be used advantageously. As shown in Figure 7-1, the hand auger can penetrate only partly into the zone of weathered rock.

Wash Boring Drilling a Hole with Casing and with a Chopping Bit Where pressured water is available it is sometimes expedient and efficient to use the simple pipe jet with casing of 2- to 4-in diameter, depending on the size of the flushed particles. Materials are judged by the washings and consolidation is appraised by the driller. With the ordinary two-person crew depths to 75 ft are possible, providing materials and water pressure are favorable. Again, drilling can reach only upper levels of the weathered rock zone.

Wash Boring Drilling a Hole and Sampling with Split-Tube Sampler Dry sampling, in the sense of securing an undisturbed sample, may be used in connection with wash boring. The chopping bit of Figure 7-1 is removed from the bottom of the wash pipe and a split-tube core sampler is substituted. This tube is forced into the earth-rock at the bottom of the hole and the core or sample is brought to the surface.

Manual tools have three desirable features, simplicity, portability, and economy. They have two limitations: (1) the limiting depth is 100 ft under the most favorable circumstances and generally it is considerably less; and (2) penetration is confined to maximum-weathered rock, as indicated in Figure 7-1.

Figure 7-2 displays a variety of manual tools as furnished in a soil sampling kit intended for up to 25-ft depths. It weighs 180 lb and contains 43 items, including 15 soil sampling tools. Samples are 1½-in diameter.

Among the important items are: (1) handle; (2) drive head; (3) nine 2½-ft drill rods with couplings; (4) ship auger; (5) closed-spiral auger; (6) open-spiral auger; (7) Iwan post-hole or sampling auger; (8) probing rod; (9) split-tube sampler; (10) saw-tooth and pocket shoes for split-tube sampler.

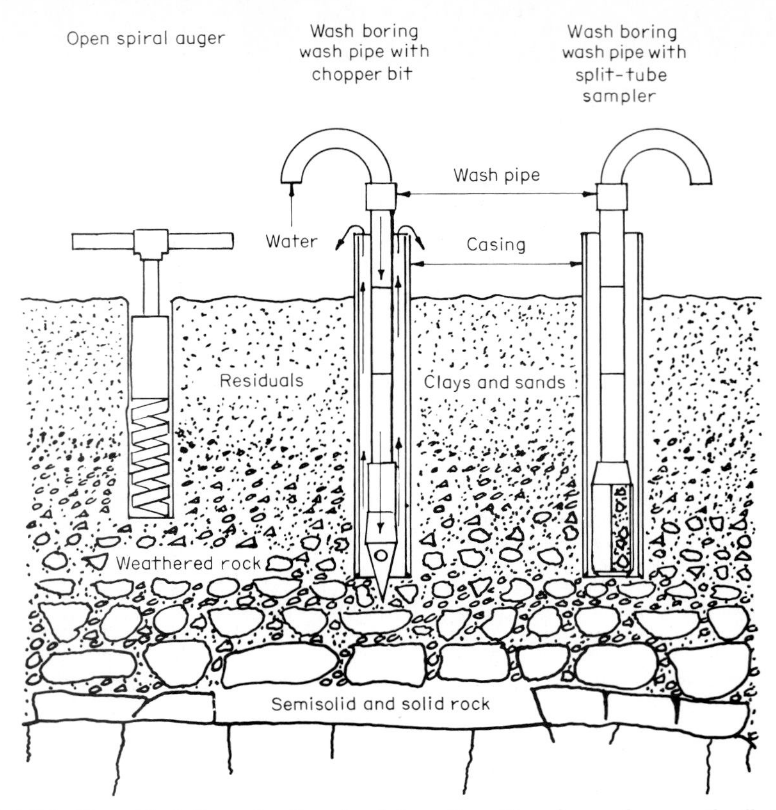

Figure 7-1 **Manual tools for exploration of rock formations.** No scales are shown, as the illustrative drawing is not in proportion. Spiral augers average 3-in diameter. Casings average 3-in diameter and water pipes average 1-in diameter. Residuals may be taken at 10-ft thickness, weathered rock at 10-ft thickness, and the semisolid and solid rock at indefinite thickness.

Figure 7-3 shows use of the soil sampling kit. The one-person operation is ideal for simple shallow-depth exploration through residuals and well-weathered rock to bedrock. The split-tube sampler is used for qualitative analyses of the earth-rock.

For more than one-person operation, affording up to 6-in-diameter holes and up to 100-ft depths under ideal conditions, there are larger-scale tools than those of the pictured soil sampling kit. These include longer handles for two- or four-person operation, 5-ft drill rods, augers of up to 6-in diameter, heavy chopping bits for weathered rock, and large-diameter split-tube samplers.

MECHANICAL MEANS

Backhoes

Small and medium-size backhoes, either wheel- or crawler-type, are efficient exploratory tools. Such machines can excavate trenches up to 20 ft deep, giving stratigraphic cross sections of residuals, weathered rock, and sometimes even semisolid rock.

Figure 7-4 illustrates exploratory work and initial excavation for a 3000-ft trench down the steep side of Slide Mountain, Nevada. The 5-ft-deep, 2-ft-wide trench is for utilities for a radio communication center atop the nearly 9000-ft-high mountain. The combination exploratory-excavation work was done to ascertain the suitability of the dual-purpose crawler-tractor-mounted-backhoe-bulldozer for the difficult work. The machine was ade-

Figure 7-2 Sampling kit for soil exploration. Manual soil sampling kit, offering a variety of hand augers, soil samplers, and split-tube sampler with accessories. Weight of 180 lb and depth capacity of 25-ft with the furnished drill rods. Simplicity, portability, and economy are the advantages of this assembly for one-person operation. *(Acker Drill Company.)*

Figure 7-3 Spiral auger and split-tube sampler for soil investigation. Use of open-spiral auger and split-tube sampler in the exploration of excavation in connection with highway building. The simple split-tube sampler gives a compact, corelike sample from a given depth of the drilled hole. *(Acker Drill Company.)*

quate for the rough terrain and the highly variable earth-rock, which consisted of weathered granite boulders embedded in sands of decomposed granite.

Hourly production was 15 lineal feet of trench, or 6 yd^3 neat trench measurement. Actual production, with allowance for caving of the sides of the trench, was about 19 yd^3/h.

Exploratory work showed that up to ½-yd^3 boulders persisted to and below the bottom of the 5-ft-deep trench, an example of the difficulty of trench exploration in weathered granite of the Sierra Nevada.

Bulldozers and Angledozers

These machines excavate a trench or sidehill cut, affording ample width and depth to analyze and diagram the exposed rock formation. Such trenches are popular with contracting officers in the case of public works and with designing engineers in the case of private works.

Figure 7-4 **Rock exploration with tractor-backhoe-bulldozer.** Medium-weight crawler-tractor-mounted backhoe-bulldozer trenching for utilities installation on the flank of Slide Mountain, Nevada. The weight of the combination machine is about 26,000 lb, and the diesel engine is 100 hp. The backhoe bucket is 30 in wide with 6.75 ft^3 struck capacity. The earth-rock is weathered granite, consisting of large boulders embedded in a matrix of sands and decomposed granite. *(J. I. Case Company.)*

Figure 7-5 illustrates a face of rock exposed by a sidehill bulldozer cut in connection with the exploration of a 1,840,000-yd^3 sidehill cut for a freeway in San Diego County, California. The rock is schist of an igneous-metamorphic formation made up of greenstones, tuffs, schists, and granites. The face of the exploratory cut is about 12 ft deep and the rock line of the semisolid formation is at a depth of about 10 ft.

Figure 7-6 is a diagram of the wall section of a 20-ft-deep exploratory sidehill cut dug by a medium-weight bulldozer equipped with a two-shank ripper. This cut called for about 250 yd^3 of earth-rock excavation and it was about a 3-h job for the medium-weight tractor.

The face, equivalent to a 60-ft-long 20-ft-high mural, showed the designing engineers what stratigraphy might be expected in the first 20-ft depth of excavation. Several similar sidehill cuts were made in critical locations of the multimillion-cubic-yard excavation for a huge housing project.

The ease or difficulty in ripping and bulldozing the hard rock formation gave indications of probable grading methods and machinery. Correlative to this investigation, one of the interested bidding contractors made additional sidehill cuts and trenches and put down several test holes in the sedimentary rock cuts.

As a result of all these tests, the builder's engineer had excellent data for design work relating to the characteristics of the materials and for estimating the probable costs of unclassified excavation, and the interested bidders could figure closely their grading costs. In short, there resulted ultimately a low cost to the buyer of a lot for his home.

Drills

Prior to discussing various drills for exploratory work, it is well to approximate their efficiency in this special work. Several of the types of drills are used for blasthole drilling, in which, because of repetitive production work, their efficiencies are just about double that for exploratory drilling.

Efficiency may be expressed as the ratio of average drilling speed in feet of hole per hour over a work period of days, weeks, or months to *penetration rate,* or rate while

Figure 7-5 Sidehill cut by tractor-ripper-bulldozer for rock investigation. An exploratory sidehill cut by a heavyweight combination tractor-bulldozer-ripper. The rock is schist of an igneous metamorphic formation of San Diego County, California. The cut face is about 12 ft deep and the rock line, or separation between rippable weathered rock and drill-and-shoot semisolid rock, is about 10 ft deep.

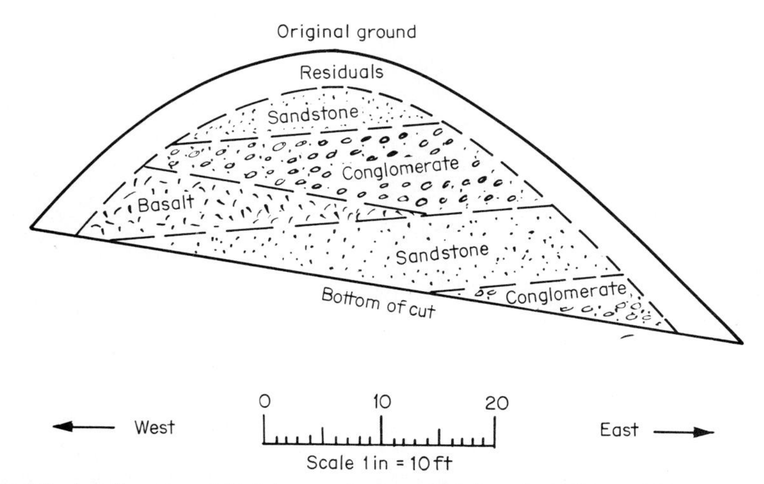

Figure 7-6 A wall section of 20-ft-deep exploratory sidehill cut by bulldozer. This cut was made in the Topanga formation of the Santa Monica Mountains, southern California, in connection with a proposed excavation for a site for homes. The 23-million-yr-old Topanga formation is made up of marine sandstones and conglomerates with volcanic intrusive and extrusive basalt rocks.

The history of the wall cross section, proceeding from bottom to top, is:

1. At least 6 ft of conglomerate was deposited on the bottom of the sea.
2. Next 7 ft of sandstone was laid down on the conglomerate.
3. At least 6 ft of basalt lava flow, tonguing to nothing at the right, was extruded on the sandstone.
4. An 8-ft layer of conglomerate was deposited on the basalt and the sandstone.
5. An unknown depth of sandstone, 4 ft within the wall section, was laid down on the conglomerate and possibly on the tongued basalt.
6. Finally all rocks have been weathered to a 3-ft depth of residuals.
7. Since deposition, the rock formation has been dipped a few degrees to the west.

actually drilling in feet of hole per hour. Thus, efficiency is a measure of the working time losses due to moving to the next hole, setting necessary casing, changing drill rods and bits, freeing stuck drill rods and bits, mechanical failures, and the like.

In blasthole drilling, efficiency may be approximated at 67 percent. In exploratory drilling, efficiency, obviously a more variable figure, may be estimated at about 33 percent. The chief reasons for the lower efficiency are the considerably more time required to move to the next hole and drill setup, deeper holes, more time in setting casing, and intermittent drilling while taking samples and making notes.

This example illustrates the relationship of these efficiencies. A combination of medium-weight track drill and a 600-ft³/min, compressor, putting down a 3½-in-diameter hole in quarry blasting of granite, requires 7.5 min to drill down an 11-ft length of drill rod. The penetration rate is 1.47 ft/min. Whence:

Penetration rate at 100% efficiency	88 ft/h
Blasthole drilling at 67% efficiency	59 ft/h
Exploration drilling at 33% efficiency	29 ft/h

Auger Drill Perhaps midway between the manual auger and the mechanical auger is the portable one-person-operated drill shown in Figure 7-7. This unit, powered by a 10-hp air-cooled gasoline engine, weighs about 300 lb with accessory equipment.

Auxiliary equipment includes augers of up to 3-in diameter and core barrels and bits for ⅞-in-diameter cores, as well as a 1½-hp water pump. A variety of tools is available, including the soil sampling kit of Figure 7-2. Depth capacity is about 50 ft.

When one considers that much exploration for rock takes place within a 50-ft depth, the portability and economy of this simple auger drill are obvious advantages.

Figure 7-7 Portable power-operated auger drill for rock exploration. Portable one-person-operated auger drill, powered by 10-hp air-cooled gasoline engine. Weight with tools about 300 lb. Depth capacity is about 50 ft for both auger and core drilling. The driller is checking the depth of soft residuals down to shales and limestones in fields of Pennsylvania. The variety of tools used includes the soil sampling kit of Figure 7-2. *(Acker Drill Company.)*

The drill of Figure 7-7 is being used in farm country with maximum depth of residuals in order to investigate the depth to shales and limestones. In such work the average drilling speeds during a complete investigation are approximately as tabulated below.

Class of rock	Speed, ft/h	
	Auger drill	Core drill
Soft (residuals)	60–90	
Medium	30–60	3–5
Hard		0–3

A truck-mounted auger drill is efficient and economical for exploratory drilling in soft to medium-hard rock formations with overburden cover of clay-silt-sand residuals. Figure 7-8 shows such a machine exploring for a shallow highway cut in Indiana. Rock formation is soft earthy residuals overlying shales and limestones.

The auger drill has a capacity of up to 200-ft depth and 5- to 36-in hole diameter. Figure 7-9 shows the principal parts of an auger drill assembly and the following five kinds of auger cutterheads and bits for the conveyor-type flight auger:

a. Earth auger for residuals with single flight and up to 24-in diameter.

b. Earth-rock auger for earth and well-weathered rock with double flight and up to 24-in diameter.

Figure 7-8 Carrier-mounted heavyweight auger drill for rock investigation. Truck-mounted auger drill putting down 12-in-diameter hole through soft residuals for exploration of depth of shales and limestones. The flight of spirals is loaded with the sticky clay residuals. When the shales and limestones are encountered in the weathered zone, the bit will be changed to the rock-type. *(Acker Drill Company.)*

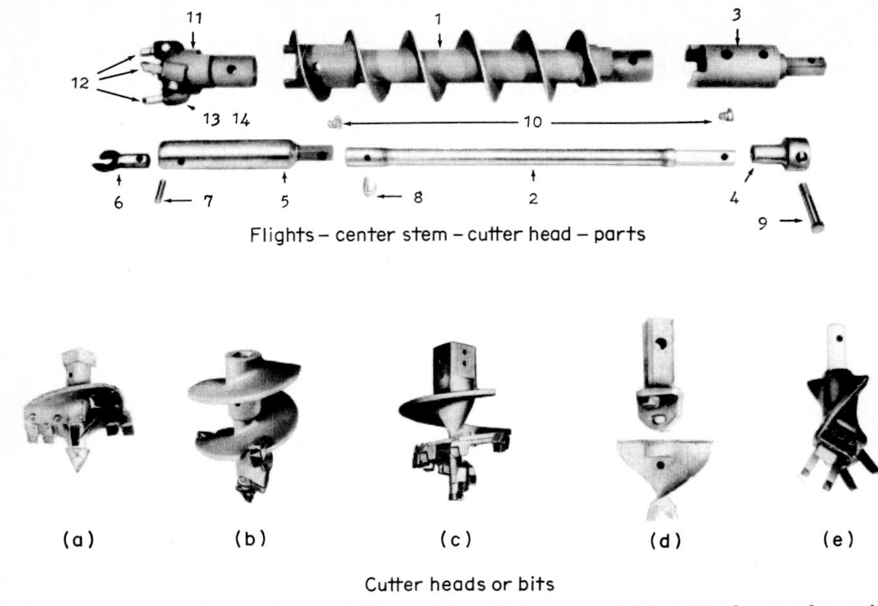

Figure 7-9 Principal parts of auger-drill assembly with five types of bit. *(a)* Earth auger for earth and residuals. *(b)* Earth-rock auger for well-weathered rock. *(c)* Rock auger for weathered rock. *(d)* Fishtail bit for loosely consolidated rocks. *(e)* Finger bit for shales and medium-hard rocks. *(Acker Drill Company.)*

c. Rock auger for weathered rock with double flight and up to 24-in diameter.

d. Fishtail bit for clean, straight holes in soft unconsolidated residuals and rocks. Up to about 8-in diameter.

e. Finger-type head for shales and medium-hard rocks. Up to about 12-in diameter.

Most heads have alloy-steel replaceable bits, either hardfaced or with carbide inserts. The conveyor flight augers are of slightly smaller diameter than the heads or bits and they are of either solid or hollow-stem construction; the latter type makes it possible to take core samples, as the hollow stem acts as a casing.

Speed of drilling for this abrasive-type drill varies from tens of feet hourly in soft earth to a few feet hourly in medium hardness of rock. The auger drill is inadequate for work in hard rocks. In the hands of an experienced driller it is a proper tool for determining the depth of the rippable rock. Ranges of hourly drilling speeds for a 5- or 6-in-diameter hole are shown in the table. Drilling efficiency is estimated at 33 percent.

Class of rock	Speed, ft/h
Soft (residuals)	60–120
Medium (well-weathered rock)	0–60

Bucket Drill The bucket drill is an abrasive drill in the auger drill family. The cutting edges of the rotating bucket function as an auger and the bucket, in addition to bringing the cuttings to the surface, serves as a large sampling device. It is a powerful drill for exploratory work, as well as for large-diameter holes for caissons, structure footings, water wells, and the like.

Figure 7-10 shows a medium-weight bucket drill discharging a bucketful of residual clays. The carrier truck is all-wheel drive, especially designed as a versatile vehicle for

Figure 7-10 Carrier-mounted bucket drill for sampling weathered rock. Bucket drill exploring for fire clay near Dunfermline, Scotland. A 30-in-diameter hole is being put down through residuals of ordinary clay to determine the depth of the desirable fire clay. Qualitative analyses are made throughout the depth of the hole to fix the interface at the top of the fire clay seam. *(Earthdrill, Incorporated.)*

the exploratory drill. The depth of overburden above a fire clay seam is being investigated near Dunfermline, Scotland. As the 30-in-diameter bucket brings up large samples, qualitative analyses may be made to determine the interface separating the ordinary clays from the desired merchantable fire clay.

Bucket drills range from a ½-yd³, 48-in-diameter bucket with 100-ft depth capacity to a 1½-yd³, 120-in-diameter bucket with 225-ft depth capacity. Specifications for a suitable bucket drill for exploratory work would include:

Capacity: ¾-yd³ bucket and 150-ft depth.

Power: Diesel engine of 100 hp.

Controls: Hydraulic with variable speeds and crowding or down pressure for the bucket.

Truck chassis: Three-axle all-wheel-drive type. Six large traction-flotation tires. Gross vehicle weight (GVW) rating 45,000 lb.

Like auger drills with somewhat similar cutting bits, the bucket drill is limited to soft to medium-hard rock formations. For example, in hard sandstones the penetration becomes a milling rather than a drilling operation, resulting in drilling speeds as low as 1 ft/h with excessive consumption of bits. Likewise, drilling speeds vary from several feet per hour in soft shales to several inches per hour in hard sedimentary rocks. A popular axiom of the experienced driller is: If the rock formation cannot be drilled by a bucket drill, it cannot be ripped, and blasting is in order.

As an approximate guide for estimating drilling speeds of a medium-weight bucket drill with 24-in-diameter bucket and capacity for drilling to 150-ft depth, the following table is suggested:

Class of rock	Speed, ft/h
Soft (residuals)	20–40
Medium (well-weathered rock)	10–20
Hard (weathered rock)	0–10

Figure 7-11 portrays combined bucket-drill and seismic explorations for Santa Susana Pass Freeway in southern California. The massive Chico sandstone formation of this 8.5-million-yd^3 rock job was discussed in Chapter 2.

The heavy-duty bucket drill of the California Division of Highways, called "Big Mo," is shown putting down test holes up against the ledge sandstone in order to determine if a near vertical face separates the residuals from the ledge rock or if there is a pediment or inclined surface of sandstone beneath the residuals. On striking the sandstone pediment drilling speed was down to 1 ft/h of 18-in-diameter hole.

In the foreground is the state seismic crew checking out the Chico formation and securing up to 13,000 ft/s shock-wave velocities a few feet below the original ground. The seismic study is being made in a test area where a heavyweight tractor-ripper with one shank attempted without success to rip economically the heavily bedded sandstone. Only a prohibitively costly production of 80 yd^3/h bank measurement was possible below a 6-ft depth.

The combined drilling, seismic, and ripping tests were made to acquaint both the state and the bidding contractors with the possible pitfalls in this spectacular job. The investigations, made after the first bids were rejected as too high, established the justifications for the initial bids. The bids of the second letting were about equal to the first bids, both initial and final bids being submitted by experienced rock excavators. The low bidder blasted about 68 percent of the total 8.5 million yd^3 of roadway excavation.

Big Mo, like many modern exploratory drills of the abrasive type, is able to work as a bucket drill, an auger drill, a rotary drill, and a core drill. It lacks only the facility for percussive types of drilling.

Jackhammer The jackhammer, a percussive-type drill, is a hard-rock drill and it is not suitable for exploratory work in residuals and well-weathered rock atop the drill-and-

Figure 7-11 Combined drilling and seismic analyses of rock excavation. Combined bucket-drill and seismic explorations are shown for the massive Chico formation of sandstones in the Santa Susana Pass Freeway of southern California. The California Division of Highways made the findings available to the bidders on this 8.5-million-yd^3 rock job. At the particular locations the speed of the bucket drill was reduced to 1 ft/h in the heavily bedded marine formation, and seismic shock-wave velocities up to 13,000 ft/s were secured within a few feet below original ground.

shoot rock at the rock line. This is because the drilled rock or cuttings are forced up and out of the hole by the same compressed air which actuates the drill. Earthy, claylike residuals cannot be forced up the hole because of the tendency to ball up, although this limitation can be partially overcome by placing casing in the hole to the depth of the hard rock. Accordingly, the overburden must be removed down to the semisolid rock before efficient drilling is possible. After this, satisfactory exploratory work may proceed through the semisolid rock into the solid rock.

Equipment for use of a 60-lb jackhammer generally includes: a truck-mounted or pull-type compressor of about 165-ft^3/min capacity; a set of drill steel in 2 ft-multiple lengths up to 20 ft; a supply of 1½-in-diameter tungsten carbide bits; several 20-ft lengths of 1-in-diameter air hose; and miscellaneous air hose fittings.

Such an exploratory drill will put down the following hourly footages of 1½-in-diameter hole with drilling efficiency at 33 percent:

Class of rock	Speed, ft/h
Soft	14–20
Medium	8–14
Hard	0–8

In most blasting work the drilling is done by percussive-type drills mounted on crawlers, particularly in the field of public works construction. In drilling there is a correlation between the drilling speeds of jackhammers and track drills. Track drills put down holes of about 2-in to 6-in diameter in systematic blasting, and in exploratory work the holes are usually of 3½- or 4-in diameters.

Correlation for exploratory drilling by jackhammers and by track drills gives the following table of average drilling speeds at 33 percent drilling efficiency:

	Speed, ft/h	
Class of rock	Jackhammer 1½-in hole	Track drill 4-in hole
Soft	17	62
Medium	11	38
Hard	4	12

The table indicates that the drilling speed of the jackhammer averages about 12 percent that of the track drill.

A distinct advantage of using the jackhammer or track drill for exploratory work is that the operation gives a good estimate or guide for the anticipated blasthole drilling speeds and for the drilling characteristics of the rock formation. It goes without saying that in such appraisals it is necessary to make full and complete notes, and preferably time studies, of the drilling operation.

Figure 7-12 displays the variety of bits used for jackhammers and for small and large track drills. The bits vary in size from 1¼-in diameter for jackhammers to 6-in diameter for large track drills. Different configurations and compressed air outlets are shown in examples *(a), (b), (c), (d),* and *(e).* Example *(f)* illustrates a bit with tungsten-carbide-steel inserts, which increase the life of the bit by a factor of 30 or more as compared with an alloy-steel bit.

Figure 7-13 shows the use of a 57-lb jackhammer for drilling hard granite. Residuals and weathered rock have been removed so as to provide efficient drilling conditions for exploratory work. The semisolid and solid granite is fairly uniform and drilling speed averages about 7 ft/h. The heavy-weight jackhammer has about a 20-ft depth capacity. The driller is checking for the presence of joints or seams and for drilling speeds. These factors will cast light on methods to be used for the drilling and blasting program.

If metamorphic rocks such as schists were being explored, the goals would be the same. However, if sedimentary rocks such as alternate beddings of sandstone and shale were being drilled, then thicknesses of beddings and angles of dips as well as the respective

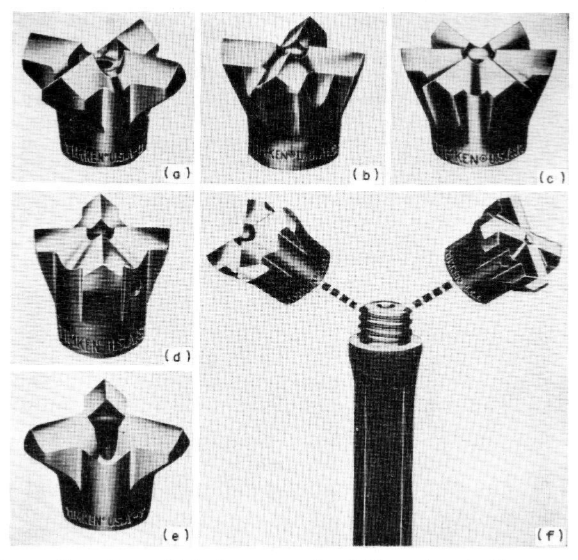

Figure 7-12 Some configurations of bits for percussive drills such as jackhammers and track drills. *(a)* Four-point bit, standard for most drilling, with center air hole for hard rock. *(b)* Four-point bit with modified airways. *(c)* Six-point bit with center air hole. *(d)* Four-point bit with side air hole for soft rock. *(e)* Three-point bit with center air hole. *(f)* Four-point alloy-steel and carbide-steel insert bits with drill steel. Not shown is the single-chisel bit for short holes in soft to medium rock. *(Timken Roller Bearing Company.)*

drilling speeds would be significant for making drilling and blasting plans for the job under investigation.

Track Drill The track drill is perhaps the most popular exploratory drill for the rock excavator because of its availability, portability, depth range, and indications of drilling speeds to be encountered in blast-hole drilling. The last capability is most important because drilling and blasting techniques and costs can be anticipated by exploratory drilling by the same kind of drill that probably will do the subsequent blast-hole drilling.

Figure 7-14 illustrates a track drill putting down 3½-in-diameter hole in weathered andesite of eastern New York State. The reach of the mast from the front of the crawlers is about 9 ft, affording an efficient spread of holes from the one drill stand. Thereby travel time over the rough natural ground and setup time are minimized. Average drilling speed in such a hard-rock formation is about 20 ft/h, the hour including travel and setup times. Figure 7-14 also shows a team of track drill and compressor traveling between exploratory holes for a canal job across Cactus Plain in Arizona. In spite of the soft desert alluvium and steep wadis of the dissected land, the track drill pulling the air compressor averages 2 mi/h. The traction air motors of the drill receive air from the compressor, making the combination a self-contained exploratory unit. In rather soft alluvia the team is able to traverse steep-sided dry streams whose banks have +20 percent grades. In one day the drill is able to put down several exploratory holes in alluvia and schists along the centerline and to travel 8 mi over the rough mountainous desert terrain.

Estimated ranges of hourly drilling speeds for a track drill putting down 3½-in-diameter exploratory holes are set forth in the following tabulation:

Class of rock	Speed, ft/h
Soft	50–75
Medium	25–50
Hard	0–25

Figure 7-13 **Percussive jackhammer for rock exploration.** Heavyweight jackhammer investigating hard rock excavation. Residuals and weathered rock have been removed and drilling is at rate of about 7 ft/h in semisolid and solid granite. Joints and seams are being checked, as well as drilling speeds. These factors relate to blasthole drilling methods and costs. *(Joy Manufacturing Company.)*

The practical maximum depth of hole for the track drill is about 80 ft unless the drill is of the downhole hammer type, as beyond this depth the inertia of the drill steel lowers the energy delivery to the bit and thereby lowers the drilling speed. At the same time expulsion of the cuttings by the compressed air becomes difficult and finally impossible. Generally this depth limit is adequate because most cuts are shallower than 80 ft and because below this depth the rock formation has become uniform and needs no deeper investigation.

Rotary Drill The rotary drill, equipped with either a drag bit or a tricone bit, is able to penetrate both soft and hard rocks to considerable depths. The drag bit is usually of the fishtail or finger type with removable hardfaced teeth and it is suitable for soft formations. Two- and four-blade bits are used, the two-blade bit giving faster drilling speed but being subject to stalling. At this juncture the four-blade bit of smaller diameter is substituted.

Tricone bits are similar to those used in drilling for oil and gas. Figure 7-15 shows a tricone bit, consisting of three rotating conical cutters mounted in the rotating head. Cuttings from rotary drill bits are removed by compressed air delivered through the hollow drill stem and released through the bit. It is possible, of course, to use *drilling fluid* or *mud*, as in oil-field drilling, but this method for removing cuttings is not normally used in exploratory drilling.

Figure 7-16 shows a rotary drill, equipped with a drag bit, exploring the excavation for a 180-ft-deep, 50-ft-diameter surge chamber. The chamber is part of the Los Angeles Tunnel at Castaic Dam, California. After the 5-in-diameter hole was drilled, uphole seismic studies were made in the Hungry Valley sedimentary rock formation. Shock-wave velocities were correlated with rates of drilling in the several zones of depth. Velocities

(a)

(b)

Figure 7-14 Exploratory drilling by track drill and compressor teams. *(a)* Putting down 3½-in-diameter holes in andesites of eastern New York. Large weathered rocks present traveling difficulties but these are offset partially by the extended boom facility, which makes it possible to drill several holes from the same central drill stand. *(b)* Crossing a wadi of the Arizona desert between exploratory holes for canal excavation. Rock is valley alluvia overlying hard schist. The track drill and towed compressor average 2 mi/h.

ranged from 3100 ft/s near the surface to 6200 ft/s at depth, indicating a mixture of tedious tractor ripping and blasting in the small-diameter confined area of the surge chamber.

Drilling rates for rotary drills, based on 33 percent efficiency, are given in the following tabulation of suggested hourly rate ranges for 5-in-diameter holes.

	Speed, ft/h	
Class of rock	With drag bit	With tricone bit
Soft	130–200	35–50
Medium	60–130	15–35
Hard	—	0–15

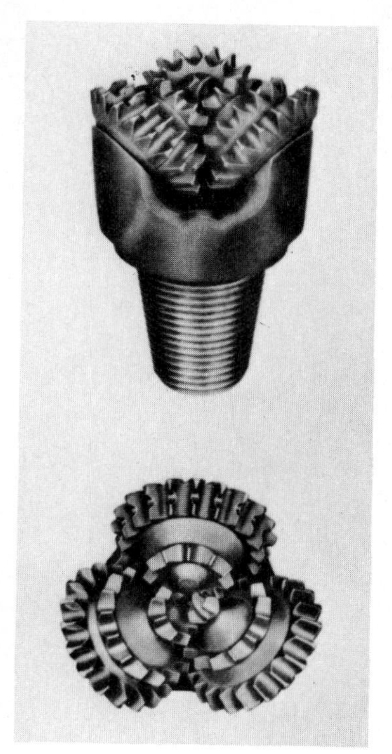

Figure 7-15 **Rotary drill bit for investigation of rock excavation.** Tricone bit for rotary drilling, showing the three rotating cones. One cone has two cutting elements and the other two have three cutting elements each. Such a bit performs well in either soft or hard rock. *(Joy Manufacturing Company.)*

As in the case of the track drill, the rotary drill using either drag bit or tricone bit furnishes valuable data anticipatory to future drilling and blasting. This is especially true if use of the rotary drill with tricone bit is planned for the blasthole drilling. Derived information is:

1. Net drilling speeds, or penetration while actually drilling.
2. Life of bits, generally a costly item of drilling.
3. Uniformity of rock formation, which usually contributes to stable drilling speeds and fixed powder factor in pounds per cubic yard or per ton of excavation.
4. Presence of joints, seams, fissures, or pockets of soft residuals, which make for expensive drilling and blasting.

In short, after exploratory drilling by the kind of drill to be used for blasthole work on the job, a wealth of data is available for a well-calculated cost estimate for production drilling and blasting. If different types of drills are used for exploratory work and anticipated blasthole drilling, either a correlation factor for drilling speeds with the two drills or data from experience in the same rock formation with the particular drill must be used.

Core Drill The core drill is an abrasive rotary type for the express purpose of extracting near continuous cores of rock. It is used almost universally by designers of public and private works involving rock excavation. It is also used less extensively by excavators in their appraisals of rock excavation.

There are two kinds of core drills, according to the nature of the abrasive or cutting material of the bit. More common is the diamond drill, in which the bits are studded with embedded diamonds and which is used to extract cores of up to 6-in diameter. Less common is the shot or calyx drill, in which chilled steel shot is fed down through the drill stem and under the serrated cutting edge of the rotating bit and which yields cores of

Figure 7-16 **Carrier-mounted rotary drill for exploration of tunnel excavation.** Rotary drill, equipped with 5-in drag bit, exploring excavation for 180-ft-deep surge chamber of the Los Angeles Tunnel of the Castaic Dam. The rock formation is the Hungry Valley, made up of average-thickness beddings of sandstones and shales. The rate of drilling in this medium rock was about 25 ft/h, which was unusually low because of delays with stuck steel and bits.

up to 6-ft diameter. Both drills have similar operating elements, including power unit, drill stem, rotating table, water pump, casing, core barrel, and bit. Such machinery for the larger-diameter shot drill is necessarily much heavier and more powerful then that for the diamond drill.

Diamond Drill. The diamond drill may be truck-mounted or skid-mounted, the truck mounting being more popular. Core diameters range from ⅞ to 6 in. Commonly used core diameters are: EX, ⅞ in; AX, 1⅛ in; BX, 1⅝ in; and NX, 2⅛ in.

Figure 7-17 shows several types of bits used in the drilling process. The *casing bit* is used to ream a hole of proper size for setting casing through porous or water-bearing formations which would impede or prevent drilling progress. The *chopping bit* has two functions. When, as sometimes happens, driving standpipe becomes impossible because of materials gathering inside the pipe, the chopping bit serves to break up the material and clear the pipe to the bottom of the drive shoe. The chopping bit also serves to break up any small portion of the core which might have broken off during extraction of the core. The *coring bit,* with diamond surface settings, drills an annular hole which permits the passage of the core barrel down over the core. The *impregnated coring bit* is applicable to most of the rock formations that are too rough for the set coring bit to handle. It is most practical in hard, badly broken, or fractured formations.

Figure 7-18 shows a typical box of core samples, type NX of 2⅛-in diameter. The cores were taken by the California Division of Highways in anticipation of a future highway in the Fortuna Mountain area of San Diego County. The highway cut being investigated is 300 ft deep. The rock of the area is granitic and metavolcanic, and these particular cores

Figure 7-17 Bits for diamond core drilling. Four of the principal kinds of bits used in diamond core drilling. *(a)* Casing bit. *(b)* Chopping bit. *(c)* Coring bit. *(d)* Impregnated coring bit. These four bits have variations to the pictured configurations. Other types of bits, usually for some special purpose, are impregnated-casing, pilot-plug, reaming, fishtail, rose, and ferrule. *(Joy Manufacturing Company.)*

are of granite. The cores were taken at depths of 27.5 to 32.5 ft in an exploratory hole of 285-ft depth.

These cores indicate that at an average 32-ft depth the rock is hard and firm and that recovery is practically 100 percent. *Recovery* is that percentage of a given length of core that is unbroken. In this example a break in the core indicates cooling joints in the granite, or perhaps breakage during core extraction. The cause can be determined by close study of any indications of old weathering or young breakage. It is probable that the granite at the horizon of the core samples will require blasting.

When this rock excavation job is up for bidding, these and all other cores will be available for the contractors' inspection. Available also will be logs of all exploratory holes, including auger and bucket drill logs. It is probable that the state will also have available seismic studies of excavation.

Figure 7-19 shows a core drill exploring overburden and rock structure at Lincoln Park in Chicago, Illinois. Test holes were drilled to a 1500-ft depth in order to check the feasibility of constructing underground tunnels for temporary storage of rainwater and sewage overflow. This drill may be tractor-mounted or truck-mounted. It is a multitype drill, affording auger drilling to a depth of 200 ft depth, rotary drilling to 1000 ft, and core drilling to 1500 ft.

Representative performance in four test holes through an average 100 ft of glacial drift made up of hard till, sands, and gravels and through an average 264 ft of limestones and shales is given in Table 7-1. The rig used was a Mobile Drilling Company Model B-50 Explorer, and the maximum hole depth for the project was 591 ft.

Figure 7-18 **Box of typical cores, taken by core drill.** Typical box of cores taken by the contracting officer in the investigation of a highway cut in granite. The cores were extracted in October 1970, and the highway had yet not been fully planned as of 1976. Sometimes a decade of rock exploration precedes a large rock excavation project. The cores show good recovery at average 32-ft depth, and they indicate that blasting will be necessary at this depth in the huge 300-ft-deep cut in granite and metavolcanics. *(California Division of Highways.)*

Figure 7-19 **Carrier-mounted core drill taking cores from deep rock strata.** Core drilling through 100 ft of overburden, consisting of sands and gravels of till, and 362 ft of limestones and shales. Rock exploration is for the feasibility of building deep tunnels for the storage of rain water and sanitary sewage beneath Lincoln Park in Chicago, Illinois. The average drilling speed in overburden by a 5⅝-in tricone bit was 5.6 ft/h and the speed in limestone and shale with a 2⅛-in core bit was 12.8 ft/h. *(Mobile Drilling Company.)*

TABLE 7-1

Date: July 1971
Project: Metropolitan Sanitary District, Chicago
Contractor: American Testing & Engineering, Indianapolis, Ind.

Job conditions: Overburden consisted of hardpan, sand and gravel, and free-running sand to average depth of 90 ft. Overburden drilled with 5⅝-in tricone bit and cased with 4-in pipe to rock, then NX casing placed 5 ft into rock in preparation for coring. Rock coring performed using wire line method with NQ rods and core barrel. Clear water was pumped by John Bean Model 435-11 (35 maximum gal/min) at pressures ranging from 200 to 500 lb/in². Coring performed through thin shale and heavy limestone formations to average depth of 360 feet.

Drill Log Data*

	Type of Drilling	Depth, ft	Total Time, h	Rate, ft/h
#1 Hole	Overburden	90	9.5	9.5
	Core	263	24.5	10.7
	Total	353	34.0	10.3
#2 Hole	Overburden	104	17.0	6.1
	Core	216	23.0	9.3
	Total	320	40.0	8.1
#3 Hole	Overburden	104	11.0	9.5
	Core	217	17.5	12.4
	Total	321	28.5	11.3
#4 Hole	Overburden	100	33.0	3.0
	Core	362	17.5	20.9
	Total	462	50.5	9.1

*Figures taken from drill log do not include setup time and are based on three 8-h shifts daily.
SOURCE: Mobile Drilling Company.

The averages for the drill log data of Table 7-1 are:

Average rate of drilling overburden with inclusion of setting of casing, tricone bit of 5⅝-in diameter.	5.6 ft/h
Average rate of drilling limestone and shale, core bit of 2⅛-in diameter	12.8 ft/h

Note that drilling time does not include setup times from hole to hole but does include all other necessary times for the drilling operation. It is probable that penetration rate or actual drilling speeds are about 50 percent faster, giving 8.4 ft/h in overburden and 19.2 ft/h for combined limestone and shale rock formation.

Actual penetration speeds, exclusive of all delays to the drilling, for diamond core drills are given approximately in the following table for all sizes of cores from ⅞- to 2⅛-in diameter. Efficiency is 100 percent. The table is adjusted for probable average drilling rates at 33 percent efficiency.

Class of rock	Speed, ft/h: 100% efficiency	Speed, ft/h: 33% efficiency
Soft, as salt and limestone	30–40	10–13
Medium, as limestone, dolomite, graywacke, and greenstone	20–30	7–10
Hard, as granite and gabbro	10–20	3– 7

SOURCE: Christensen Diamond Products Company.

Shot or Calyx Drill. The *shot* or *calyx drill* is used for large holes up to about 6 ft in diameter. Occasionally these holes are used for exploratory work, since a worker can be lowered into the hole in order to make visual inspections of the rock structures. However, the drill is used more commonly for large-diameter holes for foundations and penstocks and for communication and ventilating shafts in underground mines.

Figure 7-20 shows a cross section of the workings of a shot drill within a rock formation. Water acts as a lubricant and as a carrier for the cuttings. The cuttings and the intermingled particles of the spent shot settle in the *calyx*. The casing is equipped with a drive shoe for penetrating the residuals and weathered rock and for seating onto the semisolid or solid rock. Below this interface the cores of rock are extracted.

Standard shot drills put holes down to 600-ft depth and extract cores of 3- to 20-in diameter. Specially built drills are sometimes used for unusual rock exploration involving deeper and larger holes of up to 3-ft diameter and 1000-ft depth.

Figure 7-21 illustrates a derrick-type calyx drill exploring a dam foundation in South America. The heavy, large-diameter cores call for massive drills and for powerful drilling

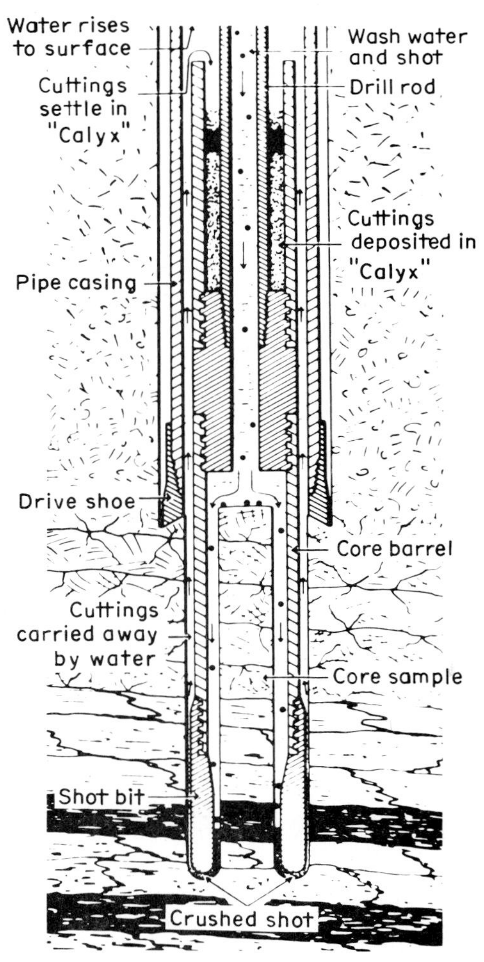

Figure 7-20 **Drill head of shot or calyx drill.** The workings of a shot or calyx drill exploring rock formation of residuals, weathered rock, and semisolid rock. Water, bearing chilled steel shot, is circulated downward between core and core barrel to the serrated shot-bit cutting edge, whence it is circulated upward with its cuttings and spent shot particles between the core barrel and casing. At the top of the core barrel, the mixture falls into the calyx for deposit. *(Ingersoll-Rand Company.)*

Figure 7-21 Head works of large-capacity shot or calyx drill. Shot or calyx drill exploring dam foundation in South America. The derrick-type drill is extracting 66-in-diameter cores from this combination hole for penstock and examination of rock formation. It is probable that the average drilling speed is about 0.4 ft/h. Such massive and powerful drills are of special construction and they are of a "one-job" nature. *(Ingersoll-Rand Company.)*

machinery. This installation is of a semipermanent, nonportable nature, except for the supporting trackage for lineal movement.

The huge size of some calyx cores is emphasized in Figure 7-22.These 36-in-diameter cores of hard sandstone were extracted for the examination of foundation work for Norris Dam, located on the Clinch River of Tennessee. In this investigation began the now established practice of lowering a geologist into the hole for minute inspection and photographing of the rock structure. It is probable that the sustained rate of drilling in this hard sedimentary rock was about 0.4 ft/h.

At 33 percent efficiency a shot drill will perform at about the drilling speeds tabulated below, according to class of rock and diameter of hole.

	Speed, ft/h, according to hole diameter, in		
Class of rock	3 to 6 in	6 to 15 in	15 to 30 in
Soft	3–6	1.5–3	0.8–1.5
Medium	1.5–3	0.8–1.5	0.4–0.8
Hard	0.8–1.5	0.4–0.8	0.2–0.4

The large-diameter calyx is used by the contracting officer in qualitative investigations of rock pertaining to design rather than by the earthmover in his appraisal of the excavation characteristics. Of course, with the use of modern camera equipment it is possible

Figure 7-22 **Large-diameter cores from calyx drill.** Large-diameter (36-in) calyx cores taken from the rock foundations for Norris Dam, one of the huge dams of the Tennessee Valley Authority. During the examination the chief geologist, Dr. Charles P. Berkey, was lowered down into the hole to make one of the first visual inspections of the walls of a test hole. Anxiety over the seamy nature of strata prompted this personal observation in addition to examination of small-diameter cores. These explorations by the contracting officer gained much valuable data for the selection of methods and machinery for the rock excavation. *(Ingersoll-Rand Company.)*

to scan the rock structure in the walls of a small-diameter hole and thus eliminate the lowering of a geologist except under some unusual investigative problem. However, when exploring variable rock formations, the unusual is sometimes the most valuable investigation.

Cable Drill The well-known and long-used *cable drill* is a gravity percussive machine, depending on the free fall of a string of drilling tools for the breakage of the rock. Over a period of about a century it has been known as a spudder, a churn drill, and a well drill.

Figure 7-23 shows this rig set up for exploratory work. It is set up on cribbing on a sidehill in order to check the depth of overburden and weathered rock overlying merchantable trap rock or basalt in northern New Jersey. The rig is drilling a 5⅝-in-diameter hole and swinging a 1350-lb string of tools, including a set of jars. The hole was cased through about 15 ft of residuals and weathered rock, at which depth solid basalt was encountered. The rate of drilling through some 20 ft of rock was 0.8 ft/h. Inasmuch as the proposed quarry floor was some 100 ft below the top of the suitable rock, the quarry site was regarded as economically feasible.

During drilling the percussive energy comes from the fall of the string of tools through an adjustable 18- to 36-in stroke of the walking beam with a frequency of 40 to 60 blows per minute. In Figure 7-23 the stroke was 36 in and the frequency was 53 blows per minute. Normally, drilling is done without jars and the string of tools is made up of swivel rope socket, drill stem, and bit. When a 5⅝-in-diameter hole is being drilled, such tools weigh about 1000 lb, and with a 30-in stroke and a frequency

Figure 7-23 Cable drill for rock exploration. An efficient crawler-mounted cable drill exploring the site for a proposed trap rock or basalt quarry in northern New Jersey. A 5⅝-in-diameter hole is being put down through residuals and weathered rock into the solid basalt. After setting casing through 15 ft of overburden, acceptable rock was encountered. Three more comparable holes were drilled and the site was judged acceptable.

of 50 blows per minute, some 125,000 ft·lb/min of energy is available for breaking the rock.

The commonly used bit is the chisel bit. As with the kindred percussive bits for jackhammer and track drill (Figure 7-12), other types are available to be used according to the nature of the rock structure. Bits may be dressed by forging and sledging on the job or, more efficiently, by work in a shop. Drilling may be with or without casing, depending on the earth-rock formation. Water is fed into the hole during drilling, forming a mixture with the cuttings. Periodically the string of tools is pulled and the mixture is removed with the dart-valve bailer. If samples are desired, a special drill bit with hollow-tube sampler may be substituted for the regular bit. Otherwise, the nature of the rock is judged by the cuttings in the bailer. Casing is driven by the stroking of the tools, the drive head being driven by drive clamps attached to the stem.

The reciprocating action of the spudder beam provides the hammerlike blows for drilling and driving casing. A bull reel carries the drilling cable, a calf reel handles the cable for the casing, and a sand reel contains the bailer cable. Different positions of the crankpin control length of stroke, usually three lengths. Engine throttle controls frequency of blows.

Such a popular-sized rig as shown in Figure 7-23 has a depth capacity of about 700 ft when swinging 900 lb of tools.

Figure 7-24 displays the common drilling tools and some of the many fishing tools used to extract stuck tools or tools which have become loose because of broken cable.

Based on 33 percent efficiency for exploratory drilling, average hourly drilling speeds for holes of 4- to 6-in diameter are as listed in the following tabulation:

Class of rock	Speed, ft/h
Soft	6–12
Medium	3–6
Hard	0–3

Wash-Boring Drill The properties of the cable drill make it an ideal machine for wash boring simply by the addition of a high-pressure water pump. Wash boring is an efficient and economical means of exploring for rock when the overburden is made up of residuals, alluvia, or extremely well-weathered rock. The principles of operation are the same as those for manual work as shown in Figure 7-1.

In the case of the mechanical cable drill a greater quantity of water is introduced into the drill rod or pipe, a cutting bit is affixed to the end of the rod, a driving head is attached to the top of the rod, and almost always the hole is cased.

The spudder beam furnishes reciprocating blows to the drill rod and to the casing pipe if necessary. Bull reel, calf reel, and sand reel suffice for all hoisting and setting work.

An inherent advantage of the use of the cable drill is in exploring through alternate layers of hard and soft rocks, such as lava and alluvium. In such cases the lava may be drilled by cable tools and the underlying alluvium may be cased and wash-bored.

Wash boring is limited to about 100-ft depth because of the abnormal water pressures needed to flush out large particles of material. Hole diameter is limited to about 6 in. In soft soils, such as residuals and alluvia, overall drilling speeds of 20 ft/h are possible. In stiff clays footage drops to about 8 ft/h.

By means of a split-tube sampler attached to the drill pipe samples may be taken at any desired depth of hole.

Although the ubiquitous veteran cable drill may lack certain sophistications of principle, it is an efficient, reliable, simple, and economical drill for much rock exploration.

Many drills are multipurpose for exploratory work, being custom-built for a variety of kinds of drilling. The machine of the California Division of Highways, Figure 7-11, can put down auger, bucket-auger, rotary, and core holes. With its 10-wheel truck chassis, it is suitable for travel over almost all terrains and for drilling in almost all kinds of rock formations.

INSTRUMENTAL MEANS—THE SEISMIC TIMER

Refraction Studies

The principles of refraction seismography have been known and applied for about 75 yr. In the decade 1920 to 1930 they were applied to oil exploration and during the past 25 yr they have been applied successfully to exploration of earth-rock excavation in the thin regolith excavated by human agency.

Principles The theory is simple. Artificially created shock waves travel through an earth-rock structure at velocities proportional to the degrees of consolidation of the residuals, weathered rock, and semisolid and solid rock. In their downward penetration of the crust they are refracted away from the normal or vertical toward the tangential or horizontal according to Snell's law of refraction. It is possible, then, to calculate the shock-wave velocities according to the degrees of consolidation and the depth to the interfaces at which velocities and degrees of consolidation change.

As will be explained later is this chapter, excavation methods and costs bear an empirical relationship to the shock-wave velocities. Expressed arithmetically, cost of excavation is proportional to degree of consolidation of earth-rock, degree of consolidation is proportional to velocity, and therefore cost of excavation is proportional to shock-wave velocity. Much work in correlating excavating machinery and costs with velocity has made this generalization just about axiomatic.

There are two types of seismic timer in common use and acceptance, the single-station and the multistation machine. The single-station unit involves a fixed position for the geophone or pickup element of the timer and variable positions for the point of excitation of the shock wave.

The multistation unit includes a fixed point of excitation and variable positions for the several, usually 12, geophones. Times for the several distances are recorded graphically

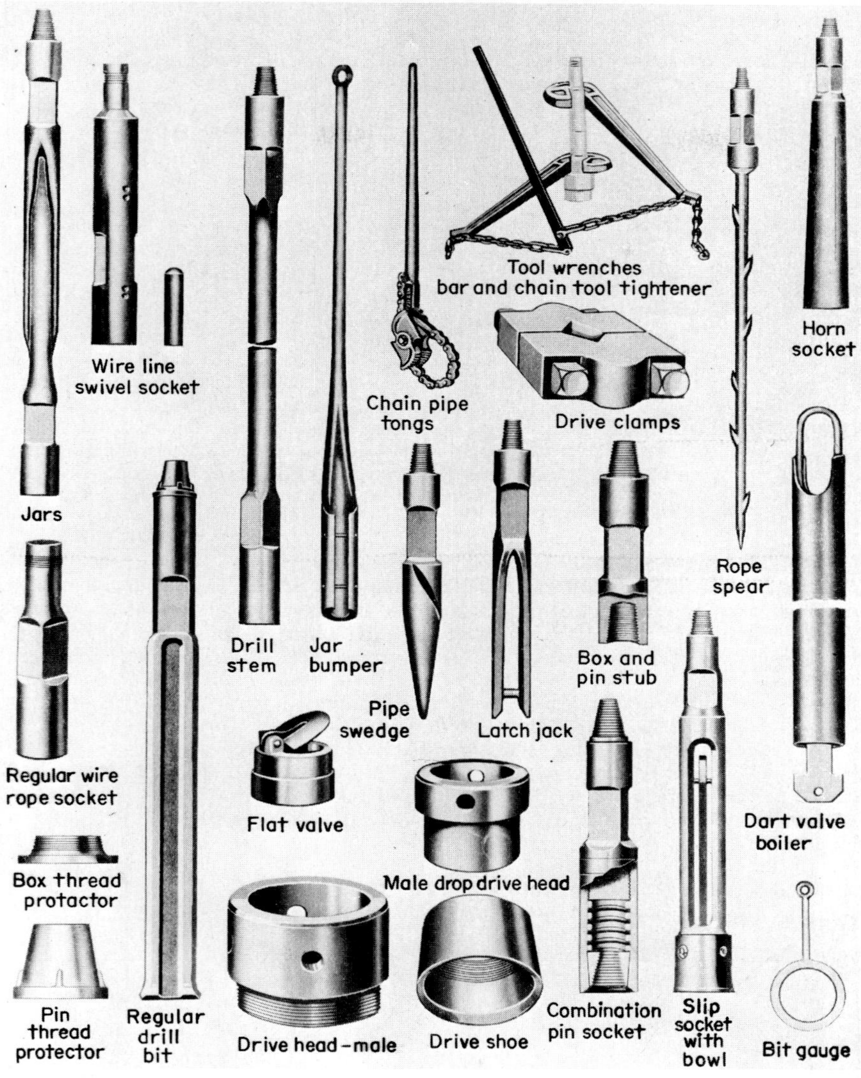

Figure 7-24 Cable drill tools. Some cable drill tools for drilling and for fishing out stuck drill steel or drill steel which is loose by reason of a broken cable. Jars are a unique device for drilling in variable rock formations in which the tools are apt to get stuck. They are set between the cable socket and the stem and the stroking of the spudder beam jars loose the tools. In characteristically difficult rock structures they are used continuously. *(Bucyrus Erie Company.)*

on a photographic negative, from the picture of which times in milliseconds are scaled.

Both types are reliable and they have inherent advantages and disadvantages. As both types produce time elements for the passage of the shock waves from the point of excitation to the geophone, this description of the seismic timer is confined to the use of the single-station unit.

In this discussion of theory, field techniques, and calculations there will be no elaboration on stratigraphic problems sometimes encountered in seismic studies or on their mathematical solution. For such professional treatment the reader is referred to the bibliography and to the instruction manuals of the makers of the several seismic timers.

Snell's law of refraction as applied to seismic work may be expressed verbally as follows: When a shock wave passes through a layer in which its velocity is V_u and strikes an interface separating this upper zone of lower velocity from the lower earth-rock zone of higher velocity V_l *at an angle* I_u with the normal, it is bent at an angle R_l with the normal in the lower layer, V_l. The relationship among these four elements is expressed by the following derived equation:

$$\frac{\sin I_u}{\sin R_l} = \frac{V_u}{V_l}$$

Figure 7-25 shows ideally the paths of travel of shock waves from points of excitation to the geophone in a three-layer earth-rock structure consisting of residuals, weathered rock, and semisolid and solid rock.

The further development of the above equation results in combination theoretical and practical equations for depths from the ground surface to the interfaces between the layers having the calculated velocities. These equations are set forth in Figure 7-26, a typical study of a 100-ft-deep cut in granite with four layers of different velocities.

Equipment and Methods Figure 7-27 illustrates the equipment needed by the instrument operator and sledge operator when the sledge is used for excitation of the shock wave. The sledge method is suitable for seismic lines of maximum 320 ft for 107-ft depth exploration. Usually explosives are used on longer lines (up to 640 ft and longer) for investigating to 217-ft or greater depths. When explosives are used, a few pounds of stick dynamite, instantaneous electric caps, blaster box, firing line, and an adze for cutting out holes for the explosive charges are added to the two packs.

Figure 7-28 illustrates the setup for a refraction seismic study. The job is a 406,000-yd³ mountain highway in the Mother Lode country of the Sierra Nevada Mountains, California. The highway is 2.04 mi long. The rock formation is the Calaveras, which, along with the Mariposa, is a common metamorphic assemblage of rocks in this fractured and faulted country. It is made up of altered sedimentary rocks consisting of schists, slates, quartzites, phyllites, and conglomerates. These two kindred formations were involved in 11 of the 509 rock excavation jobs used to develop the correlations of Figure 7-33.

Pictured in Figure 7-28 is the beginning of a 320-ft seismic line across the centerline of a 272,000-yd³ cut with maximum 69-ft centerline depth. Two other 320-ft lines and a 200-ft line were run along the 1200-ft length of the centerline. The pickup geophone is spiraled into the residuals under the right foot of the instrument operator. The 100-ft

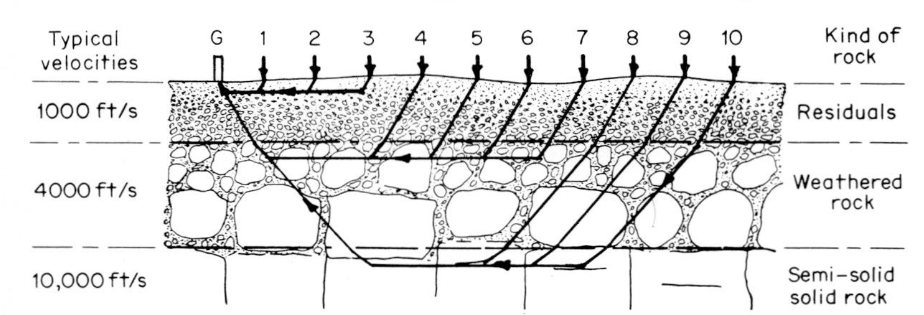

Figure 7-25 Ideal paths of travel of shock waves in an earth-rock structure according to Snell's law of refraction. The seismic timer is actuated by the earliest of the many shock waves arriving from a single point of excitation. The waves from each of the excitation points 1 through 10 are picked up by the geophone G.

Waves from 1, 2, and 3 arrive first through the upper layer of 1000 ft/s. However, of the many waves from 4, that one passing through both the upper layer and the middle layer of 4000 ft/s arrives first. Likewise, waves from 5, 6, and 7 follow the same pattern of travel through the upper and middle layers. Again, however, wave 8 reaches the geophone before any other waves through the upper and middle layers by passing through all three layers, 1000, 4000, and 10,000 ft/s. Of course, waves from 9 and 10 follow the path of wave 8.

As diagramed, a wave passing from a lower-velocity layer through the interface to a higher-velocity layer is bent farther away from the normal to the interface.

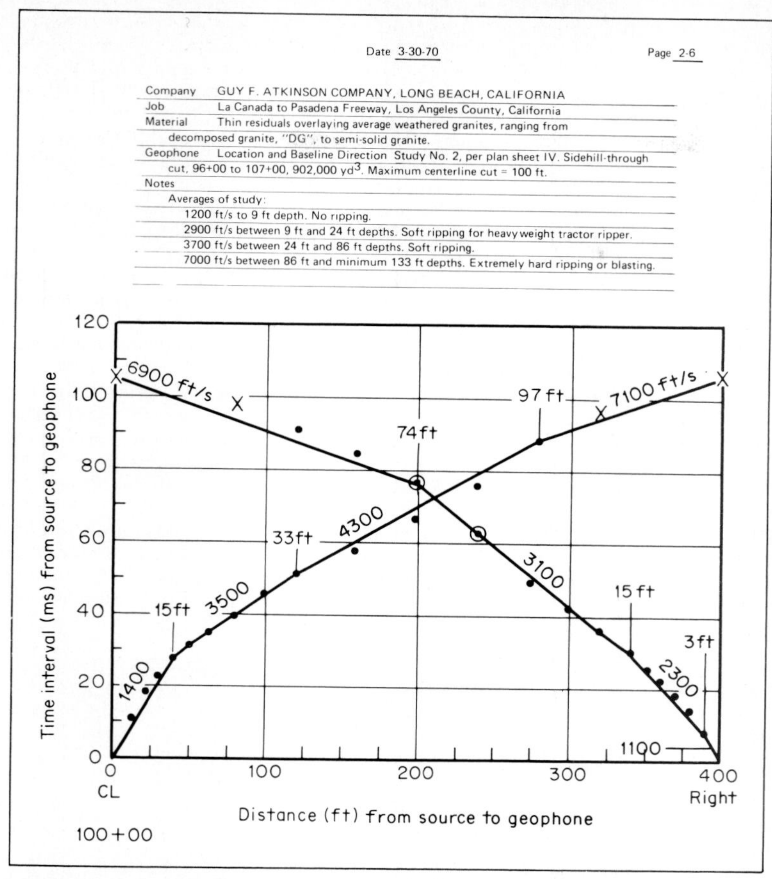
Date 3-30-70 Page 2-6

Company GUY F. ATKINSON COMPANY, LONG BEACH, CALIFORNIA
Job La Canada to Pasadena Freeway, Los Angeles County, California
Material Thin residuals overlaying average weathered granites, ranging from decomposed granite, "DG", to semi-solid granite.
Geophone Location and Baseline Direction Study No. 2, per plan sheet IV. Sidehill-through cut, 96+00 to 107+00, 902,000 yd^3. Maximum centerline cut = 100 ft.
Notes
Averages of study:
1200 ft/s to 9 ft depth. No ripping.
2900 ft/s between 9 ft and 24 ft depths. Soft ripping for heavyweight tractor ripper.
3700 ft/s between 24 ft and 86 ft depths. Soft ripping.
7000 ft/s between 86 ft and minimum 133 ft depths. Extremely hard ripping or blasting.

Figure 7-26 Seismic study of excavation. Calculations for depths and velocities below 100+00 centerline.

$$\text{D-1} = \frac{40}{2}\left(\frac{3500-1400}{3500+1400}\right)^{1/2} = 15 \text{ ft}$$

$$\text{D-2} = 0.9\times 15 + \frac{120}{2}\left(\frac{4300-3500}{4300+3500}\right)^{1/2} = 33 \text{ ft}$$

$$\text{D-3} = 0.1\times 15 + 0.8\times 33 + \frac{280}{2}\left(\frac{7100-4300}{7100+4300}\right)^{1/2} = 97 \text{ ft}$$

tape is stretched out from the geophone beyond the sledge operator. The sledge with welded-on steel disk is cable-connected to the seismic timer. The sledge operator's back is toward the geophone so that the shock wave will be driven downward toward the geophone. At the instant of impact a leaf switch, mounted on the sledge handle near the head, will close, opening up the gate of the timer, and time will start recording on the indicator-light panel. When the shock wave reaches the geophone, the gate will close and the time interval in milliseconds will be visible in the lighted bulbs. In this favorable terrain the reversed 320-ft line was run in 35 min. The averages for this study, probing to a depth of 107 ft minimum in this 69-ft deep cut, are listed below:

1300 ft/s to 6-ft depth. No ripping.
3000 ft/s between 6- and 28 ft depth. Soft ripping for heavyweight tractor-ripper.

7500 ft/s between 28- and 48-ft depths. Blasting.

9600 ft/s between 48- and minimum 107-ft depth. Blasting.

The averages for all four studies in the cut fixed the rock line at an average 29-ft depth, below which blasting would be necessary. A quantity takeoff established that 106,000 yd³, or 39 percent of the yardage of the cut, would be blasted.

Eight seismic studies were made in the three principal cuts, and the averages of the studies compared favorably with those for an adjacent section of the highway in the same Calaveras formation which was analyzed several years ago. The history of excavation in the adjacent job, along with the studies of both jobs, provided the bidding contractor with a wealth of knowledge for his bid preparation for the unclassified excavation.

The correlation of the studies of these two adjacent jobs, backed up by job experience on one of them, is a good example of the correlative approach to exploration of rock excavation which was discussed in Chapter 2 and charted in Table 2-3. When such methods are used, bidding on questionable rock excavation becomes less hazardous and approaches more closely an exact science.

Figure 7-29 illustrates shock-wave excitation by blasting. A 400-ft reversed line is being run in a 1,484,000-yd³ through cut, examining the rock excavation to 133-ft minimum depth. Excitation by sledge was carried out to 280 ft of line and thereafter explosives were used at 340- and 400-ft distances. The formation is the Panoche, which was excavated for the Pacheco Pass Freeway in Merced County, California. The formation was discussed in Chapter 2.

The rock line in this cut was established at 30-ft depth. Calculations for estimating unit bid price on the unclassified roadway excavation called for 371,000 yd³ or 25 percent, to be ripped by heavyweight tractor-rippers and 1,113,000 yd³, or 75 percent, to be blasted. The rock excavator blasted about 65 percent of a total 11 million yd³ of the Panoche and Franciscan formations of the job.

Figure 7-27 Equipment for seismic investigation of rock-earth by two-person crew. *(Left)* Pack of sledge operator. Total weight, including sledge, is about 30 lb. Equipment includes: 14-lb sledge with welded-on striking disk; 220- and 340-ft cables on reels; sender for radio link between sledge and seismic timer in lieu of connecting cable; canteen; and two lunch pails. It is sometimes possible, if there is no interference, to use the radio link. However, the cable connection is more satisfactory.

(Right) Pack of instrument operator. Total weight of about 30 lb. Equipment includes seismic timer with radio-link receiver, geophone, 100-ft engineer's tape, geologist's hand pick, case for plans, and canteen. Extra sledge with striking plate for soft ground is shown but it is not a part of the regular gear.

Figure 7-28 **Field setup of seismic equipment for refraction study** for 320-ft line across the centerline of a 69-ft-deep mountain highway cut of 272,000 yd³. Geophone under right foot of instrument operator marks the beginning of the right-to-left leg of the line, located 160 ft to the right of centerline. This tranverse line was run in favorable topography and brush and it required 35 min. Two other 320-ft lines and a 200-ft line were made along centerline with steep gradients and much brush, requiring 2 h. Thus the time for the analysis of the rock excavation of the cut required about 2½ h, about normal for mountain work in favorable weather.

Figure 7-29 **Seismic studies using explosives for excitation of shock wave.** Excitation of shock wave by blasting. The line is 400 ft long, probing to a minimum 133-ft-depth in a 95-ft deep cut of 1,484,000 yd³. The cut is in the Panoche formation, made up of thin residuals overlying weathered, semisolid, and solid heavily bedded sandstones. The rock line was established at 30-ft depth.

The shot of three sticks of 1¼-in by 8-in cartridges of dynamite is 280 ft from the geophone near the instrument operator. The worker on the blaster box has just pressed the firing button and simultaneously fired the shot and started the seismic timer to record the time interval before the shock wave is picked up by the geophone. By such means depth probes may be extended to 200 ft and more, but generally topography and other limitations prevent examinations of more than 250-ft depth.

Ripping and Blasting Zones According to Seismic Velocities

Commencing some 25 yr ago with the use of the seismic timer, it was necessary to express results in terms of shock-wave velocities through earth-rock structures with respect to two methods of rock fragmentation. Concurrent with the introduction of the seismic timer was the development of medium-weight and heavyweight tractor-rippers as economical tools for rock fragmentation. It was then natural to express rippability of tractor-rippers in terms of shock-wave velocities until a velocity was reached beyond which blasting was necessary. In the ensuing years many correlative studies have been made. Suggested guides for rippability and blasting are listed below.

1. For medium-weight tractor-rippers in the 200- to 300-engine-hp and 60,000- to 90,000-lb working-weight specification ranges:

0 to 1500 ft/s	No ripping
1500 to 3000 ft/s	Soft ripping
3000 to 4000 ft/s	Medium ripping
4000 to 5000 ft/s	Hard ripping
5000 to 6000 ft/s	Extremely hard ripping or blasting
6000 ft/s and higher	Blasting

2. For heavyweight tractor-rippers in the 300- to 525-engine-hp and 100,000- to 160,000-lb working-weight specification ranges:

0 to 1500 ft/s	No ripping
1500 to 4000 ft/s	Soft ripping
4000 to 5000 ft/s	Medium ripping
5000 to 6000 ft/s	Hard ripping
6000 to 7000 ft/s	Extremely hard ripping or blasting
7000 ft/s and higher	Blasting

Uphole Studies

Another kind of seismic study, more precise than the refraction study, is the uphole study. An exploratory drill hole is necessary for this vertical exploration, but, most unfortunately, these are rarely available. And when they are available, they are generally caved, capped permanently at the collar, or backfilled by the one making the exploration. Of some 6000 seismic studies of 1 billion yd^3 of excavation, it has been possible to make only 95 studies of the uphole kind.

The uphole study is based simply on the determination of the time for a shock wave to travel upward along the wall of the hole between two different depths of the hole. The difference between the depths, or the thickness of the layer of the rock formation, is divided by the difference between two seismic travel times for the two elevations and the quotient is the shock-wave velocity. The results are precise.

Experience with up to 340-ft-deep uphole studies indicates definitely better results than with the refraction technique, especially when one realizes that the baseline for a 340-ft-deep refraction study should be at least 1020 ft long, practically impossible in average topography.

Figure 7-30 shows the field layout for an 80-ft deep uphole study in volcanic rock of San Diego County, California. Explosive charges were set at intervals of depth in the hole and detonated by seismic electric caps, and the seismic timer recorded the time from the depth of the charge to the collar of the hole. A stick of dynamite, or a fraction of a stick, was suspended from the two leg wires of the cap, which was embedded conventionally in the stick of dynamite. Deep-hole seismic caps are spooled with leg wires of up to 400-ft length and the wires are marked at 5-ft intervals.

Usually holes are wet, sometimes with clear water and sometimes with light mud. Under such conditions the charges are weighted with 6-in spikes so as to settle to the desired depth. The long legs of the cap may be used for successively shallower shots.

Figure 7-31 is the seismic study sheet for one of 10 uphole studies in granite of the San Raphael Hills, Glendale, California. The study was made in the same tract excavation shown in Figure 5-2. The drilling and blasting contractor put down the exploratory holes by track drill, had uphole studies made, and then bid on the hard-rock drilling and blasting on a lump-sum basis.

Figure 7-30 **Field setup of seismic timer equipment for uphole study.** Firing cable with reel from blaster box to leg wires of seismic cap of charge. Blaster box connected to seismic cap and to gate which opens the inlet of the seismic timer. Firing button simultaneously fires charge and opens gate of the timer. Engineer's tape, 100-ft length, with weight for measuring depth of hole. If depth of hole, usually unknown, is greater than 100 ft, additional cord is used. Seismic timer with connected geophone located near collar of hole (foreground, left to right).

Hole, with light board from which are suspended leg wires of the cap with explosives charge. In wet holes the charge is weighted with 6-in spikes. Geophone with cable stabilizing geologist's pick (middle ground, left to right).

As a result of these combined drilling and seismic exploratory data, the contractor determined:

1. Ten depths to and beyond the rock line separating rippable from drill-and-shoot rock.
2. Rates for drilling in the hard rock.
3. Seismic data for the hard rock, from which hole diameter and hole spacing pattern were estimated.

From these data, quantities and unit costs for blasting hard rock, necessary for a venturesome lump-sum bid, were determined.

The seismic study sheet shows a slowly bending velocity versus depth curve, typical of weathered granite. The rock line is at 55-ft depth in the 100-ft-deep cut. In the prism of rock represented by the study, a quantity takeoff based on the assumption that the rock line paralleled the original ground revealed the quantity of drill-and-shoot rock.

When making seismic studies it is sometimes expedient to take advantage of an existing exploratory trench, as pictured in Figure 7-32. This trench in weathered granite was excavated by the contracting officer for the New Lake Arrowhead Dam, San Bernardino County, California. Under this fortuitous circumstance there are two advantages.

First, in the walls of this 150-ft-long, 18-ft-deep excavation one may see the weathering of the rock structure. Second, one may run a "below-original-ground" study to corroborate the findings of the nearby parallel study on the original ground. In this case a 120-ft line was run, with depth penetration of 40 ft below the bottom of the trench to a total 58-ft depth. The results of the study agreed with the results of the other four studies of the site 3 excavation of 1,100,000 yd^3 to a maximum 80-ft depth. The rock line separating rippable weathered granite from drill-and-shoot semisolid granite averaged 77-ft depth.

Such exploratory trenches are not uncommon. However, they are characteristically short and of shallow depth, the shortness necessarily controlling the length of the seismic line and therefore the depth penetration.

Velocity—Depth Relationships for the Three Classes of Rock

Figure 7-33 summarizes the velocity-depth relationships of igneous, sedimentary, and metamorphic rock formations. One may cross-refer this figure to Table 5-1, which shows

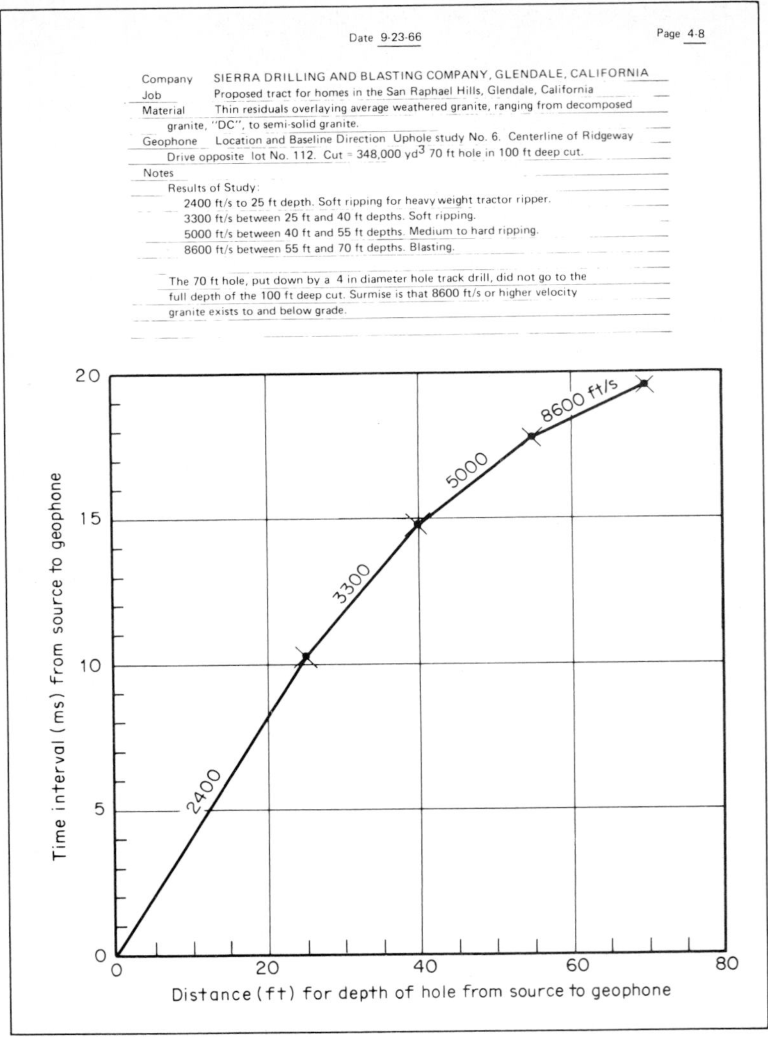

Date 9-23-66 Page 4-8

Company SIERRA DRILLING AND BLASTING COMPANY, GLENDALE, CALIFORNIA

Job Proposed tract for homes in the San Raphael Hills, Glendale, California

Material Thin residuals overlaying average weathered granite, ranging from decomposed granite, "DC", to semi-solid granite.

Geophone Location and Baseline Direction Uphole study No. 6. Centerline of Ridgeway Drive opposite lot No. 112. Cut = 348,000 yd^3 70 ft hole in 100 ft deep cut.

Notes

Results of Study:

2400 ft/s to 25 ft depth. Soft ripping for heavy weight tractor ripper.
3300 ft/s between 25 ft and 40 ft depths. Soft ripping.
5000 ft/s between 40 ft and 55 ft depths. Medium to hard ripping.
8600 ft/s between 55 ft and 70 ft depths. Blasting.

The 70 ft hole, put down by a 4 in diameter hole track drill, did not go to the full depth of the 100 ft deep cut. Surmise is that 8600 ft/s or higher velocity granite exists to and below grade.

Figure 7-31 Seismic study of excavation. Calculations for average velocities within four zones of depth:

Original ground to 25-ft depth:	25 ft/10¼ ms	=2400 ft/s
25-ft to 45-ft depth:	15 ft/4½ ms	=3300 ft/s
40-ft to 55-ft depth:	15 ft/3 ms	=5000 ft/s
55-ft to 70-ft depth:	15 ft/1¾ ms	=8600 ft/s

the relative percentages of the three basic kinds of rock in some 1 billion yd^3 of rock excavation in public and private works.

Chapter 3 contains several figures illustrative of the use of Figure 7-33.

Figure 3-4 pictures minimum-weathered limestone. Entering Figure 7-33 at the ordinate 6000 ft/s, one finds for the curve "Sedimentary—Min W" an abscissa of 31-ft depth where the rock excavator might commence to blast the rock.

Figure 3-6 shows average-weathered igneous intrusive granite. If a highway were

Figure 7-32 Ideal seismic study at bottom of exploratory trench. Refraction seismic study in an exploration trench of borrow pit for dam building. Excavation is in tonalite granite, made up of thin residuals with some surface boulders atop weathered granite with embedded boulders, atop semisolid and solid granite. The findings of the 120-ft seismic line corroborated the findings of a nearby parallel 320-ft line on natural ground. These two studies, together with three other studies in site 3, set the rock line at 77-ft depth in the 1,100,000-yd³, 80-ft maximum depth rock borrow pit. Trenches present excellent opportunities to inspect rock formations in the weathered zone, as in this case, where the method reveals weakening dikes, cooling joints, and the embedded boulders in the matrix of decomposed granite.

projected through the ridge with a cut of 50-ft depth, one would enter the graph at the abscissa 50-ft depth, intersect the curve "Igneous—Av W," and find the ordinate 6700 ft/s within the questionable zone. The canny rock mover would anticipate some costly drill-and-shoot rock near grade.

Figure 3-12 illustrates well-weathered quartzite. If one were considering a dam-abutment excavation of 80-ft depth in this metamorphic rock formation, one would enter the graph at the abscissa 80 ft, intersect the curve "Metamorphic-Max W," and read the ordinate of 5800 ft/s for the probable shock-wave velocity at grade. The velocity approaches the zone of questionable ripping or blasting, and so ripping near grade would be difficult.

One appreciates that Figure 7-33 is "average" and is merely an indicator to be proved by localized tests and studies of the particular rock formation. In a cut of the same rock formation with the same degree of weathering, two seismic studies may differ materially because of the vagaries of nature.

The advantages of the single-station seismic timer for both refraction and uphole studies are many, as summarized below:

1. *Portability.* As Figure 7-27 emphasizes, the two-person crew carries two packs and can go anywhere they can crawl.

2. *Speed.* It is normal to run an average of eight studies for depth penetrations in the 60- to 100-ft range in eight working hours.

3. *Precision.* Seismic examination being a form of sampling, it is axiomatic that "the more samples, the greater the precision." Much experience based on thousands of studies indicates that errors in velocity and depth determinations are about ± 5 percent for the individual study. In the some 1 billion yd³ of rock explored, the average was 170,000 yd³ per study.

4. *Versatility.* Any rock formation may be explored, ranging from alluvium of 1000 ft/s to andesite of 24,000 ft/s and to practical depths over 200 ft.

5. *Economy.* A typical 5-million-yd³ rock excavation with less than five different rock

formations calls for about 25 studies. Four field days and four office days, along with perhaps two travel days, are required, giving a total of 20 labor-days. In 1976 the cost of such an investigation amounted to less than 0.1 cent/yd³.

COMPARISON OF METHODS FOR EXPLORATION OF EXCAVATIONS

Choice of any or several of the manual, mechanical, and instrumental means for exploration depends on the purpose of the investigation. The contracting officer, the rock mover, and the miner may have different or the same objectives. Table 7-2 is suggested as a guide to the chief capabilities of the several means of rock exploration.

The vast array of exploratory tools offers just about every facility to the inquisitive earthmover. Excavators vary greatly in their completeness of rock investigations.

Some big companies do little exploratory work, relying on vast experience and the data furnished by the contracting officer. Others, more inquisitive, go to great lengths to learn everything possible about the invisible hard rock.

When one considers the great disparity in unit prices for unclassified excavation submitted by experienced bidders, one must conclude that more exploratory work should be done. In April 1975 bids were taken for a budgeted $24 million freeway in granite excavation in San Diego County, California. All seven bidders had worked in the rather common granite formation. Bids ranged from $0.50 to $1.50 per cubic yard of roadway excavation in rather well balanced bids. Admittedly, this is an unusual percentage difference in a unit price bid, but it suggests that more exploratory work would have resulted in more logical conformity of unit prices for the unclassified excavation. There was 2,300,000 yd³ of roadway excavation, giving a $2.3

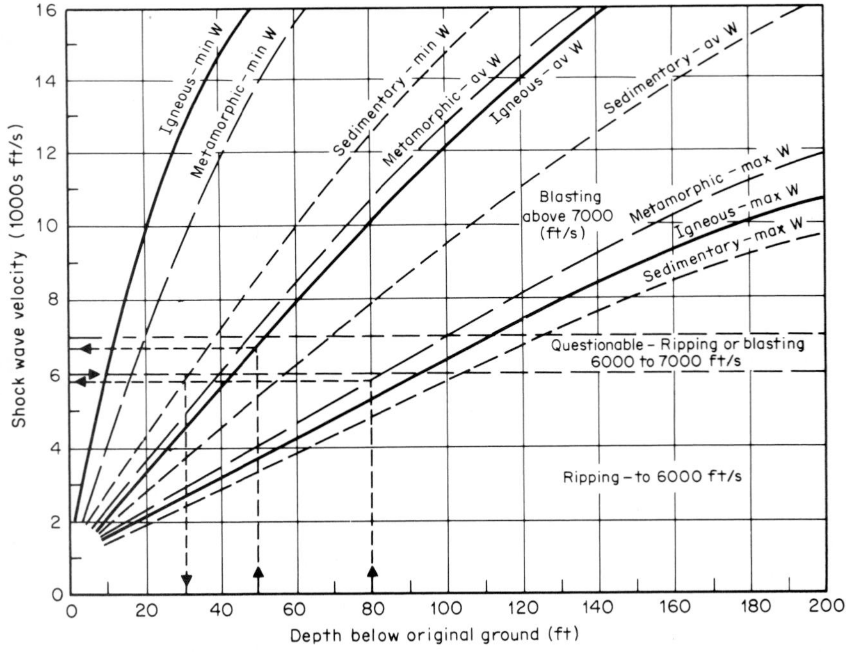

Figure 7-33 **Relationship between seismic shock-wave velocities and depths below ground surface** for the three basic kinds of rocks with minimum, average, and maximum degrees of weathering. The data for the curves are based on some 6100 seismic studies made in 509 rock excavation jobs totaling about 1 billion yd³.

TABLE 7-2

Method of exploration	Limitation of rock hardness	Limitation of depth	Purpose of exploration: Qualitative	Purpose of exploration: Quantitative
Manual:				
Pick and shovel	Hard	Limited	Yes	Yes
Sounding rod	Medium	Limited	No	Yes
Auger	Medium	Limited	Yes	Yes
Wash boring	Soft	Limited	Yes	Yes
Mechanical:				
Backhoe	Medium	Limited	Yes	Yes
Bulldozer	Medium	Limited	Yes	Yes
Auger drill	Medium	Limited	Yes	Yes
Bucket drill	Medium	Limited	Yes	Yes
Jackhammer	None	Limited	Limited	Yes
Track drill	None	Limited	Limited	Yes
Rotary drill	None	Unlimited	Yes	Yes
Core drill	None	Unlimited	Yes	Yes
Cable drill	None	Unlimited	Yes	Yes
Wash boring	Soft	Limited	Yes	Yes
Instrumental:				
Seismic timer:				
Refraction study	None	Limited	Limited	Yes
Uphole study	None	Unlimited	Limited	Yes

million difference in bid total due to the item. The low bid was $16,729,332 and the high bid was $19,921,313.

SUMMARY

As was explained in the summary of Chapter 6, exploration of excavation is a correlative to examination of excavation. Accordingly, good management regards examination and exploration as of equal importance. In so doing ownership is assured of a good appraisal of the excavation.

CHAPTER 8

Costs of Machinery and Facilities

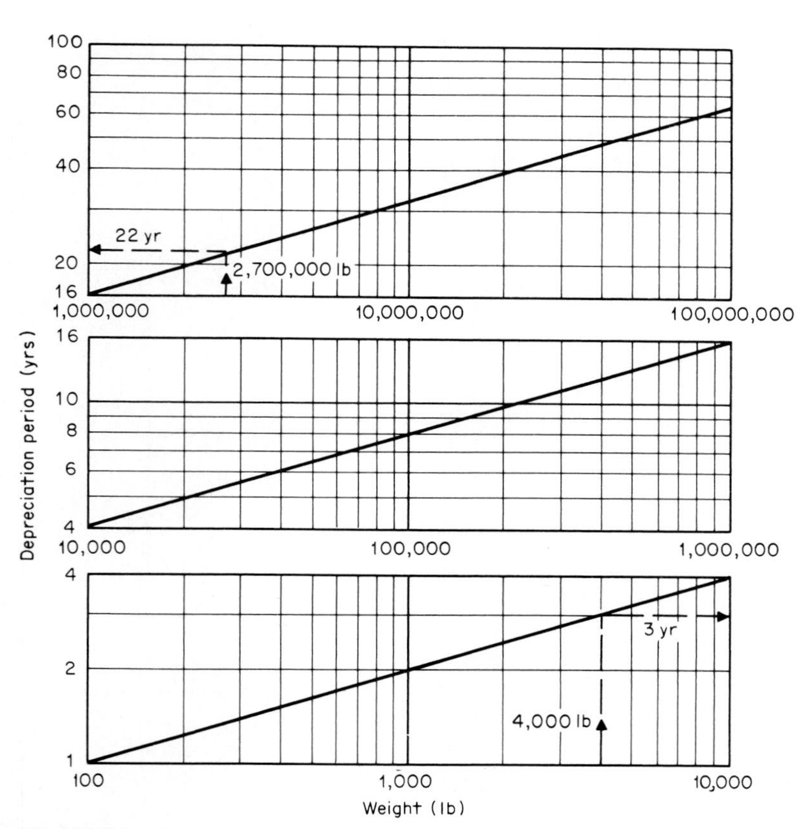

PLATE 8-1

Depreciation period in years according to weight in pounds for excavating machinery powered by engines and motors. The idealized graph is based on the formula:

$$DY = 2^{(\log WP - \log 100)}$$

where DY=depreciation period in years and WP=weight of machine in pounds.

COST OF OWNERSHIP AND OPERATION OF MACHINERY AND FACILITIES

The hourly cost of ownership and operation of rock excavation machinery varies greatly for a given machine or facility because of the range of the following factors:

1. Total cost of job, depending on discount buying, cost of financing or borrowing, sales and use taxes, and freight to job location.
2. Salvage or trade-in value.
3. Period or rate of depreciation in years and hours of use yearly.
4. Severity of work, as expressed in consumption of parts and replacements of both mechanical equipment and tires.
5. Fuel and lubricant consumption.
6. Labor rates, including fringe benefits, for operators, oilers, and maintenance mechanics.

These items are also functions of location, latitude, altitude, weather, and physical characteristics of the job. In the face of these many influencing factors, the estimator uses one or more of the following methods for estimating costs of ownership and operation.

1. Company records of machinery costs, tempered by adjustment for the particular job conditions.
2. In the case of a new variation or type of machine, a basic buildup of estimated costs founded on experience with a similar machine.
3. Local rental schedules for machines, from which costs of ownership and operation may be deduced.
4. Fundamental buildup in accordance with the material furnished by:
   a. Books and magazines of the construction and mining industries.
   b. Data from the manufacturers of the machines.
   c. Guides and tables for the costs of ownership and operation, as furnished by associations such as the Associated General Contractors of America.

A buildup of costs based on diverse information is illustrated in Tables 8-1 and 8-2, the one for a heavyweight tractor-ripper and the other for an average-size shovel for massive excavation.

In these tables, under "Hourly Cost of Ownership and Operation" is one column headed "Dollars" and one headed "% of $1000s of Cost." The second column translates the dollars of the first column into terms of the cost of the machine. In the case of the tractor ripper depreciation is $17.10 hourly and is 7.1 percent of 240, the number of thousands of dollars of cost.

This method of translating all hourly costs of ownership and operation, except the wages and fringe benefits of operator(s), is used in Table 8-3. The percentage method automatically takes care of inflationary increases in all elements of machinery costs. It is simply based on the principle that all costs are approximately proportional to the total cost of the machine. Errors are in the order of ± 5 percent.

The percentage figures of Table 8-3 are suggested averages for the United States and they are based on:

1. Average cost of machine, assembled and erected, fob job in 1978.
2. Average unit prices for diesel fuel, gasoline, electrical energy, lubricants, hydraulic oils, and filters in 1978.
3. Average unit prices for consumed items such as tires, cutting edges, bits, cables, and the like in 1978.
4. *Contractors' Equipment Manual, 1974,* issued by the Associated General Contractors of America.
5. Cost data of construction and mining companies.
6. Cost estimates of manufacturers of machinery.
7. Cost data of magazines, books, and other publications.
8. Cost data gathered by the author over 50 years.

Elements of Hourly Ownership and Operating Cost Tabulation

The elements of Table 8-3 are arranged in working sequence from left to right.

Machine with Description Machines, with auxiliary equipment when necessary for work, are arranged alphabetically with notations of capacities and ratings. Capacities and

TABLE 8-1 Estimated Hourly Cost of Ownership and Operation of Heavyweight Tractor-Ripper-Bulldozer of 410-hp Engine and 110,000-lb Weight (1978)

Cost of tractor-ripper-bulldozer, fob job site	$240,000
Less trade-in or salvage value at 20% of cost	48,000
Cost to be depreciated	192,000
Depreciation period of 8 yr of 1400 h/yr or total of 11,200 h of life.	

	Hourly cost of ownership and operation	
	Dollars	% of number of $1000s of cost
Depreciation, $192,000/11,200	17.10	7.1
Interest, 8.0%; taxes, 2.0%; insurance, 2.0%; storage, 1.0%. Total 13% of average investment of $135,000 or $17,600 yearly. $17,600/1400	12.60	5.2
Replacement cost escalation 7% of $240,000/1400	12.00	5.0
Total fixed charges	41.70	17.4
Repairs, replacements, and labor:		
Major and minor items	28.80	
Consumed parts, ripper and bulldozer	7.20	
Total	36.00	15.0
Operation, less operator:		
Diesel fuel, 16 gal at $0.40/gal	6.40	
Lubricants, hydraulic oil, filters	1.30	
Oilers and grease truck	2.50	
Total	10.20	4.2
Total hourly cost of ownership and operation, less operator	87.90	36.6

TABLE 8-2 Estimated Hourly Cost of Ownership and Operation of Electric Mining Shovel with 18-yd³ Capacity and 1600-hp Motors (1978)

Cost of shovel, erected, fob job site	$2,400,000
Less trade-in or salvage value at 20% of cost	480,000
Cost to be depreciated	1,920,000
Depreciation period of 20 yr of 4000 h/yr, or total of 80,000 h of life.	

	Hourly cost of ownership and operation	
	Dollars	% of number of $1000s of cost
Depreciation, $1,920,000/80,000	24.00	1.0
Interest, 8.0%; taxes 2.0%; insurance 2.0%. Total 12.0% of average investment of $1,260,000, or $151,200 yearly. $151,200/4000	37.80	1.6
Replacement cost escalation 7% of $2,400,000/4000	42.00	1.8
Total fixed charges	103.80	4.4
Repairs, replacements, and labor	66.00	2.7
Operation, less operator(s):		
Electrical energy, 1069 hp, average consumption 790 kWh @ $0.045	35.60	
Lubricants	3.60	
Total	39.20	1.6
Total hourly cost of ownership and operation, less operator(s)	209.00	8.7

TABLE 8-3 Cost of Ownership and Operation of Excavating Machinery and Facilities

	Weight and power		Depreciation factors			Yearly fixed charges as % of cost				Hourly costs as % of number of $1000s of cost			
Machinery with description	Weight of machine, lb	Delivered horsepower of prime mover(s)	Depreciation period, yr	Hours of use yearly	Depreciation period, total hours	Depreciation	Interest, taxes, insurance, storage	Replacement cost escalation	Total fixed charge	Fixed charges	Repairs, replacements, labor thereto	Operating expenses	Total hourly costs
Air compressors, portable, wheel-tire-mounted:													
Diesel engine:													
125 ft^3/min	3,300	40	6	1200	7,200	13.3	7.6	7.0	27.9	23	18	6	47
175 ft^3/min	3,800	50	6	1200	7,200	13.3	7.6	7.0	27.9	23	18	7	48
250 ft^3/min	4,200	75	6	1600	9,600	13.3	7.6	7.0	27.9	17	18	10	45
365 ft^3/min	4,500	110	6	1600	9,600	13.3	7.6	7.0	27.9	17	18	13	48
600 ft^3/min	6,800	180	8	1600	12,800	10.0	7.3	7.0	24.3	15	16	14	45
900 ft^3/min	10,000	300	8	1600	12,800	10.0	7.3	7.0	24.3	15	16	14	45
1200 ft^3/min	14,000	360	8	1600	12,800	10.0	7.3	7.0	24.3	15	16	14	45
1750 ft^3/min	23,000	525	8	1600	12,800	10.0	7.3	7.0	24.3	15	16	12	43
2000 ft^3/min	28,000	600	8	1600	12,800	10.0	7.3	7.0	24.3	15	16	12	43
Gasoline engine:													
85 ft^3/min	2,400	30	6	1100	6,600	13.3	7.6	7.0	27.9	25	20	11	56
125 ft^3/min	2,900	40	6	1100	6,600	13.3	7.6	7.0	27.9	25	20	13	58
175 ft^3/min	3,400	50	6	1100	6,600	13.3	7.6	7.0	27.9	25	20	16	61
Electric motor:													
900 ft^3/min	6,800	300	8	1600	12,800	10.0	7.3	7.0	24.3	15	12	40	67
1200 ft^3/min	9,500	360	8	1600	12,800	10.0	7.3	7.0	24.3	15	12	37	64
1750 ft^3/min	15,600	525	8	1600	12,800	10.0	7.3	7.0	24.3	15	12	34	61
2000 ft^3/min	19,000	600	8	1600	12,800	10.0	7.3	7.0	24.3	15	12	31	58
Air compressors, stationary, skid-mounted:													
Diesel engine:													
900 ft^3/min	9,000	300	8	1600	12,800	10.0	7.3	7.0	24.3	15	16	18	49
1200 ft^3/min	12,600	360	8	1600	12,800	10.0	7.3	7.0	24.3	15	16	16	47
1750 ft^3/min	20,700	525	8	1600	12,800	10.0	7.3	7.0	24.3	15	16	14	45
2000 ft^3/min	24,300	600	8	1600	12,800	10.0	7.3	7.0	24.3	15	16	13	44

TABLE 8-3 Cost of Ownership and Operation of Excavating Machinery and Facilities *(Continued)*

	Weight and power		Depreciation factors			Yearly fixed charges as % of cost				Hourly costs as % of number of $1000s of cost			
Machinery with description	Weight of machine, lb	Delivered horsepower of prime mover(s)	Depreciation period, yr	Hours of use yearly	Depreciation period, total hours	Depreciation	Interest, taxes, insurance, storage	Replacement cost escalation	Total fixed charge	Fixed charges	Repairs, replacements, labor thereto	Operating expenses	Total hourly costs
Air compressors, stationary, skid-mounted:													
Electric motor:													
900 ft^3/min	6,100	300	8	1600	12,800	10.0	7.3	7.0	24.3	15	12	44	71
1200 ft^3/min	8,500	360	8	1600	12,800	10.0	7.3	7.0	24.3	15	12	38	65
1750 ft^3/min	14,000	525	8	1600	12,800	10.0	7.3	7.0	24.3	15	12	32	59
2000 ft^3/min	21,800	600	8	1600	12,800	10.0	7.3	7.0	24.3	15	12	26	52
Air drills, complete with tools:													
Crawler-mounted:													
3½-in-diameter hole	12,000		6	1600	9,600	13.3	7.6	7.0	27.9	17	35	1	53
4½-in-diameter hole	13,000		6	1600	9,600	13.3	7.6	7.0	27.9	17	35	1	53
5½-in-diameter hole	14,000		6	1600	9,600	13.3	7.6	7.0	27.9	17	35	1	53
Crawler-mounted, with compressor, downhole hammer, and rotary drilling:													
Diesel engine:													
6½-in-diameter hole, 330 ft^3/min	43,000	200	8	1600	12,800	10.0	7.3	7.0	24.3	15	22	3	40
7½-in-diameter hole, 330 ft^3/min	48,000	300	8	1600	12,800	10.0	7.3	7.0	24.3	15	22	4	41
8½-in-diameter hole, 600 ft^3/min	53,000	400	10	1600	16,000	8.0	7.2	7.0	22.2	14	20	5	39
9½-in-diameter hole, 600 ft^3/min	58,000	500	10	1600	16,000	8.0	7.2	7.0	22.2	14	20	6	40
Electric motor:													
8½-in-diameter hole, 600 ft^3/min	39,700	400	10	1600	16,000	8.0	7.2	7.0	22.2	14	20	12	46
9½-in-diameter hole, 600 ft^3/min	43,400	500	10	1600	16,000	8.0	7.2	7.0	22.2	14	20	13	47
Drifters, with feed shell and support column:													
Lightweight	400		3	1200	3,600	26.7	8.6	7.0	42.3	35	50	1	86
Medium-weight	500		3	1200	3,600	26.7	8.6	7.0	42.3	35	50	1	86
Heavyweight	600		3	1200	3,600	26.7	8.6	7.0	42.3	35	50	1	86

Jackhammers:													
Lightweight	30		3	1200	3,600	26.7	8.6	7.0	42.3	35	50	1	86
Medium-weight	45		3	1200	3,600	26.7	8.6	7.0	42.3	35	50	1	86
Heavyweight	60		3	1200	3,600	26.7	8.6	7.0	42.3	35	50	1	86
Air equipment:													
Pipe or hose, with valves and fittings			3	1400	4,200	26.7	8.6	7.0	42.3	30	10	1	41
Bit sharpener and forge			7	1400	9,800	11.4	7.4	7.0	24.8	18	12	1	31
Auger drills, spiral-type, diesel engine, complete with tools:													
Vertical and inclined holes, 3- to 36-in-diameter holes, to 200-ft depth, 140 hp, truck-mounted	35,900	350	10	1200	12,000	8.0	7.2	7.0	22.2	18	15	5	38
Horizontal and inclined holes, 3- to 24-in-diameter holes, to 100-ft length, 150 hp, crawler-mounted	22,700	230	10	1200	12,000	8.0	7.2	7.0	22.2	18	15	5	38
Automobiles and pickup trucks, gasoline engines:													
Lightweight	3,000	100	2	2000	4,000	40.0	9.8	7.0	56.8	28	18	36	82
Medium-weight	4,000	150	3	2000	6,000	26.7	8.6	7.0	42.3	21	15	36	72
Heavyweight	6,000	200	4	2000	8,000	20.0	8.1	7.0	35.1	18	12	36	66
Backhoe excavators, diesel engine:													
Wheel-tire- and crawler-tractor-mounted, with front-end loader:													
½ yd^3, 100 hp	26,000	100	8	1400	11,200	10.0	7.3	7.0	24.3	17	14	4	35
1 yd^3, 150 hp	40,000	150	8	1400	11,200	10.0	7.3	7.0	24.3	17	14	4	35
Truck-mounted, all diesel engines:													
½ yd^3, 250 hp	36,000	250	8	1400	11,200	10.0	7.3	7.0	24.3	17	12	6	35
1 yd^3, 350 hp	52,000	350	8	1400	11,200	10.0	7.3	7.0	24.3	17	12	5	34
2 yd^3, 450 hp	74,000	450	8	1400	11,200	10.0	7.3	7.0	24.3	17	12	4	33
Crawler-mounted, full revolving type:													
1 yd^3, 100 hp	69,000	100	10	1400	14,000	8.0	7.2	7.0	22.2	16	5	3	24
2 yd^3, 150 hp	113,000	150	10	1400	14,000	8.0	7.2	7.0	22.2	16	4	2	21
3 yd^3, 200 hp	174,000	200	12	1400	16,800	6.7	7.0	7.0	20.7	15	4	2	21
4 yd^3, 250 hp	201,000	250	14	1400	19,600	5.7	7.0	7.0	19.7	14	3	1	18
5 yd^3, 300 hp	263,000	300	16	1400	22,400	5.0	6.9	7.0	18.9	14	3	1	18

TABLE 8-3 Cost of Ownership and Operation of Excavating Machinery and Facilities *(Continued)*

Machinery with description	Weight and power		Depreciation factors			Yearly fixed charges as % of cost				Hourly costs as % of number of $1000s of cost			
	Weight of machine, lb	Delivered horsepower of prime mover(s)	Depreciation period, yr	Hours of use yearly	Depreciation period, total hours	Depreciation	Interest, taxes, insurance, storage	Replacement cost escalation	Total fixed charge	Fixed charges	Repairs, replacements, labor thereto	Operating expenses	Total hourly costs
Belt conveyors, with rollers, frame, supports, and power:													
Portable, stackers, 100-ft long units. Belt speed 400 ft/min, belt angle 15–26°:													
Diesel engine:													
36-in belt width, 45 hp	12,400	45	8	1400	11,200	10.0	7.3	7.0	24.3	17	14	3	34
48-in belt width, 80 hp	24,300	80	8	1400	11,200	10.0	7.3	7.0	24.3	17	14	3	34
60-in belt width, 125 hp	43,100	125	8	1400	11,200	10.0	7.3	7.0	24.3	17	14	2	33
72-in belt width, 180 hp	61,900	180	8	1400	11,200	10.0	7.3	7.0	24.3	17	14	2	33
Electric motor:													
36-in belt width, 45 hp	12,000	45	10	1400	14,000	8.0	7.2	7.0	22.2	16	12	7	35
48-in belt width, 80 hp	23,000	80	10	1400	14,000	8.0	7.2	7.0	22.2	16	12	6	34
60-in belt width, 125 hp	41,000	125	10	1400	14,000	8.0	7.2	7.0	22.2	16	12	5	33
72-in belt width, 180 hp	59,000	180	10	1400	14,000	8.0	7.2	7.0	22.2	16	12	5	33
Stationary, 100-ft-long units. Belt speed 400 ft/min, belt angle 3°:													
Diesel engine:													
36-in belt width, 15 hp	11,000	15	8	1400	11,200	10.0	7.3	7.0	24.3	17	14	2	33
48-in belt width, 30 hp	20,000	30	8	1400	11,200	10.0	7.3	7.0	24.3	17	14	2	33
60-in belt width, 45 hp	37,000	45	8	1400	11,200	10.0	7.3	7.0	24.3	17	14	1	32
72-in belt width, 60 hp	53,000	60	8	1400	11,200	10.0	7.3	7.0	24.3	17	14	1	32
Electric motor:													
36-in belt width, 15 hp	10,000	15	10	1400	14,000	8.0	7.2	7.0	22.2	16	12	3	31
48-in belt width, 30 hp	18,000	30	10	1400	14,000	8.0	7.2	7.0	22.2	16	12	3	31
60-in belt width, 45 hp	35,000	45	10	1400	14,000	8.0	7.2	7.0	22.2	16	12	2	30
72-in belt width, 60 hp	50,000	60	10	1400	14,000	8.0	7.2	7.0	22.2	16	12	2	30

Belt loaders, diesel engines:													
Movable, stub-type:													
Belt 48 in wide and 60 ft long. 120 hp	52,000	120	10	1200	12,000	8.0	7.2	7.0	22.2	18	16	3	37
Belt 60 in wide and 60 ft long. 180 hp	60,000	180	10	1200	12,000	8.0	7.2	7.0	22.2	18	16	3	37
Belt 72 in wide and 60 ft long. 260 hp	70,000	260	10	1200	12,000	8.0	7.2	7.0	22.2	18	16	4	38
Mobile:													
Pull type, by crawler-tractor:													
Wheel-tire-mounted. Belt 48 in wide and 40 ft long. 120 hp.	27,000	120	10	1200	12,000	8.0	7.2	7.0	22.2	18	16	5	39
Crawler-mounted. Belt 60 in wide and 40 ft long. 300 hp.	90,000	300	10	1200	12,000	8.0	7.2	7.0	22.2	18	16	5	39
Self-propelled type, crawler-tractor-mounted:													
Belt 72 in wide and 40 ft long. Complete with two tractors. Loader hp 550. Tractor hp 820. Total hp 1370.	24,2000	1370	10	1200	12,000	8.0	7.2	7.0	22.2	18	16	5	39
Brush burners, portable, wheel-tire-mounted, gasoline engine:													
Medium-duty, 3 hp, 30-gal/h fuel consumption	500	3	4	1400	5,600	20.0	8.1	7.0	35.1	25	16	285	326
Heavy-duty, 12 hp, 60-gal/h fuel consumption	1,500	12	4	1400	5,600	20.0	8.1	7.0	35.1	25	16	233	274
Bucket drills, truck-mounted, with tools, diesel engines:													
½-yd³ bucket, to 48-in-diameter and 100-ft-deep hole, 250 total combined hp	23,000	250	10	1200	12,000	8.0	7.2	7.0	22.2	18	14	5	37
1-yd³ bucket, to 120-in-diameter and 150-ft-deep hole, 350 total combined hp	36,000	350	10	1200	12,000	8.0	7.2	7.0	22.2	18	14	5	37
1½-yd³ bucket, to 120-in-diameter and 200-ft-deep hole, 450 total combined hp	48,000	450	10	1200	12,000	8.0	7.2	7.0	22.2	18	14	4	36
Buckets:													
Clamshell, heavy-duty, and grapple, heavy-duty:													
2 yd³	11,500		6	1600	9,600	13.3	7.6	7.0	27.9	17	16	1	34
3 yd³	16,000		6	1600	9,600	13.3	7.6	7.0	27.9	17	16	1	34
4 yd³	21,000		6	1600	9,600	13.3	7.6	7.0	27.9	17	16	1	34
5 yd³	27,000		6	1600	9,600	13.3	7.6	7.0	27.9	17	16	1	34
6 yd³	34,000		6	1600	9,600	13.3	7.6	7.0	27.9	17	16	1	34

TABLE 8-3 Cost of Ownership and Operation of Excavating Machinery and Facilities *(Continued)*

Machinery with description	Weight and power		Depreciation factors			Yearly fixed charges as % of cost				Hourly costs as % of number of $1000s of cost			
	Weight of machine, lb	Delivered horsepower of prime mover(s)	Depreciation period, yr	Hours of use yearly	Depreciation period, total hours	Depreciation	Interest, taxes, insurance, storage	Replacement cost escalation	Total fixed charge	Fixed charges	Repairs, replacements, labor thereto	Operating expenses	Total hourly costs
Buckets:													
Dragline, heavy-duty:													
2 yd^3	6,000		8	1600	12,800	10.0	7.3	7.0	24.3	15	14	1	30
3 yd^3	8,000		8	1600	12,800	10.0	7.3	7.0	24.3	15	14	1	30
4 yd^3	10,000		8	1600	12,800	10.0	7.3	7.0	24.3	15	14	1	30
5 yd^3	12,000		8	1600	12,800	10.0	7.3	7.0	24.3	15	14	1	30
6 yd^3	15,000		8	1600	12,800	10.0	7.3	7.0	24.3	15	14	1	30
Buildings, portable, wheels-tires-mounted:													
Office, 60 × 12 ft	7,000		5	2000	10,000	16.0	7.8	7.0	30.8	15	2	5	22
Bunk house, 60 × 24 ft	14,000		5	3800	19,000	16.0	7.8	7.0	30.8	8	2	5	15
Mess hall, 60 × 24 ft	22,000		5	3800	19,000	16.0	7.8	7.0	30.8	8	2	5	15
Machine shop, 60 × 24 ft	10,000		5	2900	14,500	16.0	7.8	7.0	30.8	11	2	5	18
Tool house, 60 × 12 ft	5,000		5	2900	14,500	16.0	7.8	7.0	30.8	11	2	1	14
Powder magazine, 12 × 12 ft	2,000		5	2000	10,000	16.0	7.8	7.0	30.8	15	1	1	17
Cable drill, complete with tools:													
Truck-mounted, diesel engines, 3- to 6-in-diameter, up to 700-ft-deep holes, 225 total combined hp	17,000		10	1200	12,000	8.0	7.2	7.0	22.2	18	12	6	36
Cars, steel, rail, side-dump, quarry-type:													
10 cysm, 14 tons	14,000		15	2000	30,000	5.3	6.9	7.0	19.2	10	12	1	23
20 cysm, 27 tons	27,000		15	2000	30,000	5.3	6.9	7.0	19.2	10	12	1	23
40 cysm, 54 tons	54,000		15	2000	30,000	5.3	6.9	7.0	19.2	10	12	1	23
60 cysm, 81 tons	81,000		15	2000	30,000	5.3	6.9	7.0	19.2	10	12	1	23

Chippers, log, portable, wheel-tire-mounted, diesel engine:													
8 tons/h of chips, 130 hp	16,000	130	8	1400	11,200	10.0	7.3	7.0	24.3	18	12	6	36
20 tons/h of chips, 290 hp	45,000	290	8	1400	11,200	10.0	7.3	7.0	24.3	18	12	5	35
Cleaning equipment, steam, portable, wheel—tire-mounted, gasoline engine:													
100 gal/h, water to steam, 3 hp, 2 gal/h fuel oil	500	3	5	1800	9,000	16.0	7.8	7.0	30.8	17	15	75	107
200 gal/h, water to steam, 4 hp, 4 gal/h fuel oil	1,100	4	5	1800	9,000	16.0	7.8	7.0	30.8	17	15	47	79
400 gal/h, water to steam, 5 hp, 8 gal/h fuel oil	2,400	5	6	1800	10,800	13.3	7.6	7.0	27.9	16	15	46	77
600 gal/h, water to steam, 10 hp, 12 gal/h fuel oil	3,600	10	6	1800	10,800	13.3	7.6	7.0	27.9	16	15	43	74
Core drills, complete with tools, diesel engine:													
Diamond-type, 1- to 6-in-diameter and up to 1500-ft-deep hole:													
Crawler-mounted, 50 hp	18,000	50	8	1400	11,200	10.0	7.3	7.0	24.3	18	22	2	42
Truck-mounted, 200 total combined hp	15,000	200	8	1400	11,200	10.0	7.3	7.0	24.3	18	22	5	45
Skid-mounted, 50 hp	4,000	50	8	1400	11,200	10.0	7.3	7.0	24.3	18	35	5	58
Shot or calyx type, 2- to 20-in diameter and up to 600-ft-deep hole:													
Crawler-mounted, 60 hp	21,000	60	8	1400	11,200	10.0	7.3	7.0	24.3	18	22	2	42
Truck-mounted, 210 total combined hp	18,000	210	8	1400	11,200	10.0	7.3	7.0	24.3	18	22	4	44
Skid-mounted, 60 hp	7,000	60	8	1400	11,200	10.0	7.3	7.0	24.3	18	35	6	59
Crane excavators, complete with bucket or grapple, diesel engine:													
Crawler-mounted:													
2 yd^3, 40 tons, 150 hp	102,000	150	10	1600	16,000	8.0	7.2	7.0	22.2	14	6	2	22
3 yd^3, 50 tons, 200 hp	154,000	200	12	1600	19,200	6.7	7.0	7.0	20.7	13	5	2	20
4 yd^3, 60 tons, 250 hp	178,000	250	14	1700	23,800	5.7	7.0	7.0	19.7	12	4	2	18
5 yd^3, 70 tons, 300 hp	215,000	300	16	1700	27,200	5.0	6.9	7.0	18.9	11	3	2	16
Truck-mounted:													
2 yd^3, 30 tons, 10-ft radius, 400 total hp	62,000	400	10	1400	14,000	8.0	7.2	7.0	22.2	16	6	3	25
3 yd^3, 60 tons 12 ft radius, 500 total hp	90,000	500	12	1400	16,800	6.7	7.0	7.0	20.7	15	5	3	23
4 yd^3, 90 tons 12 ft radius, 600 total hp	118,000	600	14	1600	22,400	5.7	7.0	7.0	19.7	12	4	3	19
5 yd^3, 120 tons 16 ft radius, 700 total hp	147,000	700	16	1600	25,600	5.0	6.9	7.0	18.9	12	3	3	18

TABLE 8-3 Cost of Ownership and Operation of Excavating Machinery and Facilities *(Continued)*

	Weight and power		Depreciation factors			Yearly fixed charges as % of cost				Hourly costs as % of number of $1000s of cost			
Machinery with description	Weight of machine, lb	Delivered horsepower of prime mover(s)	Depreciation period, yr	Hours of use yearly	Depreciation period, total hours	Depreciation	Interest, taxes, insurance, storage	Replacement cost escalation	Total fixed charge	Fixed charges	Repairs, replacements, labor thereto	Operating expenses	Total hourly costs
Crushers, primary, complete, electric motors:													
Jaw-type, with pan-feed:													
48 × 42 in, 220 tons/h, 200 total hp	207,000	200	10	1600	16,000	8.0	7.2	7.0	22.2	14	11	3	28
60 × 48 in, 340 tons/h, 270 total hp	361,000	270	10	1600	16,000	8.0	7.2	7.0	22.2	14	11	3	28
84 × 66 in, 500 tons/h, 330 total hp	469,000	330	10	1600	16,000	8.0	7.2	7.0	22.2	14	11	3	28
Gyratory type:													
36 in, 480 tons/h, 225 hp	180,000	225	15	1600	24,000	5.3	6.9	7.0	19.2	12	15	4	31
54 in, 1250 tons/h, 350 hp	280,000	350	15	1600	24,000	5.3	6.9	7.0	19.2	12	15	4	31
72 in, 2500 tons/h, 500 hp	540,000	500	15	1600	24,000	5.3	6.9	7.0	19.2	12	15	3	30
Cutters, stump, portable, wheels—tires-mounted, gasoline engine:													
10-in depth, 37 hp	3,200	37	8	1400	11,200	10.0	7.3	7.0	24.3	18	12	16	46
24-in depth, 65 hp	5,200	65	8	1400	11,200	10.0	7.3	7.0	24.3	18	12	19	49
Dragline excavators, complete with bucket:													
Crawler-mounted:													
Diesel engine:													
2 yd^3, 150 hp	100,000	150	10	1600	16,000	8.0	7.2	7.0	22.2	14	8	2	24
3 yd^3, 200 hp	140,000	200	12	1600	19,200	6.7	7.0	7.0	20.7	13	7	2	22
4 yd^3, 250 hp	185,000	250	14	1700	23,800	5.7	7.0	7.0	19.7	12	6	2	20
5 yd^3, 300 hp	225,000	300	16	1700	27,200	5.0	6.9	7.0	18.9	11	5	2	18
6 yd^3, 350 hp	270,000	350	18	1700	30,600	4.4	6.9	7.0	18.3	11	4	2	17
Electric motors:													
2 yd^3, 150 hp	98,000	150	10	1600	16,000	8.0	7.2	7.0	22.2	14	7	3	24
3 yd^3, 200 hp	137,000	200	12	1600	19,200	6.7	7.0	7.0	20.7	13	6	3	22
4 yd^3, 250 hp	181,000	250	14	1700	23,800	5.7	7.0	7.0	19.7	12	5	3	20
5 yd^3, 300 hp	221,000	300	16	1700	27,200	5.0	6.9	7.0	18.9	11	4	3	18
6 yd^3, 350 hp	265,000	350	18	1700	30,600	4.4	6.9	7.0	18.3	11	3	3	17

Pontoon-mounted, walker-type:													
Diesel engine, electric motors, or diesel-electric drive:													
8 yd³, 700 hp	750,000	700	20	4000	80,000	4.0	6.8	7.0	17.8	4	3	1	8
12 yd³, 1100 hp	1,150,000	1150	20	4000	80,000	4.0	6.8	7.0	17.8	4	3	1	8
18 yd³, 1600 hp	1,700,000	1600	20	4000	80,000	4.0	6.8	7.0	17.8	4	3	1	8
Electric motors-generators-motors, ac to dc:													
18 yd³, 1600 hp	1,700,000	1600	20	7000	140,000	4.0	6.8	7.0	17.8	3	3	2	8
27 yd³, 4200 hp	2,600,000	4200	20	7000	140,000	4.0	6.8	7.0	17.8	3	3	2	8
36 yd³, 5600 hp	3,400,000	5600	20	7000	140,000	4.0	6.8	7.0	17.8	3	3	2	8
54 yd³, 8400 hp	5,100,000	8400	30	7000	210,000	2.7	6.7	7.0	16.4	3	2	2	7
72 yd³, 11,000 hp	6,800,000	11000	30	7000	210,000	2.7	6.7	7.0	16.4	3	2	2	7
108 yd³, 16,800 hp	10,900,000	16800	30	7000	210,000	2.7	6.7	7.0	16.4	3	2	2	7
Dredges, hydraulic, diesel engines, complete:													
12-in pump, 725 hp; auxiliary engine 335 hp	140,000	1060	12	2800	33,600	6.7	7.0	7.0	20.7	7	6	4	17
18-in pump, 1125 hp; auxiliary engine 335 hp	460,000	1460	16	2800	44,800	5.0	6.9	7.0	18.9	7	6	4	17
24-in pump, 2875 hp; auxiliary engine 800 hp	670,000	3680	20	2800	56,000	4.0	6.8	7.0	17.8	6	6	4	16
Engines, clutch, skid-mounted:													
Diesel:													
10 hp	150	10	5	1200	6,000	16.0	7.8	7.0	30.8	26	24	34	84
20 hp	300	20	5	1200	6,000	16.0	7.8	7.0	30.8	26	24	34	84
40 hp	600	40	5	1200	6,000	16.0	7.8	7.0	30.8	26	18	34	78
60 hp	900	60	5	1200	6,000	16.0	7.8	7.0	30.8	26	18	34	78
100 hp	1,400	100	6	1200	7,200	13.3	7.6	7.0	27.9	23	16	34	73
200 hp	2,900	200	6	1200	7,200	13.3	7.6	7.0	27.9	23	16	34	73
300 hp	4,400	300	6	1200	7,200	13.3	7.6	7.0	27.9	23	16	34	73
400 hp	5,800	400	6	1200	7,200	13.3	7.6	7.0	27.9	23	16	34	73
600 hp	8,700	600	8	1200	9,600	10.0	7.3	7.0	24.3	20	14	34	68
800 hp	11,600	800	8	1200	9,600	10.0	7.3	7.0	24.3	20	14	34	68
1000 hp	14,500	1000	8	1200	9,600	10.0	7.3	7.0	24.3	20	14	34	68
1200 hp	17,400	1200	10	1200	12,000	8.0	7.2	7.0	22.2	18	12	34	64
1600 hp	23,200	1600	10	1200	12,000	8.0	7.2	7.0	22.2	18	12	34	64

TABLE 8-3 Cost of Ownership and Operation of Excavating Machinery and Facilities *(Continued)*

Machinery with description	Weight and power		Depreciation factors			Yearly fixed charges as % of cost				Hourly costs as % of number of $1000s of cost			
	Weight of machine, lb	Delivered horsepower of prime mover(s)	Depreciation period, yr	Hours of use yearly	Depreciation period, total hours	Depreciation	Interest, taxes, insurance, storage	Replacement cost escalation	Total fixed charge	Fixed charges	Repairs, replacements, labor thereto	Operating expenses	Total hourly costs
Engines, clutch, skid-mounted:													
Gasoline or gas:													
10 hp	100	10	5	1200	6,000	16.0	7.8	7.0	30.8	26	26	121	173
20 hp	200	20	5	1200	6,000	16.0	7.8	7.0	30.8	26	26	121	173
40 hp	400	40	5	1200	6,000	16.0	7.8	7.0	30.8	26	20	121	167
60 hp	600	60	5	1200	6,000	16.0	7.8	7.0	30.8	26	20	121	167
100 hp	1,000	100	6	1200	7,200	13.3	7.6	7.0	27.9	23	18	121	162
200 hp	2,000	200	6	1200	7,200	13.3	7.6	7.0	27.9	23	18	121	162
300 hp	3,000	300	6	1200	7,200	13.3	7.6	7.0	27.9	23	18	121	162
Explosives loading truck, bulk ammonium nitrate and fuel oil mixture, pneumatic- or auger-type, diesel engine:													
5 tons, 2 axles, 150 hp	14,000	150	8	1600	12,800	10.0	7.3	7.0	24.3	15	12	5	32
12 tons, 3 axles, 250 hp	22,000	250	8	1600	12,800	10.0	7.3	7.0	24.3	15	12	5	32
Floodlights, portable, wheel-tire mounting, 30-ft tower:													
Diesel engine-generator:													
5 kW, 10 hp	2,200	10	5	1200	6,000	16.0	7.8	7.0	30.8	26	20	2	48
10 kW, 20 hp	2,600	20	5	1200	6,000	16.0	7.8	7.0	30.8	26	20	3	49
15 kW, 30 hp	3,100	30	5	1200	6,000	16.0	7.8	7.0	30.8	26	20	4	50
Gasoline engine-generator:													
5 kW, 10 hp	2,200	10	5	1200	6,000	16.0	7.8	7.0	30.8	26	20	5	51
10 kW, 20 hp	2,600	20	5	1200	6,000	16.0	7.8	7.0	30.8	26	20	9	55
15 kW, 30 hp	3,700	30	5	1200	6,000	16.0	7.8	7.0	30.8	26	20	13	59

Fusion piercing drill, diesel engine, complete with delivered oxygen, fuel oil, and water:													
12,000 ft³/h of oxygen, 200 hp	120,000	200	20	4000	80,000	4.0	6.8	7.0	17.8	4	3	5	12
Generators, electric, skid-mounted:													
Diesel engine:													
5 kW, 8 hp	150	8	5	1200	6,000	16.0	7.8	7.0	30.8	26	20	30	76
10 kW, 15 hp	250	15	5	1200	6,000	16.0	7.8	7.0	30.8	26	20	30	76
20 kW, 30 hp	500	30	5	1200	6,000	16.0	7.8	7.0	30.8	26	20	30	76
40 kW, 60 hp	1,000	40	5	1200	6,000	16.0	7.8	7.0	30.8	26	20	30	76
60 kW, 90 hp	1,500	90	5	1200	6,000	16.0	7.8	7.0	30.8	26	20	30	76
100 kW, 150 hp	2,500	150	6	1400	8,400	13.3	7.6	7.0	27.9	20	18	30	68
200 kW, 300 hp	5,000	300	6	1400	8,400	13.3	7.6	7.0	27.9	20	18	30	68
300 kW, 450 hp	7,000	450	6	1400	8,400	13.3	7.6	7.0	27.9	20	18	30	68
400 kW, 600 hp	9,500	600	8	1600	10,800	10.0	7.3	7.0	24.3	15	16	30	61
500 kW, 750 hp	12,000	750	8	1600	10,800	10.0	7.3	7.0	24.3	15	16	30	61
600 kW, 900 hp	14,400	900	8	1600	10,800	10.0	7.3	7.0	24.3	15	16	30	61
800 kW, 1200 hp	19,100	1200	10	1600	16,000	8.0	7.2	7.0	22.2	14	14	30	58
1000 kW, 1500 hp	24,000	1500	10	1600	16,000	8.0	7.2	7.0	22.2	14	14	30	58
1200 kW, 1800 hp	28,900	1800	10	1600	16,000	8.0	7.2	7.0	22.2	14	14	30	58
Gasoline engine:													
5 kW, 8 hp	100	8	5	1200	6,000	16.0	7.8	7.0	30.8	26	22	87	135
10 kW, 15 hp	200	15	5	1200	6,000	16.0	7.8	7.0	30.8	26	22	87	135
20 kW, 30 hp	450	30	5	1200	6,000	16.0	7.8	7.0	30.8	26	22	87	135
40 kW, 60 hp	950	60	5	1200	6,000	16.0	7.8	7.0	30.8	26	22	87	135
60 kW, 90 hp	1,400	90	5	1200	6,000	16.0	7.8	7.0	30.8	26	22	87	135
100 kW, 150 hp	2,200	150	6	1400	8,400	13.3	7.6	7.0	27.9	20	20	87	127
Graders:													
Drag, tow-type, by tractor, wheel-tire-mounted:													
12-ft width, 6 cysm.	12,000		8	1400	11,200	10.0	7.3	7.0	24.3	17	8	2	27
16-ft width, 8 cysm.	13,000		8	1400	11,200	10.0	7.3	7.0	24.3	17	8	2	27
20-ft width, 10 cysm	14,500		8	1400	11,200	10.0	7.3	7.0	24.3	17	8	2	27
Motor grader, 3 axles, with scarifier, diesel engine:													
12-ft blade length, 180 hp	36,000	180	8	1400	11,200	10.0	7.3	7.0	24.3	17	10	7	34
14-ft blade length, 210 hp	45,000	210	8	1400	11,200	10.0	7.3	7.0	24.3	17	10	6	33
16-ft blade length, 240 hp	60,000	240	8	1400	11,200	10.0	7.3	7.0	24.3	17	10	5	32

TABLE 8-3 Cost of Ownership and Operation of Excavating Machinery and Facilities *(Continued)*

Machinery with description	Weight and power		Depreciation factors			Yearly fixed charges as % of cost				Hourly costs as % of number of $1000s of cost			
	Weight of machine, lb	Delivered horsepower of prime mover(s)	Depreciation period, yr	Hours of use yearly	Depreciation period, total hours	Depreciation	Interest, taxes, insurance, storage	Replacement cost escalation	Total fixed charge	Fixed charges	Repairs, replacements, labor thereto	Operating expenses	Total hourly costs
Loaders, bucket type, diesel engine													
Crawler-mounted, with ripper:													
2 cysm, 140 hp	34,000	140	8	1400	11,200	10.0	7.3	7.0	24.3	17	12	5	34
3 cysm, 190 hp	54,000	190	8	1400	11,200	10.0	7.3	7.0	24.3	17	12	5	34
4 cysm, 280 hp	82,000	280	8	1400	11,200	10.0	7.3	7.0	24.3	17	12	5	34
Wheel-tiremounted:													
2 cysm, 100 hp	21,000	100	8	1400	11,200	10.0	7.3	7.0	24.3	17	12	6	35
3 cysm, 130 hp	27,000	130	8	1400	11,200	10.0	7.3	7.0	24.3	17	12	6	35
4 cysm, 170 hp	36,000	170	8	1400	11,200	10.0	7.3	7.0	24.3	17	12	6	35
5 cysm, 260 hp	50,000	260	8	1400	11,200	10.0	7.3	7.0	24.3	17	12	6	35
6 cysm, 325 hp	74,000	325	8	1400	11,200	10.0	7.3	7.0	24.3	17	12	5	34
8 cysm, 410 hp	106,000	410	8	1400	11,200	10.0	7.3	7.0	24.3	17	12	5	34
10 cysm, 550 hp	139,000	550	8	1400	11,200	10.0	7.3	7.0	24.3	17	12	5	34
Locomotives, rail, diesel engine:													
15 tons, 180 hp	30,000	180	12	2000	24,000	6.7	7.0	7.0	20.7	15	10	6	31
25 tons, 300 hp	50,000	300	12	2000	24,000	6.7	7.0	7.0	20.7	15	10	6	31
45 tons, 540 hp	90,000	540	16	2000	32,000	5.0	6.9	7.0	18.9	14	8	6	28
65 tons, 780 hp	130,000	780	16	2000	32,000	5.0	6.9	7.0	18.9	14	8	6	28
120 tons, 1400 hp	240,000	1400	16	2000	32,000	5.0	6.9	7.0	18.9	14	8	6	28
Monitors, hydraulicking, complete with stands:													
1½-in discharge-nozzle diameter, 3-in intake pipe diameter	4,000		8	1400	11,200	10.0	7.3	7.0	24.3	17	8	1	26
2-in discharge-nozzle diameter, 4-in intake pipe diameter	5,000		8	1400	11,200	10.0	7.3	7.0	24.3	17	7	1	25
3-in discharge-nozzle diameter, 6-in intake pipe diameter	7,000		8	1400	11,200	10.0	7.3	7.0	24.3	17	6	1	24

Motors, electric, with controls, 1800 r/min:													
10 hp	200	10	6	1400	8,400	13.3	7.6	7.0	27.9	20	12	119	151
20 hp	400	20	6	1400	8,400	13.3	7.6	7.0	27.9	20	12	119	151
40 hp	700	40	6	1400	8,400	13.3	7.6	7.0	27.9	20	12	119	151
60 hp	1,100	60	6	1400	8,400	13.3	7.6	7.0	27.9	20	12	119	151
100 hp	1,800	100	8	1400	11,200	10.0	7.3	7.0	24.3	17	10	109	136
200 hp	2,600	200	8	1400	11,200	10.0	7.3	7.0	24.3	17	10	109	136
300 hp	3,400	300	8	1400	11,200	10.0	7.3	7.0	24.3	17	10	109	136
400 hp	3,800	400	8	1400	11,200	10.0	7.3	7.0	24.3	17	10	109	136
600 hp	5,700	600	10	1400	14,000	8.0	7.2	7.0	22.2	16	8	99	123
800 hp	7,600	800	10	1400	14,000	8.0	7.2	7.0	22.2	16	8	99	123
1000 hp	9,600	1000	10	1400	14,000	8.0	7.2	7.0	22.2	16	8	99	123
1200 hp	11,500	1200	10	1400	14,000	8.0	7.2	7.0	22.2	16	8	99	123
Pipe, steel, with fittings, 10 gauge, 100-ft units of length:													
2-in diameter	300		3	1400	4200	26.7	8.6	7.0	42.3	30	5	0	35
4-in diameter	600		3	1400	4200	26.7	8.6	7.0	42.3	30	5	0	35
6-in diameter	900		3	1400	4200	26.7	8.6	7.0	42.3	30	5	0	35
8-in diameter	1200		4	1400	5600	20.0	8.1	7.0	35.1	25	5	0	30
10-in diameter	1500		4	1400	5600	20.0	8.1	7.0	35.1	25	5	0	30
12-in diameter	1800		4	1400	5600	20.0	8.1	7.0	35.1	25	5	0	30
Pumps, complete, with power:													
Centrifugal, sump, portable, wheels-tires-mounted:													
Diesel engine:													
2 in, 10M, 7 hp	250	7	6	1200	7200	13.3	7.6	7.0	27.9	23	20	6	49
3 in, 20M, 10 hp	400	10	6	1200	7200	13.3	7.6	7.0	27.9	23	20	7	50
4 in, 40M, 20 hp	1000	20	6	1200	7200	13.3	7.6	7.0	27.9	23	20	8	51
6 in, 90M, 40 hp	1800	40	8	1200	9600	10.0	7.3	7.0	24.3	20	16	9	45
8 in, 125M, 80 hp	4200	80	8	1200	9600	10.0	7.3	7.0	24.3	20	16	10	46
10 in, 240M, 120 hp	4900	120	8	1200	9600	10.0	7.3	7.0	24.3	20	16	12	48
Gasoline engine:													
2 in, 10M, 7 hp	200	7	6	1200	7200	13.3	7.6	7.0	27.9	23	20	17	60
3 in, 20M, 10 hp	350	10	6	1200	7200	13.3	7.6	7.0	27.9	23	20	20	63
4 in, 40M, 20 hp	850	20	6	1200	7200	13.3	7.6	7.0	27.9	23	20	22	65
6 in, 90M, 40 hp	1700	40	8	1200	9600	10.0	7.3	7.0	24.3	20	16	25	61

TABLE 8-3 Cost of Ownership and Operation of Excavating Machinery and Facilities *(Continued)*

	Weight and power		Depreciation factors			Yearly fixed charges as % of cost				Hourly costs as % of number of $1000s of cost			
Machinery with description	Weight of machine, lb	Delivered horsepower of prime mover(s)	Depreciation period, yr	Hours of use yearly	Depreciation period, total hours	Depreciation	Interest, taxes, insurance, storage	Replacement cost escalation	Total fixed charge	Fixed charges	Repairs, replacements, labor thereto	Operating expenses	Total hourly costs
Pumps, complete, with power:													
Centrifugal, sump, portable, wheels-tires-mounted:													
Electric motor:													
2 in, 10M, 7 hp	150	7	6	1200	7200	13.3	7.6	7.0	27.9	23	16	23	62
3 in, 20M, 10 hp	250	10	6	1200	7200	13.3	7.6	7.0	27.9	23	16	26	65
4 in, 40M, 20 hp	500	20	6	1200	7200	13.3	7.6	7.0	27.9	23	16	29	68
6 in, 90M, 40 hp	900	40	8	1200	9600	10.0	7.3	7.0	24.3	20	12	32	64
8 in, 125M, 80 hp	2100	80	8	1200	9600	10.0	7.3	7.0	24.3	20	12	35	67
10 in, 240M, 120 hp	2500	120	8	1200	9600	10.0	7.3	7.0	24.3	20	12	38	70
Centrifugal, submersible:													
Electric motor:													
2 in, 10M, 4 hp	50	4	6	1200	7200	13.3	7.6	7.0	27.9	23	20	17	60
3 in, 20M, 10 hp	100	10	6	1200	7200	13.3	7.6	7.0	17.9	23	20	19	62
4 in, 40M, 20 hp	200	20	6	1200	7200	13.3	7.6	7.0	17.9	23	20	21	64
6 in, 90M, 40 hp	400	40	8	1200	9600	10.0	7.3	7.0	24.3	20	16	23	59
Diaphragm, portable, wheel-tire-mounted:													
Diesel engine:													
2 in, 2M, 3 hp	200	3	6	1200	7200	13.3	7.6	7.0	27.9	23	20	3	46
3 in, 3M, 6 hp	350	6	6	1200	7200	13.3	7.6	7.0	27.9	23	20	5	48
4 in, 6M, 10 hp	500	10	6	1200	7200	13.3	7.6	7.0	27.9	23	20	6	49
Gasoline engine:													
2 in, 2M, 3 hp	150	3	6	1200	7200	13.3	7.6	7.0	27.9	23	20	9	52
3 in, 3M, 6 hp	250	6	6	1200	7200	13.3	7.6	7.0	27.9	23	20	16	59
4 in, 6M, 10 hp	350	10	6	1200	7200	13.3	7.6	7.0	27.9	23	20	18	61
Electric motor:													
2 in, 2M, 3 hp	100	1	6	1200	7200	13.3	7.6	7.0	27.9	23	16	13	52
3 in, 3M, 6 hp	200	2	6	1200	7,200	13.3	7.6	7.0	27.9	23	16	16	55
4 in, 6M, 10 hp	350	3	6	1200	7,200	13.3	7.6	7.0	17.9	23	16	18	57

Turbine, well-type, stationary, complete, head of 500 ft:													
Diesel engine:													
8 in, 500 gal/min 100 hp	4,000	100	8	1200	9,600	10.0	7.3	7.0	24.3	20	12	9	41
10 in, 1000 gal/min 200 hp	8,000	200	8	1200	9,600	10.0	7.3	7.0	24.3	20	12	9	41
12 in, 1500 gal/min 300 hp	15,000	300	8	1200	9,600	10.0	7.3	7.0	24.3	20	10	7	37
14 in, 2000 gal/min 400 hp	22,000	400	8	1200	9,600	10.0	7.3	7.0	24.3	20	10	7	37
16 in, 2500 gal/min 500 hp	30,000	500	8	1200	9,600	10.0	7.3	7.0	24.3	20	10	6	36
Electric motor:													
8 in, 500 gal/min 100 hp	3,000	100	10	1200	12,000	8.0	7.2	7.0	22.2	18	8	24	50
10 in, 1000 gal/min 200 hp	6,000	200	10	1200	12,000	8.0	7.2	7.0	22.2	18	8	20	46
12 in, 1500 gal/min 300 hp	12,000	300	10	1200	12,000	8.0	7.2	7.0	22.2	18	8	17	43
14 in, 2000 gal/min 400 hp	18,000	400	10	1200	12,000	8.0	7.2	7.0	22.2	18	8	15	41
16 in, 2500 gal/min 500 hp	24,000	500	10	1200	12,000	8.0	7.2	7.0	22.2	18	8	14	40
Turbine, booster-type, stationary, skid-mounted, complete, head of 500 ft:													
Diesel engine:													
8 in, 500 gal/min 100 hp	3,000	100	8	1200	9,600	10.0	7.3	7.0	24.3	20	12	12	44
10 in, 1000 gal/min 200 hp	6,000	200	8	1200	9,600	10.0	7.3	7.0	24.3	20	12	12	44
12 in, 1500 gal/min 300 hp	10,000	300	8	1200	9,600	10.0	7.3	7.0	24.3	20	11	11	42
14 in, 2000 gal/min 400 hp	13,000	400	8	1200	9,600	10.0	7.3	7.0	24.3	20	11	11	42
16 in, 2500 gal/min 500 hp	17,000	500	8	1200	9,600	10.0	7.3	7.0	24.3	20	11	11	42
Electric motor:													
8 in, 500 gal/min 100 hp	2,000	100	10	1200	12,000	8.0	7.2	7.0	22.2	18	8	34	60
10 in, 1000 gal/min 200 hp	4,000	200	10	1200	12,000	8.0	7.2	7.0	22.2	18	8	33	59
12 in, 1500 gal/min 300 hp	6,000	300	10	1200	12,000	8.0	7.2	7.0	22.2	18	8	32	58
14 in, 2000 gal/min 400 hp	9,000	400	10	1200	12,000	8.0	7.2	7.0	22.2	18	8	31	57
16 in, 2500 gal/min 500 hp	11,000	500	10	1200	12,000	8.0	7.2	7.0	22.2	18	8	30	56
Rippers, pull-type, by crawler-tractor, with three shanks, steel, wheel-mounted:													
8,000 lb weight	8,000		8	1400	11,200	10.0	7.3	7.0	24.3	17	6	1	24
14,000 lb weight	14,000		8	1400	11,200	10.0	7.3	7.0	24.3	17	6	1	24
28,000 lb weight	28,000		8	1400	11,200	10.0	7.3	7.0	24.3	17	6	1	24
Rollers or compactors, heavy-duty, unballasted weights:													
Sheepsfoot, tow-type, by tractor:													
Double-drum, 5 ft by 5 ft, 24,000 lb	24,000		8	1400	11,200	10.0	7.3	7.0	24.3	17	8	1	26
Triple drum, 5-ft width by 5-ft diameter, 37,000 lb	37,000		8	1400	11,200	10.0	7.3	7.0	24.3	17	8	1	26

TABLE 8-3 Cost of Ownership and Operation of Excavating Machinery and Facilities *(Continued)*

Machinery with description	Weight and power		Depreciation factors			Yearly fixed charges as % of cost				Hourly costs as % of number of $1000s of cost			
	Weight of machine, lb	Delivered horsepower of prime mover(s)	Depreciation period, yr	Hours of use yearly	Depreciation period, total hours	Depreciation	Interest, taxes, insurance, storage	Replacement cost escalation	Total fixed charge	Fixed charges	Repairs, replacements, labor thereto	Operating expenses	Total hourly costs
Rollers or compactors, heavy-duty, unballasted weights:													
Sheepsfoot, vibratory, tow-type, by tractor, diesel engine:													
Single drum, 6-ft width by 5-ft diameter, 75 hp, 23,000 lb	23,000	75	8	1400	11,200	10.0	7.3	7.0	24.3	17	10	4	31
Sheepsfoot, self-propelled, two-drum or four-wheel, four-drum, with bulldozer, diesel engine:													
Two-drum or four-wheel, 6-ft width by 5-ft diameter, 180 hp, 37,000 lb	37,000	180	8	1400	11,200	10.0	7.3	7.0	24.3	17	12	6	35
Two-drum or four-wheel, 6-ft width by 6-ft diameter, 320 hp, 62,000 lb	62,000	320	8	1400	11,200	10.0	7.3	7.0	24.3	17	12	6	35
Four-drum, 6-ft width by 6-ft diameter, 400 hp, 80,000 lb	80,000	400	8	1400	11,200	10.0	7.3	7.0	24.3	17	12	6	35
Pneumatic, tow-type, by tractor, multiple box, wheels-tires:													
25 tons, ballasted, 22,000 lb	22,000		8	1400	11,200	10.0	7.3	7.0	24.3	17	10	1	28
50 tons, ballasted, 32,000 lb	32,000		8	1400	11,200	10.0	7.3	7.0	24.3	17	10	1	28
75 tons, ballasted, 36,000 lb	36,000		8	1400	11,200	10.0	7.3	7.0	24.3	17	10	1	28
100 tons, ballasted, 45,000 lb	45,000		8	1400	11,200	10.0	7.3	7.0	24.3	17	10	1	28
125 tons, ballasted, 55,000 lb	55,000		8	1400	11,200	10.0	7.3	7.0	24.3	17	10	1	28
Rolling choppers, tow-type, by crawler-tractor, unballasted weights:													
Single-drum:													
8-ft width, 14,000 lb	14,000		8	1400	11,200	10.0	7.3	7.0	24.3	17	8	1	26
12-ft width, 22,000 lb weight	22,000		8	1400	11,200	10.0	7.3	7.0	24.3	17	8	1	26
16-ft width, 46,000 lb weight	46,000		8	1400	11,200	10.0	7.3	7.0	24.3	17	8	1	26

Tandem drums:													
7-ft width, 19,000 lb weight	19,000		8	1400	11,200	10.0	7.3	7.0	24.3	17	8	1	26
8-ft width, 27,000 lb weight	27,000		8	1400	11,200	10.0	7.3	7.0	24.3	17	8	1	26
10-ft width, 37,000 lb weight	37,000		8	1400	11,200	10.0	7.3	7.0	24.3	17	8	1	26
Rotary drills, complete with tools:													
Crawler-mounted:													
Diesel engine:													
3- to 6-in-diameter hole, to 2500-ft depth	33,000	100	8	1400	11,200	10.0	7.3	7.0	24.3	17	12	2	31
5- to 8-in-diameter hole, to 2500-ft depth	62,000	150	8	1400	11,200	10.0	7.3	7.0	24.3	17	12	2	31
6- to 9-in-diameter hole, to 2500-ft depth	90,000	200	12	4000	48,000	6.7	7.0	7.0	20.7	5	12	2	19
7- to 11-in-diameter hole, to 2500-ft depth	150,000	300	12	4000	48,000	6.7	7.0	7.0	20.7	5	12	2	19
9- to 15-in-diameter hole, to 2500-ft depth	220,000	400	12	4000	48,000	6.7	7.0	7.0	20.7	5	12	2	19
Electric motors:													
3- to 6-in-diameter hole, to 2500-ft depth	32,000	100	8	1400	11,200	10.0	7.3	7.0	24.3	17	10	5	32
5- to 8-in-diameter hole, to 2500-ft depth	59,000	150	8	1400	11,200	10.0	7.3	7.0	24.3	17	10	4	31
6- to 9-in-diameter hole, to 2500-ft depth	86,000	200	12	4000	48,000	6.7	7.0	7.0	20.7	5	10	3	18
7- to 11-in-diameter hole, to 2500-ft depth	152,000	300	12	4000	48,000	6.7	7.0	7.0	20.7	5	10	3	18
9- to 15-in-diameter hole, to 2500-ft depth	209,000	400	12	4000	48,000	6.7	7.0	7.0	20.7	5	10	3	18
Truck-mounted, diesel engines:													
3- to 6-in-diameter hole, to 2500-ft depth, 2 axles, 250 total combined hp	40,000	250	8	1400	11,200	10.0	7.3	7.0	24.3	17	12	3	32
5- to 8-in-diameter hole, to 2500-ft depth, 3 axles, 400 total combined hp	60,000	400	8	1400	11,200	10.0	7.3	7.0	24.3	17	12	3	32
Saws, logging, gasoline engine:													
Chain type, manual:													
Medium weight, 30-in bar length, 2 hp	20	2	3	1400	4200	26.7	8.6	7.0	42.3	30	27	20	77
Heavyweight, 45-in bar length, 4 hp	30	3	3	1400	4200	26.7	8.6	7.0	42.3	30	27	11	58
Circular type, portable, single axle, wheel-tire-mounted 24-in diameter saw, 15 hp	250	10	5	1400	7000	16.0	7.8	7.0	30.8	22	20	42	84

TABLE 8-3 Cost of Ownership and Operation of Excavating Machinery and Facilities *(Continued)*

Machinery with description	Weight and power		Depreciation factors			Yearly fixed charges as % of cost				Hourly costs as % of number of $1000s of cost			
	Weight of machine, lb	Delivered horsepower of prime mover(s)	Depreciation period, yr	Hours of use yearly	Depreciation period, total hours	Depreciation	Interest, taxes, insurance, storage	Replacement cost escalation	Total fixed charge	Fixed charges	Repairs, replacements, labor thereto	Operating expenses	Total hourly costs
Scales, platform, complete with scale house, erected:													
50 tons	33,000		12	1400	16,800	6.7	7.0	7.0	20.7	15	5	1	21
100 tons	45,000		12	1400	16,800	6.7	7.0	7.0	20.7	15	5	1	21
150 tons	67,000		12	1400	16,800	6.7	7.0	7.0	20.7	15	5	1	21
200 tons	76,000		12	1400	16,800	6.7	7.0	7.0	20.7	15	5	1	21
Scrapers, wheels-tires:													
Tow-type, by crawler-tractor, 2 axles:													
15 cysm, 22 tons	25,000		8	1400	11,200	10.0	7.3	7.0	24.3	17	10	1	28
20 cysm, 30 tons	35,000		8	1400	11,200	10.0	7.3	7.0	24.3	17	10	1	28
25 cysm, 38 tons	48,000		8	1400	11,200	10.0	7.3	7.0	24.3	17	10	1	28
Self-propelled type, push-loaded by crawler-tractor(s), two or three axles, diesel engine(s):													
Single engine:													
15 cysm, 22 tons, 350 hp	67,000	350	8	1400	11,200	10.0	7.3	7.0	24.3	17	12	6	35
20 cysm, 30 tons, 400 hp	74,000	400	8	1400	11,200	10.0	7.3	7.0	24.3	17	12	5	34
25 cysm, 38 tons, 450 hp	96,000	450	8	1400	11,200	10.0	7.3	7.0	24.3	17	12	5	34
30 cysm, 45 tons, 500 hp	115,000	500	8	1400	11,200	10.0	7.3	7.0	24.3	17	12	5	34
35 cysm, 52 tons, 550 hp	126,000	550	8	1400	11,200	10.0	7.3	7.0	24.3	17	12	5	34
40 cysm, 60 tons, 550 hp	138,000	550	8	1400	11,200	10.0	7.3	7.0	24.3	17	12	4	33
Twin engines:													
15 cysm, 22 tons, 480 hp	71,000	480	8	1400	11,200	10.0	7.3	7.0	24.3	17	15	7	39
20 cysm, 30 tons, 600 hp	81,000	600	8	1400	11,200	10.0	7.3	7.0	24.3	17	15	7	39
25 cysm, 38 tons, 720 hp	98,000	720	8	1400	11,200	10.0	7.3	7.0	24.3	17	15	7	39
30 cysm, 45 tons, 840 hp	132,000	840	8	1400	11,200	10.0	7.3	7.0	24.3	17	15	7	39
35 cysm, 52 tons, 960 hp	153,000	960	8	1400	11,200	10.0	7.3	7.0	24.3	17	15	7	39
40 cysm, 60 tons, 960 hp	174,000	960	8	1400	11,200	10.0	7.3	7.0	24.3	17	15	6	38

Self-propelled type, self-loaded by integral elevator, 2 or 3 axles, diesel engine:													
Single engine:													
15 cysm, 22 tons, 350 hp	52,000	350	8	1400	11,200	10.0	7.3	7.0	24.3	17	13	7	37
20 cysm, 30 tons, 400 hp	74,000	400	8	1400	11,200	10.0	7.3	7.0	24.3	17	13	6	36
25 cysm, 38 tons, 450 hp	83,000	450	8	1400	11,200	10.0	7.3	7.0	24.3	17	13	6	36
Shovels, full revolving, crawler-mounted:													
Diesel engine:													
2 yd³, 150 hp	117,000	150	10	1600	16,000	8.0	7.2	7.0	22.2	14	8	1	23
3 yd³, 200 hp	168,000	200	12	1600	19,200	6.7	7.0	7.0	20.7	13	7	1	21
4 yd³, 250 hp	210,000	250	14	1700	23,800	5.7	7.0	7.0	19.7	12	6	1	19
5 yd³, 300 hp	275,000	300	16	1700	27,200	5.0	6.9	7.0	18.9	11	5	1	17
6 yd³, 350 hp	330,000	350	18	1700	30,600	4.4	6.9	7.0	18.3	11	4	1	16
Electric motor:													
2 yd³, 150 hp	115,000	150	10	1600	16,000	8.0	7.2	7.0	22.2	14	7	2	23
3 yd³, 200 hp	165,000	200	12	1600	19,200	6.7	7.0	7.0	20.7	13	6	2	21
4 yd³, 250 hp	206,000	250	14	1700	23,800	5.7	7.0	7.0	19.7	12	5	2	19
5 yd³, 300 hp	271,000	300	16	1700	27,200	5.0	6.9	7.0	18.9	11	4	2	17
6 yd³, 350 hp	325,000	350	18	1700	30,600	4.4	6.9	7.0	18.3	11	3	2	16
Electric motor-generator-motor, ac to dc:													
8 yd³, 650 hp	500,000	650	20	4000	80,000	4.0	6.8	7.0	17.8	4	3	2	9
12 yd³, 950 hp	800,000	950	20	4000	80,000	4.0	6.8	7.0	17.8	4	3	2	9
18 yd³, 1400 hp	1,200,000	1400	20	4000	80,000	4.0	6.8	7.0	17.8	4	3	2	9
27 yd³, 2400 hp	1,900,000	2400	20	7000	140,000	4.0	6.8	7.0	17.8	3	3	2	8
36 yd³, 3500 hp	2,700,000	3500	20	7000	140,000	4.0	6.8	7.0	17.8	3	3	2	8
54 yd³, 5900 hp	4,500,000	5900	30	7000	210,000	2.7	6.7	7.0	16.4	3	3	2	8
72 yd³, 9000 hp	6,900,000	9000	30	7000	210,000	2.7	6.7	7.0	16.4	3	3	2	8
108 yd³, 13,000 hp	12,000,000	13000	30	7000	210,000	2.7	6.7	7.0	16.4	3	3	2	8
Slackline excavators:													
Dragscraper-type, complete with head and tail towers:													
Diesel engine:													
3 yd³, 300-ft span, 225 hp	54,000	225	12	1600	19,200	6.7	7.0	7.0	20.7	13	10	3	26
4 yd³, 400-ft span, 300 hp	62,000	300	12	1600	19,200	6.7	7.0	7.0	20.7	13	10	3	26
5 yd³, 500-ft soan, 375 hp	79,000	375	12	1600	19,200	6.7	7.0	7.0	20.7	13	10	3	26
Electric motor:													
3 yd³, 300-ft span, 225 hp	50,000	225	12	1600	19,200	6.7	7.0	7.0	20.7	13	8	5	26
4 yd³, 400-ft span, 300 hp	60,000	300	12	1600	19,200	6.7	7.0	7.0	20.7	13	8	5	26
5 yd³, 500-ft span, 375 hp	80,000	375	12	1600	19,200	6.7	7.0	7.0	20.7	13	8	5	26

TABLE 8-3 Cost of Ownership and Operation of Excavating Machinery and Facilities *(Continued)*

Machinery with description	Weight and power		Depreciation factors			Yearly fixed charges as % of cost				Hourly costs as % of number of $1000s of cost			
	Weight of machine, lb	Delivered horsepower of prime mover(s)	Depreciation period, yr	Hours of use yearly	Depreciation period, total hours	Depreciation	Interest, taxes, insurance, storage	Replacement cost escalation	Total fixed charge	Fixed charges	Repairs, replacements, labor thereto	Operating expenses	Total hourly costs
Slackline excavators:													
Track-cable scraper-type, complete with crawler-mounted head and tail towers, 1000-ft span:													
Diesel engine:													
6 yd^3, 700 total combined hp	418,000	700	12	1600	19,200	6.7	7.0	7.0	20.7	13	8	2	23
8 yd^3, 1000 total combined hp	515,000	1000	12	1600	19,200	6.7	7.0	7.0	20.7	13	8	2	23
10 yd^3, 1200 total combined hp	770,000	1200	12	1600	19,200	6.7	7.0	7.0	20.7	13	8	2	23
12 yd^3, 1400 total combined hp	1,030,000	1400	12	1600	19,200	6.7	7.0	7.0	20.7	13	8	2	23
15 yd^3, 1700 total combined hp	1,280,000	1700	12	1600	19,200	6.7	7.0	7.0	20.7	13	8	2	23
Electric motor:													
6 yd^3, 700 total combined hp	406,000	700	12	1600	19,200	6.7	7.0	7.0	20.7	13	7	3	23
8 yd^3, 1000 total combined hp	500,000	1000	12	1600	19,200	6.7	7.0	7.0	20.7	13	7	3	23
10 yd^3, 1200 total combined hp	750,000	1200	12	1600	19,200	6.7	7.0	7.0	20.7	13	7	3	23
12 yd^3, 1400 total combined hp	1,000,000	1400	12	1600	19,200	6.7	7.0	7.0	20.7	13	7	3	23
15 yd^3, 1700 total combined hp	1,250,000	1700	12	1600	19,200	6.7	7.0	7.0	20.7	13	7	3	23
Tractors:													
Crawler-type, diesel engine, with bulldozer or angledozer, with push block for loading scrapers, or with attached other equipment for land clearing and grubbing:													
50 hp	13,000	50	8	1400	11,200	10.0	7.3	7.0	24.3	17	12	5	34
100 hp	25,000	100	8	1400	11,200	10.0	7.3	7.0	24.3	17	12	5	34
200 hp	44,000	200	8	1400	11,200	10.0	7.3	7.0	24.3	17	12	5	34
300 hp	74,000	300	8	1400	11,200	10.0	7.3	7.0	24.3	17	12	4	33
400 hp	95,000	400	8	1400	11,200	10.0	7.3	7.0	24.3	17	12	4	33
500 hp	135,000	500	8	1400	11,200	10.0	7.3	7.0	24.3	17	12	4	33
600 hp	154,000	600	8	1400	11,200	10.0	7.3	7.0	24.3	17	12	4	33
700 hp	173,100	700	8	1400	11,200	10.0	7.3	7.0	24.3	17	12	4	33

Crawler-type, diesel engine, with bulldozer and three-shank ripper:													
50 hp	14,000	50	8	1400	11,200	10.0	7.3	7.0	24.3	17	15	4	36
100 hp	29,000	100	8	1400	11,200	10.0	7.3	7.0	24.3	17	15	4	36
200 hp	51,000	200	8	1400	11,200	10.0	7.3	7.0	24.3	17	15	4	36
300 hp	86,000	300	8	1400	11,200	10.0	7.3	7.0	24.3	17	15	4	36
400 hp	108,000	400	8	1400	11,200	10.0	7.3	7.0	24.3	17	15	4	36
500 hp	155,000	500	8	1400	11,200	10.0	7.3	7.0	24.3	17	15	4	36
600 hp	172,000	600	8	1400	11,200	10.0	7.3	7.0	24.3	17	15	4	36
700 hp	190,000	700	8	1400	11,200	10.0	7.3	7.0	24.3	17	15	4	36
Wheel-tire type, with bulldozer or push block for loading scrapers, diesel engine:													
50 hp	12,000	50	8	1400	11,200	10.0	7.3	7.0	24.3	17	12	5	34
100 hp	17,000	100	8	1400	11,200	10.0	7.3	7.0	24.3	17	12	5	34
200 hp	37,000	200	8	1400	11,200	10.0	7.3	7.0	24.3	17	12	5	34
300 hp	55,000	300	8	1400	11,200	10.0	7.3	7.0	24.3	17	12	5	34
400 hp	83,000	400	8	1400	11,200	10.0	7.3	7.0	24.3	17	12	5	34
500 hp	109,000	500	8	1400	11,200	10.0	7.3	7.0	24.3	17	12	5	34
Tractor-trailer(s), wheel-tire-type, diesel engine:													
Brush haulers, on-the-road type:													
Rear dumper, 3 axles, 12 tons, 24 cysm, 250 hp	25,000	250	8	1400	11,200	10.0	7.3	7.0	24.3	17	12	12	41
Log haulers, bunks, on-the-road type:													
12 tons, 3 axles, 250 hp	22,000	250	8	1400	11,200	10.0	7.3	7.0	24.3	17	12	12	41
24 tons, 5 axles, 400 hp	35,000	400	8	1400	11,200	10.0	7.3	7.0	24.3	17	12	12	41
Machinery haulers, on-the-road type:													
35 tons, 5 axles, single gooseneck, 250 hp	30,000	250	8	1400	11,200	10.0	7.3	7.0	24.3	17	8	9	34
60 tons, 9 axles, double gooseneck, 400 hp	56,000	400	8	1400	11,200	10.0	7.3	7.0	24.3	17	8	9	34
Rock haulers, off-the-road type:													
Bottom dumpers:													
Single unit, 2 or 3 axles:													
20 cysm, 30 tons, 300 hp	49,000	300	8	1400	11,200	10.0	7.3	7.0	24.3	17	10	7	34
30 cysm, 45 tons, 360 hp	71,000	360	8	1400	11,200	10.0	7.3	7.0	24.3	17	10	7	34
40 cysm, 60 tons, 420 hp	83,000	420	8	1400	11,200	10.0	7.3	7.0	24.3	17	10	7	34
50 cysm, 75 tons, 480 hp	102,000	480	8	1400	11,200	10.0	7.3	7.0	24.3	17	10	6	33
60 cysm, 90 tons, 540 hp	109,000	540	8	1400	11,200	10.0	7.3	7.0	24.3	17	10	6	33
70 cysm, 105 tons, 600 hp	116,000	600	8	1400	11,200	10.0	7.3	7.0	24.3	17	10	6	33

TABLE 8-3 Cost of Ownership and Operation of Excavating Machinery and Facilities *(Continued)*

	Weight and power		Depreciation factors			Yearly fixed charges as % of cost				Hourly costs as % of number of $1000s of cost			
Machinery with description	Weight of machine, lb	Delivered horsepower of prime mover(s)	Depreciation period, yr	Hours of use yearly	Depreciation period, total hours	Depreciation	Interest, taxes, insurance, storage	Replacement cost escalation	Total fixed charge	Fixed charges	Repairs, replacements, labor thereto	Operating expenses	Total hourly costs
Tractor-trailer(s), wheel-tire-type, diesel engine:													
Rock haulers, off-the-road type:													
Bottom dumpers:													
Double units, 4 or 5 axles:													
80 cysm, 120 tons, 420 hp	118,000	420	8	1400	11,200	10.0	7.3	7.0	24.3	17	8	4	29
100 cysm, 150 tons, 480 hp	143,000	480	8	1400	11,200	10.0	7.3	7.0	24.3	17	8	4	29
120 cysm, 180 tons, 540 hp	156,000	540	8	1400	11,200	10.0	7.3	7.0	24.3	17	8	4	29
140 cysm, 210 tons, 600 hp	161,000	600	8	1400	11,200	10.0	7.3	7.0	24.3	17	8	4	29
Rear dumpers:													
Single unit, 2 or 3 axles:													
20 cysm, 30 tons, 300 hp	57,000	300	8	1400	11,200	10.0	7.3	7.0	24.3	17	12	6	35
30 cysm, 45 tons, 360 hp	71,000	360	8	1400	11,200	10.0	7.3	7.0	24.3	17	12	5	34
40 cysm, 60 tons, 420 hp	96,000	420	8	1400	11,200	10.0	7.3	7.0	24.3	17	12	5	34
50 cysm, 75 tons, 480 hp	119,000	480	8	1400	11,200	10.0	7.3	7.0	24.3	17	12	4	33
Side dumpers:													
Single unit, 3 axles:													
14 cysm, 20 tons, 240 hp	50,000	240	8	1400	11,200	10.0	7.3	7.0	24.3	17	12	5	34
20 cysm, 30 tons 300 hp	78,000	300	8	1400	11,200	10.0	7.3	7.0	24.3	17	12	4	33
30 cysm, 45 tons, 360 hp	106,000	360	8	1400	11,200	10.0	7.3	7.0	24.3	17	12	4	33
Double units, 5 axles:													
28 cysm, 40 tons, 240 hp	75,000	240	8	1400	11,200	10.0	7.3	7.0	24.3	17	12	3	32
40 cysm, 60 tons, 300 hp	117,000	300	8	1400	11,200	10.0	7.3	7.0	24.3	17	12	3	32
60 cysm, 90 tons, 360 hp	159,000	360	8	1400	11,200	10.0	7.3	7.0	24.3	17	12	2	31
Rock haulers, on-the-road type:													
Bottom dumpers:													
Single unit:													
10 cysm, 12 tons, 3 axles, 250 hp	22,000	250	8	1400	11,200	10.0	7.3	7.0	24.3	17	12	13	42
20 cysm, 24 tons, 5 axles, 400 hp	35,000	400	8	1400	11,200	10.0	7.3	7.0	24.3	17	12	13	42

Double units:													
20 cysm, 24 tons, 5 axles, 250 hp	35,000	250	8	1400	11,200	10.0	7.3	7.0	24.3	17	12	8	37
40 cysm, 48 tons, 9 axles, 400 hp	55,000	400	8	1400	11,200	10.0	7.3	7.0	24.3	17	12	8	37
Rear dumpers:													
Single unit:													
10 cysm, 12 tons, 3 axles, 250 hp	25,000	250	8	1400	11,200	10.0	7.3	7.0	24.3	17	15	12	44
20 cysm, 24 tons, 5 axles, 400 hp	40,000	400	8	1400	11,200	10.0	7.3	7.0	24.3	17	15	12	44
Side dumpers:													
Single unit:													
10 cysm, 12 tons, 3 axles, 250 hp	25,000	250	8	1400	11,200	10.0	7.3	7.0	24.3	17	15	12	44
20 cysm, 24 tons, 5 axles, 400 hp	40,000	400	8	1400	11,200	10.0	7.3	7.0	24.3	17	15	12	44
Double units:													
20 cysm, 24 tons, 5 axles, 250 hp	41,000	250	8	1400	11,200	10.0	7.3	7.0	24.3	17	15	7	39
40 cysm, 48 tons, 9 axles, 400 hp	65,000	400	8	1400	11,200	10.0	7.3	7.0	24.3	17	15	7	39
Water haulers, tankers, sprayers, off-the-road type:													
6,000 gal, 410 total combined hp	63,000	410	8	1400	11,200	10.0	7.3	7.0	24.3	17	10	8	35
8,000 gal, 480 total combined hp	78,000	480	8	1400	11,200	10.0	7.3	7.0	24.3	17	10	8	35
10,000 gal, 550 total combined hp	92,000	550	8	1400	11,200	10.0	7.3	7.0	24.3	17	10	8	35
12,000 gal, 620 total combined hp	107,000	620	8	1400	11,200	10.0	7.3	7.0	24.3	17	10	8	35
14,000 gal, 690 total combined hp	116,000	690	8	1400	11,200	10.0	7.3	7.0	24.3	17	10	8	35
Trenchers, crawler-mounted, diesel engine:													
Ladder type:													
16- to 24-in width, to 11-ft depth, 80 hp	17,000	80	8	1400	11,200	10.0	7.3	7.0	24.3	17	15	4	36
24- to 36-in width, to 15-ft depth, 120 hp	31,000	120	8	1400	11,200	10.0	7.3	7.0	24.3	17	15	4	36
24- to 72-in width, to 25-ft depth, 160 hp	55,000	160	8	1400	11,200	10.0	7.3	7.0	24.3	17	15	3	35
Wheel type:													
16- to 24-inch width, up to 6-ft depth, 60 hp	20,000	60	8	1400	11,200	10.0	7.3	7.0	24.3	17	12	3	32
20- to 36-inch width, up to 7-ft depth, 100 hp	29,000	100	8	1400	11,200	10.0	7.3	7.0	24.3	17	12	3	32
30- to 60-in width, to 8-ft depth, 150 hp	50,000	150	8	1400	11,200	10.0	7.3	7.0	24.3	17	12	3	32
Trucks, diesel engine:													
Brush haulers, on-the-road type:													
Rear dumper, 2 axles, 8 tons, 16 cysm, 250 hp	12,000	250	8	1400	11,200	10.0	7.3	7.0	24.3	17	12	24	53
Rear dumper, 3 axles, 12 tons, 24 cysm, 400 hp	20,000	400	8	1400	11,200	10.0	7.3	7.0	24.3	17	12	24	53

TABLE 8-3 Cost of Ownership and Operation of Excavating Machinery and Facilities *(Continued)*

	Weight and power		Depreciation factors			Yearly fixed charges as % of cost				Hourly costs as % of number of $1000s of cost			
Machinery with description	Weight of machine, lb	Delivered horsepower of prime mover(s)	Depreciation period, yr	Hours of use yearly	Depreciation period, total hours	Depreciation	Interest, taxes, insurance, storage	Replacement cost escalation	Total fixed charge	Fixed charges	Repairs, replacements, labor thereto	Operating expenses	Total hourly costs
Trucks, diesel engine:													
Lubricating trucks and refueling trucks, on-the-road type:													
2 axles, 60 ft^3/min compressor, 5 tanks and pumps, 8 reels	18,000	150	8	1400	11,200	10.0	7.3	7.0	24.3	17	12	3	32
3 axles, 90 ft^3/min compressor, 11 tanks and pumps, 18 reels	30,000	250	8	1400	11,200	10.0	7.3	7.0	24.3	17	12	3	32
Machinery haulers, flat-bed type, on-the-road type:													
2 axles, 8 tons, 250 hp	11,000	250	8	1400	11,200	10.0	7.3	7.0	24.3	17	10	26	53
3 axles, 12 tons, 400 hp	18,000	400	8	1400	11,200	10.0	7.3	7.0	24.3	17	10	26	53
Rock haulers, off-the-road type:													
Rear dumpers, 2 or 3 axles:													
10 cysm, 15 tons, 200 hp	30,000	200	8	1400	11,200	10.0	7.3	7.0	24.3	17	10	7	34
14 cysm, 20 tons, 250 hp	37,000	250	8	1400	11,200	10.0	7.3	7.0	24.3	17	10	7	34
20 cysm, 30 tons, 350 hp	50,000	350	8	1400	11,200	10.0	7.3	7.0	24.3	17	10	7	34
27 cysm, 40 tons, 450 hp	66,000	450	8	1400	11,200	10.0	7.3	7.0	24.3	17	10	7	34
40 cysm, 60 tons, 650 hp	95,000	650	8	1400	11,200	10.0	7.3	7.0	24.3	17	10	7	34
54 cysm, 80 tons, 850 hp	120,000	850	12	4000	48,000	6.7	7.0	7.0	20.7	5	8	7	20
67 cysm, 100 tons, 1000 hp	143,000	1000	12	4000	48,000	6.7	7.0	7.0	20.7	5	8	7	20
94 cysm, 140 tons, 1400 hp	190,000	1400	12	4000	48,000	6.7	7.0	7.0	20.7	5	8	7	20
120 cysm, 180 tons, 1700 hp	230,000	1700	12	4000	48,000	6.7	7.0	7.0	20.7	5	8	7	20
174 cysm, 260 tons, 2300 hp	310,000	2300	12	4000	48,000	6.7	7.0	7.0	20.7	5	8	7	20
Side dumpers:													
10 cysm, 15 tons, 200 hp	32,000	200	8	1400	11,200	10.0	7.3	7.0	24.3	17	10	6	33
14 cysm, 20 tons, 250 hp	39,000	250	8	1400	11,200	10.0	7.3	7.0	24.3	17	10	6	33
20 cysm, 30 tons, 350 hp	53,000	350	8	1400	11,200	10.0	7.3	7.0	24.3	17	10	6	33

Rock haulers, on-the-road type:													
Rear dumpers:													
6 cysm, 8 tons, 2 axles, 250 hp	13,000	250	8	1400	11,200	10.0	7.3	7.0	24.3	17	10	29	56
10 cysm, 12 tons, 3 axles, 400 hp	24,000	400	8	1400	11,200	10.0	7.3	7.0	24.3	17	10	23	50
Side dumpers:													
6 cysm, 8 tons, 2 axles, 250 hp	15,000	250	8	1400	11,200	10.0	7.3	7.0	24.3	17	10	24	51
10 cysm, 12 tons, 3 axles, 400 hp	26,000	400	8	1400	11,200	10.0	7.3	7.0	24.3	17	10	19	46
Water haulers, tankers, sprayers, 2 axles, off-the-road type:													
4000 gal, 200 hp	30,000	200	8	1400	11,200	10.0	7.3	7.0	24.3	17	10	7	34
6000 gal, 300 hp	45,000	300	8	1400	11,200	10.0	7.3	7.0	24.3	17	10	7	34
8000 gal, 400 hp	68,000	400	8	1400	11,200	10.0	7.3	7.0	24.3	17	10	7	34
Water haulers, tankers, sprayers, on-the-road type:													
2000 gal, 2 axles, 250 hp	12,000	250	8	1400	11,200	10.0	7.3	7.0	24.3	17	10	26	53
4000 gal, 3 axles, 400 hp	20,000	400	8	1400	11,200	10.0	7.3	7.0	24.3	17	10	21	48
Wagons, rock, side-dump, crawler-mounted, crawler-tractor-towed:													
6 cysm	18,000		8	1400	11,200	10.0	7.3	7.0	24.3	17	12	1	30
8 cysm	27,000		8	1400	11,200	10.0	7.3	7.0	24.3	17	12	1	30
Water tanks, gravity flow:													
Portable, power-raised, 2 axles, wheel-tire-mounted, gasoline engine, 8 hp:													
6,000 gal	8,000	8	8	1400	11,200	10.0	7.3	7.0	24.3	17	8	1	26
8,000 gal	10,000	8	8	1400	11,200	10.0	7.3	7.0	24.3	17	8	1	26
10,000 gal	13,000	8	8	1400	11,200	10.0	7.3	7.0	24.3	17	8	1	26
12,000 gal	16,000	8	8	1400	11,200	10.0	7.3	7.0	24.3	17	8	1	26
16,000 gal	20,000	8	8	1400	11,200	10.0	7.3	7.0	24.3	17	8	1	26
20,000 gal	26,000	8	8	1400	11,200	10.0	7.3	7.0	24.3	17	8	1	26
Movable, stand-skid-mounted:													
6,000 gal	6,000	8	8	1400	11,200	10.0	7.3	7.0	24.3	17	5	0	22
8,000 gal	8,000	8	8	1400	11,200	10.0	7.3	7.0	24.3	17	5	0	22
10,000 gal	10,000	8	8	1400	11,200	10.0	7.3	7.0	24.3	17	5	0	22
12,000 gal	13,000	8	8	1400	11,200	10.0	7.3	7.0	24.3	17	5	0	22
16,000 gal	16,000	8	8	1400	11,200	10.0	7.3	7.0	24.3	17	5	0	22
20,000 gal	22,000	8	8	1400	11,200	10.0	7.3	7.0	24.3	17	5	0	22

TABLE 8-3 Cost of Ownership and Operation of Excavating Machinery and Facilities *(Continued)*

	Weight and power		Depreciation factors			Yearly fixed charges as % of cost				Hourly costs as % of number of $1000s of cost			
Machinery with description	Weight of machine, lb	Delivered horsepower of prime mover(s)	Depreciation period, yr	Hours of use yearly	Depreciation period, total hours	Depreciation	Interest, taxes, insurance, storage	Replacement cost escalation	Total fixed charge	Fixed charges	Repairs, replacements, labor thereto	Operating expenses	Total hourly costs
Welding equipment, portable, wheel-tire-mounted:													
Diesel engine:													
100 A, 15 hp	500	15	6	1400	8,400	13.3	7.6	7.0	27.9	20	24	22	66
200 A, 25 hp	1,000	25	6	1400	8,400	13.3	7.6	7.0	27.9	20	24	22	66
300 A, 40 hp	1,500	40	6	1400	8,400	13.3	7.6	7.0	27.9	20	24	22	66
400 A, 50 hp	2,000	50	8	1400	11,200	10.0	7.3	7.0	24.3	17	24	22	63
600 A, 80 hp	3,000	80	8	1400	11,200	10.0	7.3	7.0	24.3	17	24	22	63
800 A, 100 hp	4,000	100	8	1400	11,200	10.0	7.3	7.0	24.3	17	24	22	63
Electric motor:													
100 A, 10 hp	250	10	6	1400	8,400	13.3	7.6	7.0	27.9	20	20	55	95
200 A, 15 hp	500	15	6	1400	8,400	13.3	7.6	7.0	27.9	20	20	56	96
300 A, 20 hp	700	20	6	1400	8,400	13.3	7.6	7.0	27.9	20	20	57	97
400 A, 25 hp	900	25	8	1400	11,200	10.0	7.3	7.0	24.3	17	20	58	95
600 A, 40 hp	1,400	40	8	1400	11,200	10.0	7.3	7.0	24.3	17	20	58	95
800 A, 60 hp	1,800	60	8	1400	11,200	10.0	7.3	7.0	24.3	17	20	58	95
Well-point systems, complete:													
Diesel engine:													
6-in centrifugal pump, 50 well points, 40 hp	11,000	40	5	1200	6,000	16.0	7.8	7.0	30.8	26	24	3	53
10-in centrifugal pump, 150 well points, 120 hp	24,000	120	6	1200	7,200	13.3	7.6	7.0	27.9	23	20	5	48
Electric motor:													
6-in centrifugal pump, 50 well points, 40 hp	9,000	40	5	1200	6,000	16.0	7.8	7.0	30.8	26	20	7	53
10-in centrifugal pump, 150 well points, 120 hp	19,000	120	6	1200	7,200	13.3	7.6	7.0	27.9	23	16	10	49

Wheel bucket excavators, belt-loading, crawler-mounted:													
Linear digging, linear traveling:													
Diesel engine:													
16-ft wheel, twelve 1-yd³ buckets, 500 hp	250,000	500	10	1400	14,000	8.0	7.2	7.0	22.2	16	8	2	26
Circular digging, linear traveling:													
Diesel engine-generator-motor drive:													
18-ft wheel, eight 1 yd³ buckets, 640 hp	450,000	640	20	4000	80,000	4.0	6.8	7.0	17.8	4	4	1	9
25-ft wheel, eight 2 yd³ buckets, 1600 hp	700,000	1600	20	4000	80,000	4.0	6.8	7.0	17.8	4	4	1	9
Combination ac motor-dc generator-dc motor drives and ac-motor drives:													
30-ft wheel, ten 2½ yd³ buckets, 2400 hp	1,620,000	2400	20	4000	80,000	4.0	6.8	7.0	17.8	4	4	1	9

ratings are given specifically rather than within a range, so that the user can approximate closely a given machine.

Weight and Power of Machine These are averages for a given machine as built by one or more manufacturers. In some cases they have been estimated from specifications for a machine of given capacity and rating.

Depreciation Factors These three factors, depreciation period in years, hours of use yearly, and depreciation period in total hours, are estimated for average working conditions of construction and mining in the United States. They may be varied for particular physical working conditions, shifts of work daily, and bookkeeping practices.

Depreciation Period in Years. The years of depreciation vary greatly, even for the same machines working under like conditions. The double logarithmic graph of Plate 8-1 rationalizes a value for years of depreciation according to the weight of the machine. Of all possible determinants weight is the most consistent. The graph approximates averages for excavation machines. The graph, the abscissa of which commences with 100 lb and ends with 100 million lb of weight, shows a doubling of the depreciation period in years for every tenfold increase in weight. The equation is:

$$DY = 2^{(\log WP - \log 100)}$$

where DY = depreciation period in years and WP = machine weight in pounds.

There are two examples shown in the graph, namely:

1. A 4000-lb medium-weight automobile or pickup truck would be depreciated in 3.0 yr.
2. A 2.7-million-lb, 36-yd³ strip-mining shovel would be depreciated in 22 yr.

Hours of Use Yearly. Hours of use yearly is likewise an average figure for the United States. It varies according to length of working season, which is 5 to 12 months, and to number of shifts daily, which varies from one on an average construction job to three for a vast open pit-mine like that shown in Plate 7-1. The hours of use yearly are generally in terms of single shifting, with some double and triple shifting in cases of large, expensive machines used in the massive excavations of construction and mining.

Depreciation Period in Total Hours. This summary figure is the product of depreciation period in years times hours of use yearly.

Yearly Fixed Charges as Percentage of Cost These four charges are expressed as an annual percentage of the cost of the machine. Cost is the total cost of the machine, assembled and erected, at job site and it includes sales taxes and freight.

Depreciation. Depreciation is calculated from the depreciation period in years. All machines are given a trade-in or salvage value of 20 percent of cost. Depreciation is reckoned by the uniform or straight-line method.

Yearly cost in percentage of cost *(PC)* of machine is given:

$$PC = \frac{100 - \text{percent trade-in or salvage value}}{\text{depreciation-period, yr}}$$

or

$$PC = \frac{80}{\text{depreciation period, yr}}$$

Interest, Taxes, Insurance, and Storage. These four items are functions of the average value of the machine during the depreciation period. Their annual rates, estimated on the basis of average figures for 1978, total 13.0%, broken down as follows: interest, 8.0%; property and use taxes, 2.0%; insurance, 2.0%; storage, 1.0%. Yearly cost in percentage of cost *(PC)* of machine is given:

$$PC = \frac{0.13\left(\frac{C + C/Y}{2}\right)}{C}$$

wherein C = cost and Y = depreciation period in years. The equation reduces to:

$$PC = 6.5\left(1 + \frac{1}{Y}\right)$$

Replacement Cost Escalation. This hedge against increased inflationary cost for the replacement of the machine after the depreciation period is the third element of yearly fixed charges as percentage of cost. It allows for a like-kind replacement. It is estimated at 7 percent yearly of the cost of the machine.

Total Fixed Charges. The total of depreciation, interest, taxes, insurance, storage, and replacement-cost escalation is expressed as percentage yearly of the cost of the machine.

Hourly Costs as Percentage of Thousands of Dollars of Cost These percentage elements of four cost factors are summarized in Table 8-3, which gives factors to be applied to the current cost of a machine in order to arrive at the hourly expense of ownership and operation.

Fixed Charges. For fixed charges, the hourly percentage cost (HPC) is calculated by the equation:

$$\text{HPC} = \frac{\dfrac{\text{total yearly fixed charges} \times \text{cost}}{\text{hours of use yearly}}}{\dfrac{\text{cost}}{\$1000}}$$

The equation reduces to:

$$\text{HPC} = \frac{\text{total yearly fixed charges} \times 1000}{\text{hours of use yearly}}$$

Repairs and Replacements, Including Labor. This hourly percentage cost includes both major and minor repairs of a mechanical nature and replacement of consumed items such as tires, cutting edges, bits and drill steel, cables, and the like. These costs may be subdivided approximately as follows:

Parts and replacements, 50%
Labor, mechanics, 25%
Shop and field service, 10%
Repairs by outside agencies, 15%

As tires and tire repairs account for a significant percentage of the cost of ownership and operation, these costs are sometimes set up as a separate item. However, for simplicity's sake and without real loss of precision, they may be included in repairs, replacements, and labor.

Information for the 1978 dollar costs of repairs, replacements, and labor for an hour of operation has been translated into percentages of the thousands of dollars of cost for the 1978 machine.

Operating Expenses. As emphasized, these expenses do not include the hourly wages and fringe benefits of operator(s). For diesel and gasoline engines and for electric motors, including motor-generator drives, they include all costs for fuel and electrical energy, for all lubricants for the prime movers and power trains and assemblies, for hydraulic oils, for filters, and for oilers and grease trucks when they are used.

Approximate formulas for these costs are in terms of the rated or flywheel horsepower of prime movers delivering an average 67 percent of total horsepower. In some cases changes have been made in the 67 percent factor to accommodate special engine and electric-motor configurations of drive. In terms of 1978 fuel and energy costs the general formulas are:

Prime mover	Fuel or energy cost	Hourly cost at 67% load factor
Diesel engine	0.40/gal	\$0.024 × rated horsepower
Gasoline engine	0.50/gal	\$0.045 × rated horsepower
Gas (LPG) engine	0.52/gal	\$0.047 × rated horsepower
Electric motor	0.035/kWh	\$0.033 × rated horsepower

These general equations have been modified sometimes in Table 8-3 for special conditions. Examples are:

1. Dragline excavator, pontoon-type, diesel engine, 12-yd^3 capacity, is not lubricated by a grease truck with oilers. The dragline oiler, part of a two- or three-person crew, takes care of lubrication. The hourly cost is estimated to be \$0.019 times rated engine horsepower.

2. Belt conveyor, stationary, 100-ft segment, 60 in wide, electric motor, is rarely

lubricated and needs no grease truck. Hourly cost is estimated to be $0.029 time rated horsepower.

For Table 8-3 these hourly costs, based on formulas, are translated into percentages of thousands of dollars of 1978 machine cost.

Total Hourly Costs. This fourth element summarizes fixed charges; repairs and replacements, including labor; and operating expenses. All four percentage elements are rounded off to the nearest integer.

Use of Table of Hourly Ownership and Operating Costs

Two examples are:

1. A 1986 estimate for the hourly cost of ownership and operation of a mining shovel, full revolving, crawler-mounted, electric motors, 8 yd³, 500 hp. Cost (1986): $ 1,700,000. Table 8-3 gives the hourly cost as 9 percent of the thousands of dollars of cost of shovel, exclusive of runner and oiler.

Machine cost hourly, 9% of $1700,		$153.00
Operators' cost hourly, with fringe benefits:		
Runner	$22.00	
Oiler	20.00	42.00
Total hourly costs		$195.00

2. The data of Table 8-3 can be varied at will for particular conditions. Assume that an air drill, crawler-mounted with compressor for both downhole hammer and rotary drilling, powered by electric motors, and drilling 9½-in-diameter holes, is to be used in a dolomite quarry for fluxing stone. Work is round-the-clock triple-shifting throughout the year 1981. These determinants are in order:

a. Depreciation period, yr	5
b. Hours of use yearly, based on about 85% availability	7,400
c. Depreciation period, total hours	37,000

Yearly fixed charges as percentage of cost of $240,000 in 1981 are:

Depreciation, %	16.0
Interest, taxes, insurance (no storage), % 0.060 (1 + ⅕)	7.2
Replacement cost escalation, %	7.0
Total yearly fixed charges, %	30.2

Hourly costs as percentage of thousands of dollars of cost become:

Fixed charges, $\frac{30.2\% \text{ of } 1000}{7400}$	4%
Repairs, replacements, labor	20
Operating expenses	13
Total hourly costs, exclusive of operator(s)	37%

The hourly costs, except for operator(s), are 37 percent of $ 240, or about $89, instead of 47 percent of $240, or about $113, as suggested in Table 8-3.

In conclusion, let it be emphasized that Table 8-3 is a guide for average conditions in the United States, that it does not include the cost of operator(s), and that it can and should be interpreted and modified for particular conditions. These particular conditions are typical of excavations in massive public works such as dams and in huge private works such as open-pit mines.

PRODUCTION OF MACHINERY

Production of machinery is the time rate of performing work. The period of time may be the minute, hour, shift, day, month, year, or duration of the work. Production may be continuous or intermittent.

Continuous Production

Examples of continuous work are the productions of air compressors, engines, generators, motors, and pumps. These machines operate almost without either minor delays (less than a few minutes in duration) or major delays (more than a few minutes in duration) caused usually by mechanical failures.

The efficiency of the machine is expressed as the ratio of the time rate of energy output, horsepower, to the time rate of energy input. An example is the efficiency of an electric motor, expressed as follows:

$$\text{Efficiency} = \frac{\text{output, hp}}{\text{input, hp}}$$

$$\text{Input hp} = \frac{\text{input, kW}}{0.746}$$

Intermittent Production

Examples of intermittent work are the productions of drills, excavators, haulers, and tractors. These machines operate on cycle times, the cycle being the repetitive operation resulting in production. Cycles may be exemplified by that for a shovel, consisting of loading, swinging, dumping, and returning times, and by that for a scraper, made up of loading, hauling, dumping, turning, returning, and turning times.

These cycles are interrupted by minor and major delays. Examples of the minor delays to a shovel are checking grade, handling outsize rocks, minor maintenance, moving the shovel at the face of a cut, operator's delays, and spotting of haulers at the shovel.

Major delays, such as those due to mechanical failures and weather, are not considered to be factors affecting the operating efficiency of the machine.

Efficiencies are reckoned in two ways, each convertible to the other. Both ways express the ratio of available work time for the cycling of the machine to actual work time. Available work time is that time during which the machine can operate on continuous cycles without interruptions due to minor delays.

One well-accepted way employs the available work time in terms of the number of available minutes of the hour. Examples are the 30-min, 40-min, 50-min, and 60-min hours, equivalent respectively to 50, 67, 83, and 100 percent efficiency. As most machines with intermittent productions operate on a cycle, this kind of efficiency is convenient and readily understandable. It is the method used in the production manuals of the manufacturers of machinery, by earthmovers, and in the many examples of production and costs in the text.

An example is provided by the calculations for production of a bottom-dump hauler operating on an estimated 12.5-min hauling cycle and an estimated payload of 25 yd^3 bank measurement (25 cybm), and working an estimated 45-min working hour, or at 75 percent efficiency. The estimated hourly production is given by the equation:

$$\text{Hourly production, cybm} = \frac{45.0\text{ min}}{12.5\text{ min}} \times 25\text{ cybm} = 90$$

The other way is expressed by the equation:

$$\text{Hourly production, cybm} = \frac{75\%\text{ of }60\text{ min}}{12.5\text{ min}} \times 25\text{ cybm} = 90$$

The efficiency of a well-managed job, considering carefully the unavoidable minor delays, may be based on a 30- to 50-min working hour. One extreme is the 30-min working hour for a backhoe loading rock from a small basement excavation into trucks working in a confined area. Another extreme is the 50-min working hour for scrapers removing overburden in an open-pit mine in which working conditions are ideal.

The selection of a reasonable efficiency for a contemplated operation must be based on a combination of theory, practicability, and experience. For the sake of uniformity and ability to compare different operations, most of the efficiencies of machines used in the examples of the text are the 50-min working hour, or 83 percent.

DIRECT JOB UNIT COST

The basic unit of cost used in the text is the *direct job unit cost*. As developed in Chapter 9, this unit cost is the basic one for calculating total job costs and total bid prices.

Direct job unit cost is calculated by dividing total cost of ownership and operation of machine by production of machine for the corresponding period of time. Total cost of ownership and operation of machine is the sum of the cost of ownership and operation of the machine and the cost of operator(s) with fringe benefits. The time period is generally the hour, shift, or day.

Direct job unit cost is used mostly in the text, except where it is desired to include supervision as part of the cost. In this case, the term *total direct job unit cost* is used.

The usual equation for direct job unit cost is:

$$\text{Direct job unit cost} = \frac{\text{total hourly cost of ownership and operation of machine}}{\text{hourly production of machine}}$$

SUMMARY

Calculations for cost of ownership and operation of machinery, production of machinery, and direct job unit cost are the sequential estimates for determining the economics of excavation. Upon them depend the complete cost estimate on the basis of which the feasibility of the work is decided and the bid for the work presented.

Whether the work is small or big, knowledge and thoroughness must guide the calculations, or else one of two unfortunate circumstances will plague the earthmover. Either the estimate will be too low, resulting in a loss for the job, or it will be too high, so that the contractor ends up being an unsuccessful bidder for the work.

CHAPTER

Preparation of Bid and Schedule of Work

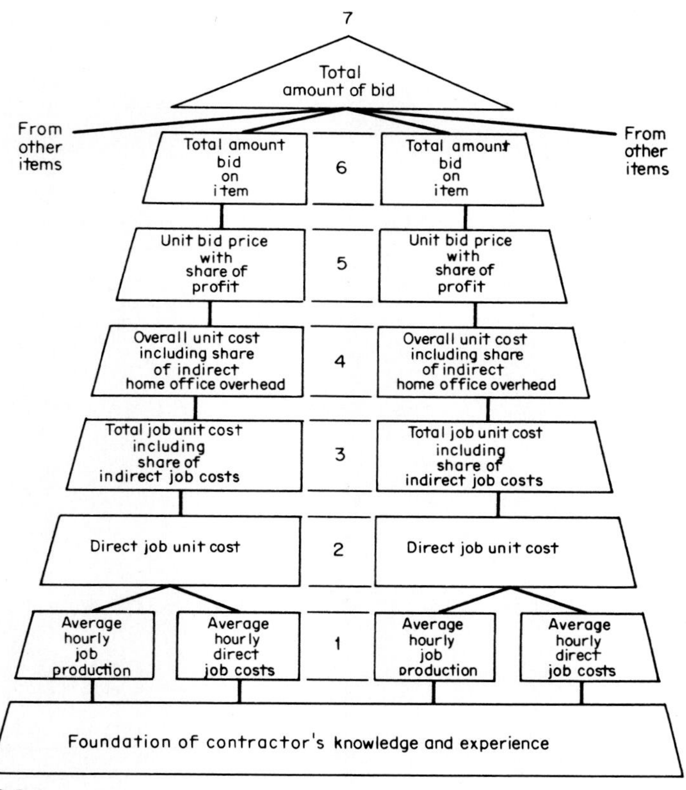

PLATE 9-1

Pyramid of the businesslike bid.

PREPARATION OF BID

There is no universally accepted plan for the preparation of a bid for excavation. Each excavator generally has a personal system, peculiar to that excavator's business methods and personal liking. The methods range from the simplest establishment of unit prices based on the experience of one person to the complex preparations of several persons for several months, as may be the case when bidding a large dam. The system suggested here for building a businesslike bid is simple and complete. It may be used in its entirety or it may be changed to suit the bookkeeping methods of the excavator, provided the fundamental procedures are carried out completely.

A bid is built like a pyramid. Like the Great Pyramid it must be based on a firm foundation. This conception is illustrated in Plate 9-1. The foundation is the excavator's knowledge and experience. The apex is the total amount of the bid. Each course of the pyramid rests squarely and solidly on the course below and supports in like manner the course above.

Hitherto in the text the discussions of cost of work have been confined largely to the direct job unit cost. This cost is secured by dividing the hourly cost of ownership and operation of the machine and of the operator(s) by the hourly production of the machine. These values make up courses 1 and 2 of the pyramid. The direct job unit cost must bear its share of the indirect job costs and the resultant total job unit cost is represented by course 3. The usual indirect job costs, chargeable to all the bid items, are set forth in Table 9-1. Likewise, the total job unit cost must bear its proportionate share of the cost of the indirect home office overhead, and the resultant overall unit cost is represented in course 4. The usual indirect home office expenses, chargeable to all the bid items, are set forth in Table 9-2. To the total job unit cost must be added a profit, either as percentages for all bid items or for certain individual bid items, and the unit bid price with its share of profit is course 5. The quantity of the bid item times the unit bid price gives the total amount bid on the item, and these amounts are course 6. Finally, the apex of the pyramid is the sum of all the amounts bid on the individual items. This is the capstone, 7.

A system of bidding must have a built-in facility for internal examination and for changes in unit bid prices at any time, and the pyramid system has this characteristic. Each bid item is calculated independently of all other bid items until it becomes part of the total amount bid at the apex or capstone. Thus it may be changed independently at any time without any necessary changes in the other bid items.

A bid is prepared around an invitation to bid a job. To illustrate the building of a

TABLE 9-1 Indirect Job Costs

1. General expenses
   - *a.* Contract surety bond
   - *b.* Special surety bond
   - *c.* Nuisance insurance
   - *d.* Protection of environment
   - *e.* Local taxes
   - *f.* Local legal charges
   - *g.* Miscellany not covered by the items listed above
   - *h.* Safety and traffic control
   - *i.* Move-in and move-out charges
   - *j.* Storage of machinery after completion of job
2. Field office
   - *a.* Ownership or rental charges
   - *b.* Light and heat
   - *c.* Telephone and telegraph
   - *d.* Equipment
   - *e.* Supplies
   - *f.* Cars
3. Office help
   - *a.* Manager
   - *b.* Bookkeepers
   - *c.* Timekeepers
   - *d.* Clerks and secretaries
   - *e.* Traveling expenses
4. Shop and yard
   - *a.* Ownership or rental charges
   - *b.* Light, power, and heat
   - *c.* Equipment
   - *d.* Supplies
   - *e.* Lubricating trucks
   - *f.* Fueling trucks
   - *g.* Maintenance trucks
   - *h.* Pickups and cars
   - *i.* Miscellany
5. Shop and yard help
   - *a.* Master mechanic
   - *b.* Mechanics
   - *c.* Greasemen
   - *d.* Guards
   - *e.* Traveling expenses
6. Supervision
   - *a.* Superintendent
   - *b.* Engineers
   - *c.* Foremen
   - *d.* Cars and pickups
   - *e.* Traveling expenses

TABLE 9-2 Indirect Home Office Overhead

1. Property taxes
2. Business taxes
3. Property and liability insurance
4. Life and special insurance
5. Legal expenses
6. Accounting expenses
7. Miscellany
8. Office
   a. Rental or fixed charges
   b. Equipment
   c. Supplies
   d. Light and heat
   e. Telephone and telegraph
   f. Miscellany
9. Managerial help
   a. Officers
   b. General superintendent
   c. Engineer
   d. Cars
   e. Traveling expenses
10. Office help
   a. Manager
   b. Bookkeeper
   c. Clerks and stenographers
   d. Cars
   e. Traveling expenses
11. Shop and yard
   a. Rental or fixed charges
   b. Equipment
   c. Supplies
   d. Light, heat, and power
   e. Trucks
   f. Cars and pickups
12. Shop help
   a. Master mechanic
   b. Mechanics
   c. Helpers
   d. Guard
   e. Traveling expenses

businesslike bid, a step-by-step example of the procedure will be given for a simple rock excavation job. All costs for the bid preparation are for 1978.

Example of Bid Preparation

The owner has property suitable as a site for a quarry, aggregate plant, asphalt plant, and concrete plant, as illustrated by the maps and pictures of Figures 9-1 to 9-4. The owner is asking for bids for the removal of overburden from the quarry site and excavation of a channel change. The overburden is to be placed in embankments for the plant site and for the access roadway and in a stockpile for materials. The excavation from the channel change is to be placed in the stockpile for materials. The owner will build several culverts for the access roadway prior to the specified time for commencing work.

The invitation for bids includes the following pertinent conditions:

1. Beginning of work is to be within 15 calendar days after receiving notice that the contract has been approved.

2. Completion of work is to be within 270 calendar days after the beginning of work.

3. Liquidated damages are $1000.00 per day for each and every calendar day of delay in finishing the work in excess of the number of calendar days allowed.

4. A contract surety bond for the total amount of the bid is required.

5. The quarry overburden is considered to be rippable. If for any reason blasting is necessary, the contingency is covered by a rock clause as part of the contract.

6. The work will be done during the dry season when there is no water in Calavera Creek, the channel of which is to be changed.

Bid Items The individual items are as follows:

1. Clearing and grubbing, 76.9 acres
2. Development of water supply, lump sum
3. Excavation of overburden for plant site embankment, 400,000 cybm
4. Excavation of overburden for roadway embankment, 42,000 cybm
5. Excavation of overburden for stockpile of material, 114,000 cybm
6. Excavation of channel change for stockpile of material, 74,000 cybm
7. Embankment of plant site, 400,000 cybm
8. Embankment of roadway, 42,000 cybm
9. Fill for stockpile from overburden excavation, 114,000 cybm
10. Fill for stockpile from channel change excavation, 74,000 cybm

The unit for all excavation, embankments, and fills is the cubic yard bank measurement (cybm). It is believed that a cubic yard in the cut will make a cubic yard in the embankment and fill, there being no swell or shrink factor.

Direct Job Costs The estimator prepares a schedule of costs for each bid item. These schedules are based on estimates for the following controlling factors.

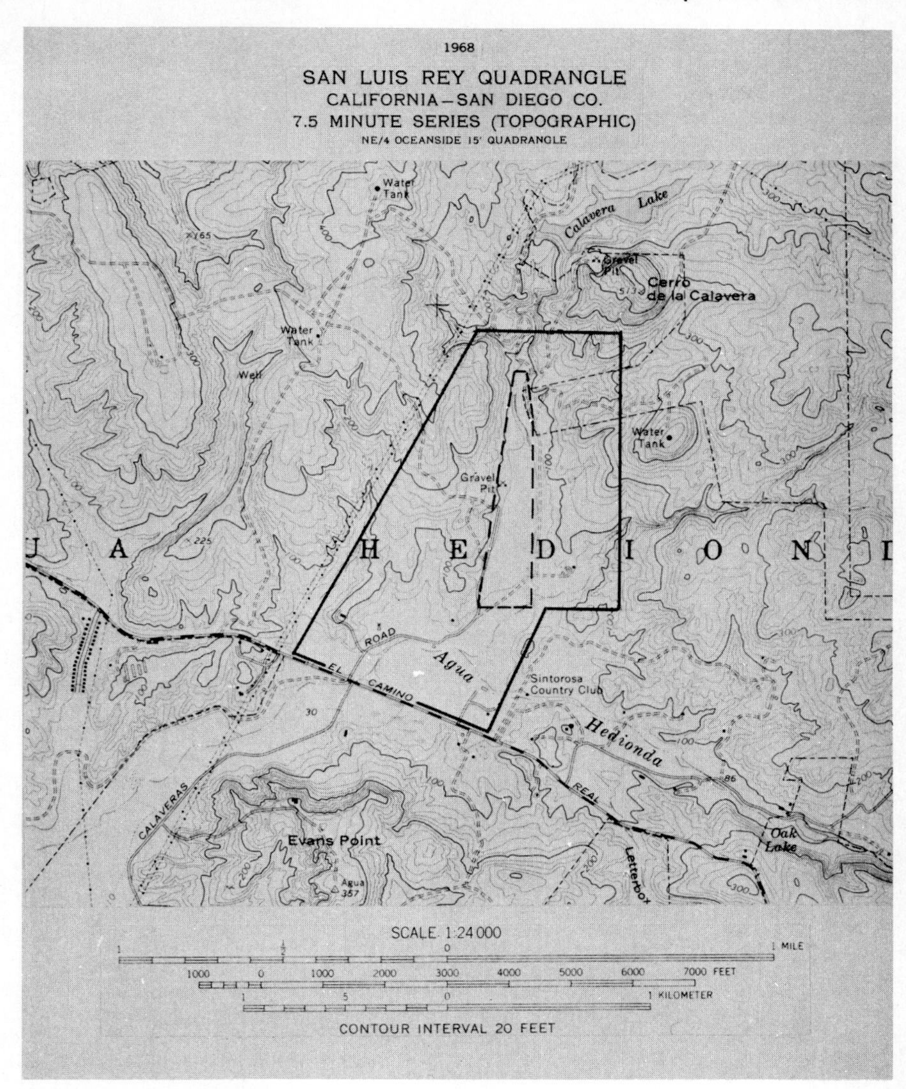

Figure 9-1 Location of property for quarry with aggregate plant, asphalt plant, and concrete plant. The property is outlined in solid lines and the location of plants is outlined in dashed lines. The rock is made up of thin residuals overlying weathered, semisolid, and solid granites. *(U.S. Geological Survey topographic map of San Luis Rey, 7½-minute quadrangle, San Diego County, California.)*

Figure 9-2 General view of west side of property. Referring to Figure 9-1, the view is to the west from the water tank, across the site for the plants, and toward the gravel pit of the quarry site. The gravel pit is in the center of the picture. The area of the quarry to be worked first is on the far right side of the picture. Here overburden to a depth of 15 ft is to be excavated from a 23-acre area and to be used for embankment for the plant site, for embankment for the access roadway, and for fill of stockpile of materials. The overburden removal is 556,000 cybm. The excavation of the channel change of Calavera Creek is to the east of the plant site. The job is set by the owner for completion in 270 calendar days after the beginning of the work.

Figure 9-3 Small abandoned quarry located at site for large quarry. The small quarry shows up as the gravel pit of Figures 9-1 and 9-4. It is located at the midpoint of the western boundary of the site for plants. At this location the rippable granite formation is about 10 ft deep. At other locations the depth of rippable rock reaches 20 ft. The average depth of the overburden over the 23 acres is estimated to be 15 ft, giving about 556,000 yd^3 of overburden to be removed.

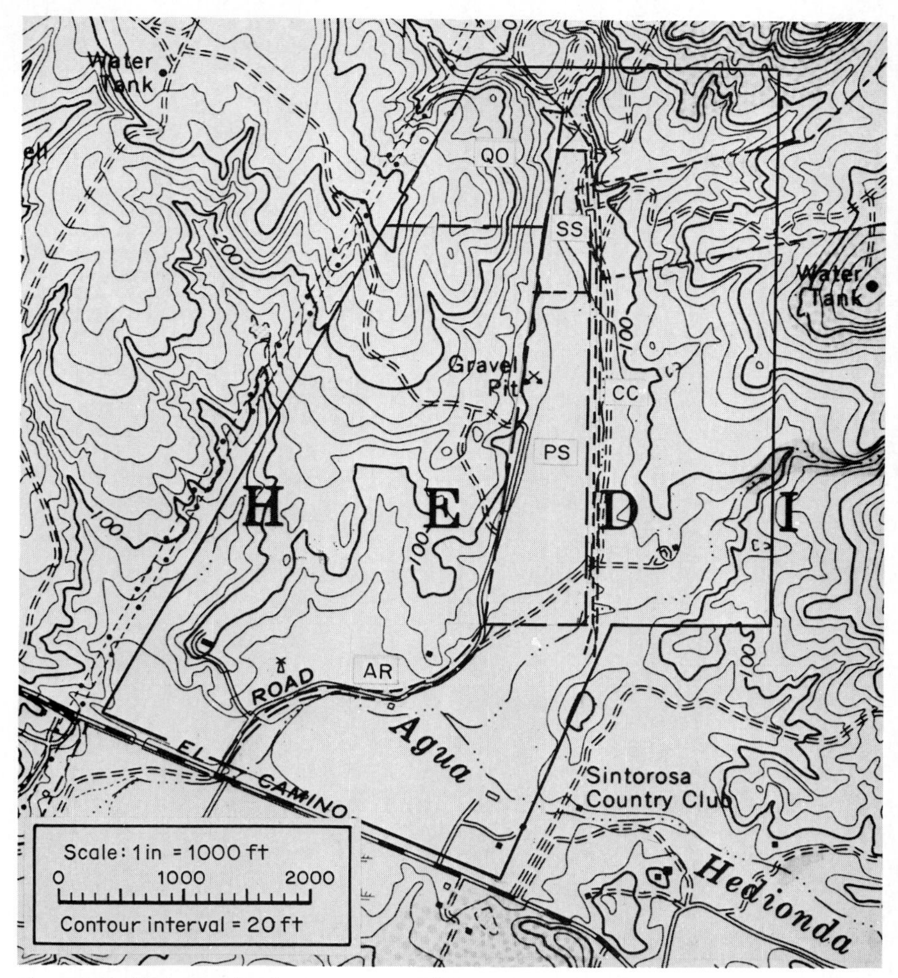

Figure 9-4 Work areas for the preparation of the property. Key: QO, quarry overburden; SS, site for stockpiles of materials; PS, site for aggregate, asphalt, and concrete plants; CC, channel change; AR, access road. Haul roads for quarry overburden are through areas QO, SS, PS, and AR; haul road for channel change excavation is through areas PS and SS. *(U.S. Geological Survey, topographic map of San Luis Rey, 7½-minute quadrangle, San Diego County, California.)*

1. Hourly production.
2. Time required in working hours for the work.
3. Machinery required for the work.
4. Hourly cost of ownership and operation of the machinery. This is sometimes called *hourly rental to the job,* but if this term is used it is not to be confused with rental of machinery from another owner.
5. Hourly cost of rental of machinery from another owner.
6. Hourly cost of labor, operator(s) and general, including all fringe benefits.

The estimates for the 10 bid items are discussed and tabulated below.

1. Clearing and Grubbing. The first work, along with development of water supply, includes removal of some 50 eucalyptus trees with up to 24-in diameters and light brush up to 4 ft in height from the 76.9-acre site. A clearing contractor will remove the trees and their stumps in exchange for the trees. The brush will be stacked and burned.

Brushing will be done by three 300-hp crawler-tractor–brush rakes. The hourly cost of ownership and operation of each machine is $60.40. Production is 1.32 acres/h and the time for the work is 59 working hours.

The summary of the direct job costs is:

Summary of Direct Job Costs — Bid Item No. 1 — Unit Acre — Quantity 76.9 — Page No 1

Working hours for completion 59 — Quantity per working hour 1.32

Kind Of Cost	Operation And Machinery Data	O/R	1 Machinery Owned Rented	2 Labor Operators	3 Labor General	4 Materials	5 Sub-Contract	6 Total
H C	Hourly Costs							
	Brushing 3-300 hp crawler tractor brush rakes	O	181 20	46 80				228 00
	Burning burn man				11 50			11 50
	Totals		181 20	46 80	11 50			239 50
T C	Total Costs 59 h		10690 80	2761 20	678 50			14130 50
U C	Unit Costs 769 acre		139 01	35 91	8 82			183 74

2. *Development of Water Supply.* This item involves a lump-sum bid. Nearby is the utility company main line from the tank to the east of the property. The line passes within 300 ft of the southeast corner of the site for plants, Figure 9-4. The hourly cost of ownership and operation of the connecting line and the portable tank of 10,000-gal capacity is $7.20. The line will be in service for the 651 working hours required for the job from move-in to move-out of machinery. The total cost is $4687.20. The estimated cost of assembly and disassembly of line and tank is $1043.40. The total cost is $5730.60.

The summary of direct job costs is:

Summary of Direct Job Costs

Page No 2

Bid Item No. 2 Unit Lump Sum Quantity 1

Working hours for completion 651 Quantity per working hour

Kind Of Cost		Operation And Machinery Data	1 Machinery O/R	1 Machinery Owned Rented	2 Labor Operators	3 Labor General	4 Materials	5 Sub-Contract	6 Total
H	C	Hourly Costs							
		Line from utility main to job	O	1.00					1.00
		Portable tank 10,000 gal capacity	O	6.20					6.20
		Totals		7.20					7.20
T	C	Total costs 651 h		4687.20					4687.20
		Estimated cost of assembly + disassembly							
		32 h of pickup truck @ $7.20	O	230.40					230.40
		64 h of pipe fitters @ $11.80				755.20	57.80		813.00
		Totals Lump sum		4917.60		755.20	57.80		5730.60

3. Excavation of Overburden for Plant Site Embankment. For the required 400,000 cybm excavation, a total of 400,000 cybm of overburden will be removed from an area of 23 acres located in the northwest corner of the property, as illustrated in Figure 9-4. The average depth of the overburden is estimated to be 15 ft. The work will commence after completion of clearing and grubbing and development of water supply. The pioneering of haul roads is minimal and it will be done by the tractor-bulldozer-ripper and the motor grader during the initial stage of work when little or no ripping is necessary. The average haul is 2500 ft one way. The maximum grade for a short distance in the overburden area is −33 percent, calling for scraper operation on a 3:1 slope. The work will proceed at an average rate of 940 cybm/h, requiring a total of 426 working hours.

The machinery for the work is:

1. A 514-hp crawler-tractor-bulldozer-ripper at a cost of $128.90/h for ownership and operation.

2. A 514-hp crawler-tractor–cushion bulldozer for push-loading scrapers at a cost of $103.30/h for ownership and operation.

3. Six 415-hp, 27-cysm push-loaded scrapers, five working and one standby. The hourly cost of ownership and operation for each machine is $84.70 when working and $42.30 when standby.

4. A 180-hp, 14-ft motor grader. The hourly cost of ownership and operation, $34.60, is divided equally between haul-road maintenance and embankment construction, as these are separate bid items. Thus, $17.30 is charged hourly to both item 3 and item 7.

5. A 250-hp, 2000-gal-capacity water wagon for haul-road maintenance. The hourly cost of ownership and operation is $15.90. Water is regarded as a material. Water consumption is estimated at 6 gal/h per cybm of excavation or 5640 gal/h. Water costs $0.225 per 100 ft³, giving a cost of $1.70/h.

The summary of direct job costs is:

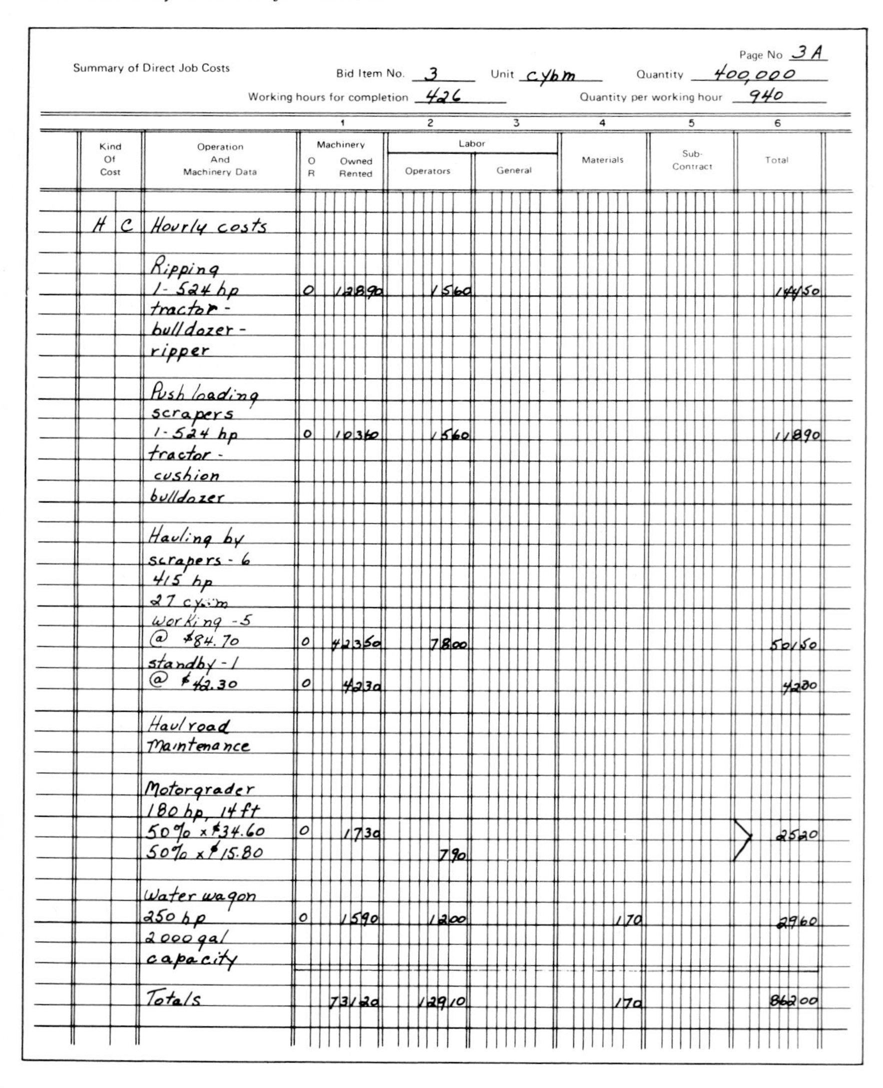

Summary of Direct Job Costs

Page No 3 A

Bid Item No. 3 Unit cybm Quantity 400,000

Working hours for completion 426 Quantity per working hour 940

Kind Of Cost	Operation And Machinery Data	O R	1 Machinery Owned Rented	2 Labor Operators	3 Labor General	4 Materials	5 Sub-Contract	6 Total
H C	Hourly costs							
	Ripping 1-524 hp tractor-bulldozer-ripper	O	128 90	15 60				144 50
	Push loading scrapers 1-524 hp tractor-cushion bulldozer	O	103 60	15 60				118 90
	Hauling by scrapers-6 415 hp 27 cysm working-5 @ $84.70	O	423 50	78 00				501 50
	standby-1 @ $42.30	O	42 30					42 30
	Haulroad Maintenance							
	Motorgrader 180 hp, 14 ft 50% x $34.60	O	17 30					25 20
	50% x $15.80			7 90				
	Water wagon 250 hp 2000 gal capacity	O	15 90	12 00		1 70		29 60
	Totals		731 20	129 10		1 70		862 00

Summary of Direct Job Costs

Page No 3B

Bid Item No. 3 Unit Cybm Quantity 400,000

Working hours for completion 426 Quantity per working hour 940

Kind Of Cost	Operation And Machinery Data	1 Machinery O/R Owned Rented	2 Labor Operators	3 Labor General	4 Materials	5 Sub-Contract	6 Total
TC	TOTAL COSTS						
	426 h	311149120	5499660		72420		36721207
UC	UNIT COSTS						
	400,000 Cybm	0.779	0.137		0.002		0.918

4. Excavation of Overburden for Roadway Embankment. The 42,000-cybm excavation will be handled similarly to that for the 400,000 cybm placed in the embankment for the plant site. It will follow the excavation for plant site. The average haul is 5600 ft one way. The work will proceed at the rate of 940 cybm per working hour, requiring 45 working hours for completion. The machinery and quantity of water for the work will be the same as those for item 3, except that a total of nine scrapers will be used because of the longer haul.

The summary of direct job costs is:

Page No 4

Summary of Direct Job Costs — Bid Item No. 4 — Unit cybm — Quantity 42,000

Working hours for completion 45 — Quantity per working hour 940

Kind Of Cost	Operation And Machinery Data	O/R	1 Machinery Owned Rented	2 Labor Operators	3 Labor General	4 Materials	5 Sub-Contract	6 Total
H C	Hourly costs							
	Ripping, as item 3	O	128.90	15.60				144.50
	Push loading scrapers, as item 3	O	103.30	15.60				118.90
	Hauling by scrapers - 9 working - 8 @ $84.70	O	677.60	124.80				802.40
	Stand by - 1 @ $42.30	O	42.30					42.30
	Haul road maintenance, as item 3							
	Motor grader	O	17.30	7.90				25.20
	Water wagon	O	15.90	12.00		1.70		29.60
	Totals		985.30	175.90		1.70		1162.90
T C	Total costs 45 h		44338.50	7916.50		76.50		52330.50
U C	Unit Costs 42,000 cybm		1.056	0.188		0.002		1.246

5. Excavation of Overburden for Stockpile of Materials. The 114,000-cybm excavation will be handled similarly to the 400,000 cybm placed in the embankment for the plant site. It will follow the excavation for the roadway. The average haul is 900 ft one way. The work will proceed at a rate of 940 cybm per working hour, requiring 121 working hours for completion. The machinery and quantity of water for the work will be the same as those for item 3, except that a total of five scrapers will be used for the shorter haul.

The summary of direct job costs is:

Summary of Direct Job Costs — Page No 5

Bid Item No. 5 — Unit cybm — Quantity 114,000

Working hours for completion 121 — Quantity per working hour 940

Kind Of Cost	Operation And Machinery Data	O/R	1 Machinery Owned Rented	2 Labor Operators	3 Labor General	4 Materials	5 Sub-Contract	6 Total
H C	Hourly costs							
	Ripping, as item 3	O	128.90	15.60				144.50
	Push loading scrapers, as item 3	O	103.40	15.60				118.90
	Hauling by scrapers,-5 working -4 @ $84.76	O	338.80	62.40				401.20
	Standby - 1 @ $42.30	O	42.30					42.30
	Haul road maintenance, as item 3							
	Motor grader	O	17.30	7.90				25.20
	Water wagon	O	15.90	12.00		1.70		29.60
	Totals		646.60	113.50		1.70		761.70
T C	Total costs 121 h		78226.50	13733.50		205.70		92165.70
U C	Unit costs 114,000 cybm		0.686	0.120		0.002		0.808

6. *Excavation of Channel Change for Stockpile of Materials.* The 74,000-cybm excavation will be handled at the same time as that for overburden excavation for the plant site, roadway, and stockpile for materials. The haul averages 1300 ft one way on an average 2.0 percent upgrade. The work will proceed at an average rate of 472 cybm per working hour and it will require 157 working hours for completion.

The machinery for this work is:

1. Four 415-hp, 29-cysm self-loading scrapers. The hourly cost of ownership and operation of each machine is $92.90.

2. A 300-hp crawler-tractor-bulldozer. The hourly cost of ownership and operation is $60.40. Of the total hourly cost, 50 percent is allocated to excavation of channel change and 50 percent is allocated to item 10, fill for stockpile from channel change excavation.

3. A 180-hp, 14-ft motor grader for haul road maintenance.

4. A 250-hp, 2000-gal water wagon for haul-road maintenance.

The hourly costs for the motor grader and the water wagon are not chargeable to this item, as they are a part of the costs for the simultaneous excavations of overburden for embankments for the plant site and roadway and for the fill of stockpile materials.

The summary of direct job costs is:

Summary of Direct Job Costs — Bid Item No. 6 — Unit cybm — Quantity 74,000 — Page No 6

Working hours for completion 157 — Quantity per working hour 472

Kind Of Cost		Operation And Machinery Data	O/R	1 Machinery Owned Rented	2 Labor Operators	3 Labor General	4 Materials	5 Sub-Contract	6 Total
H	C	Hourly costs							
		Hauling by self loading scrapers - 4 415 hp 29 cysm	O	371.60	62.40				434.00
		Utility work by 300 hp crawler tractor-bulldozer							
		50% × $60.40		30.20					38.00
		50% × $15.60			7.80				
		Totals		401.80	70.20				472.00
T	C	Total costs 157 h		63,082.60	11,021.40				74,104.00
U	C	Unit costs 74,000 cybm		0.852	0.149				1.001

7. Embankment of Plant Site. The 400,000 cybm of embankment will be placed at the rate of 940 cybm per working hour in a total of 426 working hours. The average depth of embankment is 6 ft. The machinery for this work is:

1. One 300-hp, 67,000-lb, self-propelled tamping-foot compactor. The hourly cost of ownership and operation is $58.40.
2. One 300-hp crawler-tractor-bulldozer. The hourly cost of ownership and operation is $60.40.
3. One 180-hp, 14-ft motor grader. The hourly cost of ownership and operation is $34.60, 50 percent of which, $17.30, is allocated to this item and the remaining $17.30 to the simultaneous item 3.
4. One 480-hp, 8000-gal water wagon. The hourly cost of ownership and operation is $81.90. Water is applied at the rate of 21 gal/cybm, and the hourly cost is $5.90.

The summary of direct job costs is:

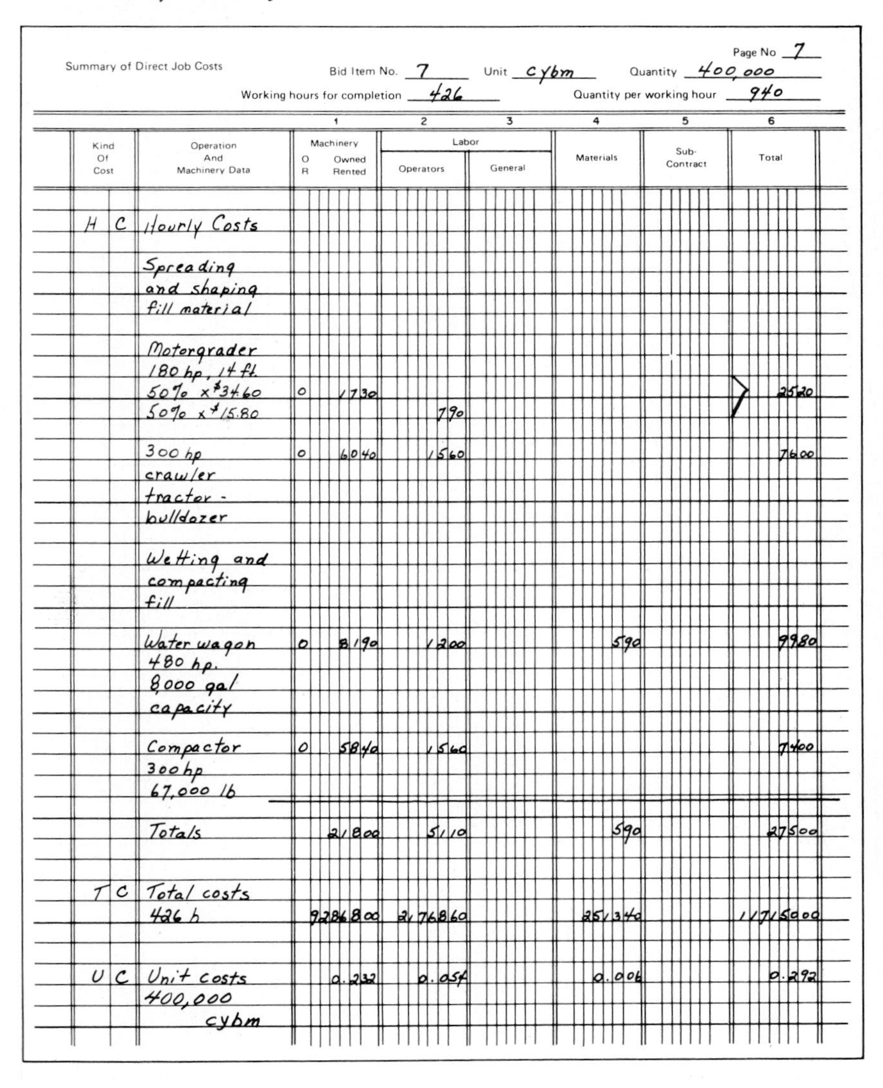

Summary of Direct Job Costs — Page No 7

Bid Item No. 7 — Unit cybm — Quantity 400,000

Working hours for completion 426 — Quantity per working hour 940

Kind Of Cost	Operation And Machinery Data	O/R	1 Machinery Owned Rented	2 Labor Operators	3 Labor General	4 Materials	5 Sub-Contract	6 Total
H C	Hourly Costs							
	Spreading and shaping fill material							
	Motorgrader 180 hp, 14 ft. 50% x $34.60	O	1730					2520
	50% x $15.80			790				
	300 hp crawler tractor-bulldozer	O	6040	1560				7600
	Wetting and compacting fill							
	Water wagon 480 hp. 8,000 gal capacity	O	8190	1200		590		9980
	Compactor 300 hp 67,000 lb	O	5840	1560				7400
	Totals		21800	5110		590		27500
T C	Total costs 426 h		9286800	2176860		251340		11715000
U C	Unit costs 400,000 cybm		0.232	0.054		0.006		0.292

8. Embankment of Roadway. The 42,000-cybm embankment will be placed at the rate of 940 cybm per working hour in a total of 45 working hours. The average depth of embankment is 6 ft. The machinery and quantity of water for this work will be the same as those for item 7, embankment for plant site. The hourly costs will be the same.

The summary of direct job costs is:

Summary of Direct Job Costs — Page No 8

Bid Item No. 8 — Unit Cybm — Quantity 42,000

Working hours for completion 45 — Quantity per working hour 940

Kind Of Cost	Operation And Machinery Data	1 Machinery O Owned R Rented	2 Labor: Operators	3 Labor: General	4 Materials	5 Sub-Contract	6 Total
HC	HOURLY COSTS, AS ITEM 7						
	TOTALS	21800	5110		590		27500
TC	TOTAL COSTS 45h	981000	229900		26560		1237500
UC	UNIT COSTS 42,000 Cybm	0.234	0.055		0.006		0.295

9. Fill for Stockpile from Overburden Excavation. The 114,000-cybm stockpile will be placed in 121 working hours at the rate of 940 cybm per working hour. The average depth of fill is 15 ft. The machinery for this work is a 300-hp crawler-tractor-bulldozer. The hourly cost of ownership and operation is $60.40.

The summary of direct job costs is:

Summary of Direct Job Costs — Page No 9

Bid Item No. 9 — Unit Cybm — Quantity 114,000

Working hours for completion 121 — Quantity per working hour 940

Kind Of Cost	Operation And Machinery Data	O/R	1 Machinery Owned Rented	2 Labor Operators	3 Labor General	4 Materials	5 Sub-Contract	6 Total
HC	Hourly Costs							
	Shaping fill							
	300 hp crawler tractor bulldozer	O	60.40	15.60				76.00
	Totals		60.40	15.60				76.00
TC	Total Costs 121 h		7308.40	1887.60				9196.00
UC	Unit Costs 114,000 cybm		0.064	0.017				0.081

10. Fill for Stockpile from Channel Change Excavation. The stockpile, amounting to 74,000 cybm, will be placed at the rate of 472 cybm per working hour in a total of 157 working hours. The average depth of fill is 15 ft. The machinery for this work is a 300-hp crawler-tractor-bulldozer. The hourly cost of ownership and operation is $60.40. Of this, 50 percent, or $30.20, is allocated to this item and the remaining $30.20 is allocated to item 6, excavation of channel change for stockpile of materials.

The summary of direct job costs is:

Summary of Direct Job Costs — Page No 10
Bid Item No. 10 Unit cybm Quantity 74,000
Working hours for completion 157 Quantity per working hour 472

Kind Of Cost	Operation And Machinery Data	O/R	1 Machinery Owned Rented	2 Labor Operators	3 Labor General	4 Materials	5 Sub-Contract	6 Total
HC	Hourly Costs							
	Shaping fill							
	300 hp crawler tractor bulldozer							
	50% x $60.40	O	30.20					38.00
	50% x $15.60			7.80				
	Totals		30.20	7.80				38.00
TC	Total Costs 157 h		4741.40	1224.60				5966.00
UC	Unit Costs 74,000 cybm		0.064	0.017				0.081

Indirect Job Costs The estimator prepares a summary of all indirect job costs, using a check list similar to Table 9-1. The summary is tabulated below. The total is $213,150, equivalent to 28.4 percent of the $749,359.80 direct job costs.

Summary Of Indirect Job Costs

Page No 1

Total Direct Job Costs $750,360.30

Calendar Days Allowed For Job 180

Item No		Kind of Cost	1 General	2 Field Office	3 Field Shop-Yard	4 Machinery	5 Salaries Wages	6 Travel Lodging
1		General expenses						
	a	Contract surety bond	11000 –					
	b	Special surety bond						
	c	Nuisance insurance						
	d	Protection of environment						
	e	Local taxes						
	f	Local legal charges						
	g	Miscellany						
	h	Safety and traffic control						
	i	Move-in & move out	14400 –					
	j	Storage of machinery						
2		Field office						
	a	Ownership or rental charges		2970 –				
	b	Light and heat		240 –				
	c	Telephone and telegraph		360 –				
	d	Equipment						
	e	Supplies		1200 –				
	f	Cars				11510		
3		Office help						
	a	Manager						
	b	Book-keepers					9000 –	
	c	Time-keepers					7200 –	
	d	Clerks and secretaries					[illegible]000 –	
	e	Travelling expenses						1800 –

Summary Of Indirect Job Costs

Page No 2

Item No		Kind of Cost	1 General	2 Field Office	3 Field Shop-Yard	4 Machinery	5 Salaries Wages	6 Travel Lodging
4		Shop-Yard						
	a	Ownership or rental charges			1080 -			
	b	Light, power, and heat			1200 -			
	c	Equipment			400 -			
	d	Supplies			6000 -			
	e	Lubricating trucks				6400 -		
	f	Fueling trucks				8380 -		
	g	Maintenance trucks				6440 -		
	h	Pickups and cars				5150 -		
	i	Miscellany						
5		Shop-Yard help						
	a	Master Mechanic					12000 -	
	b	Mechanics					8050 -	
	c	Greasemen					14750 -	
	d	Watchmen					10280 -	
	e	Travelling expenses						1200 -
6		Supervision						
	a	Superintendent					20000 -	
	b	Engineers					10000 -	
	c	Foremen					30000 -	
	d	Cars and pickups				15100 -		
	e	Travelling expenses						2000 -
		Totals	24400 -	4770 -	8680 -	53020 -	117280 -	5000 -
		Grand Totals						213150

Indirect Home Office Overhead The estimator is given a cost for a job's share of the home office overhead or expense. The cost is set by the home office and the methods for determining this cost vary with the policies of the company.

A logical method for determining the amount is to calculate the ratio of the estimated total cost of the job to the total costs of all work during an average period corresponding to the time allotted for the job. This ratio or percentage is then applied to the total job costs to determine the addition necessary to arrive at overall job costs. The cost for the home office overhead allocated to this particular job is $70,000.

The Bid Summary Sheet The bid summary sheet is generally a specially prepared sheet about 30 in wide and it contains columns similar to those of a bookkeeping sheet. It is worked horizontally bid item by bid item. In this text the sheet is divided into six pages for individual costs in order to simplify explanations. There are two repetitions of items which are eliminated in the large single sheet. They are the "bid item description" column for all but sheet A and the "percent allocation" column for all but sheet B.

Having data for direct job costs, indirect job costs, and indirect home office overhead costs, the estimator is ready to commence the bid summary sheet. The steps are explained with reference to the individual Sheets A, B, C, D, E, and F.

Sheet A. Sheet A is a recapitulation of the bid items of the invitation for bids.

Sheet B. Sheet B is a summary of the direct job cost pages for the individual bid items, together with calculated percentage distributions for the individual bid items. This is the sheet of basic data for the bid preparation.

Sheet C. Sheet C, by means of the percentage distributions of Sheet B, allocates total indirect job costs to each bid item. These indirect job costs are added to the direct job costs to give total job costs and resultant unit job costs.

Sheet D. Sheet D, also by means of the percentage distributions of Sheet B, allocates that percentage portion of home office overhead expense to each bid item. These indirect costs are added to the previous total job costs to give overall job costs and resultant unit overall costs.

Sheet E. Sheet E contains calculations for trial and revised bid prices, based on preliminary percentages of profit margins. Sheet E, for the sake of simplicity, contains only one column for revised bid prices. The large single-bid summary sheet generally contains two or more of these columns.

There are two ways of figuring profit margins. They may be calculated as a percentage addition to costs, in which case bid price equals overall job cost $\times$ (1.00 + percentage profit margin). Or they may be expressed as a percentage of bid price, in which case bid price equals

$$\frac{\text{overall job cost}}{1.00-\text{percentage profit margin}}$$

In this text the profit margin is expressed as a percentage of bid price or, equivalently, as a percentage of selling price.

Sheet F. Sheet F contains the submitted bid prices as they generally appear in the tendered bid. It represents the well-considered deliberations of the decision-making people. In this instance it is felt that a lowering of profit margin from 20 to 15 percent is necessary for the lowest total bid, and so the revised unit prices and total prices are used.

BID SUMMARY SHEET A

Bidding Schedule

Page No 1

Bid Item No.	Bid Item Description	1 Unit	2 Quantity
1	Cleaning & grubbing	Acre	169
2	Development of water supply	Lump Sum	1
3	Excavation of overburden for plant site embankment	cybm	400000
4	Excavation of overburden for roadway embankment	cybm	42000
5	Excavation of overburden for stockpile of material	cybm	114000
6	Excavation of channel change for stockpile of material	cybm	74000
7	Embankment of plant site	cybm	400000
8	Embankment of roadway	cybm	42000
9	Fill for stockpile from overburden excavation	cybm	114000
10	Fill for stockpile from channel change excavation	cybm	74000

BID SUMMARY SHEET D

Total And Unit Overall Job Costs

Share Of Total Home Office Overhead Allocated To Job $70,000.00 Page No 1

Bid Item No	Bid Item Description	Allocation (%)	Home Office Overhead Allocations	Total Job Costs	Overall Job Costs	Unit Overall Job Costs
1	C & G	1.9	1425 00	[illegible]35	1960[illegible]35	254 95
2	DOWS	0.8	600 00	7435 80	8035 80	8035 80
3	EOO FOR PSE	48.9	36675 00	471442 35	508117 35	1.270
4	EOO FOR RE	7.0	5250 00	67251 00	72501 00	1.726
5	EOO FOR SOM	12.3	9225 00	118383 15	127608 15	1.119
6	EOCC FOR SOM	9.9	7425 00	95205 85	102630 85	1.387
7	EOPS	15.6	11700 00	150401 40	162101 40	0.405
8	EOR	1.6	1200 00	15785 40	16985 40	0.404
9	FFSP FROM OE	1.2	900 00	11753 80	12653 80	0.111
10	FFSP FROM CCE	0.8	600 00	7671 20	8271 20	0.112
	Totals	100.0	75000 00	963510 30	1038510 30	

BID SUMMARY SHEET E

Trial And Revised Bid Prices

Page No 1

Bid Item No	Bid Item Description	Unit Overall Job Costs	Total Price Margin 20 %		Revised Prices Margin 15 %	
			Unit Price	Total Price	Unit Price	Total Price
1	C&G	254 95	319 00	24531 10	300 00	23070 00
2	DOWS	8035 80	10045 00	10045 00	9454 00	9454 00
3	EOO FOR PSE	1.270	1 59	636000 00	1 49	596000 00
4	EOO FOR RE	1.726	2 16	90720 00	2 03	85260 00
5	EOO FOR SOM	1.119	1 40	159600 00	1 32	150480 00
6	EOCC FOR SOM	1.387	1 73	128020 00	1 63	120620 00
7	EOPS	0.405	0 51	204000 00	0.48	192000 00
8	EOR	0.404	0 50	21000 00	0.48	20160 00
9	FFSP FROM OE	0.111	0 14	15960 00	0 13	14820 00
10	FFSP FROM CCE	0.112	0 14	10360 00	0 13	9620 00
	Totals			1300236 10		1221484 00
	Total costs			1038510 30		1038510 30
	Total profits			261725 80		182973 70
	% of profit			20.1		15.0

BID SUMMARY SHEET F

Submitted Bid Prices

Page No 1

Bid Item No	Bid Item Description	1 Unit	2 Quantity	3 Unit Bid Price	4 Total Bid Price
1	C & G	Acre	769	300 00	230700 00
2	DOWS	Lump Sum	1	945400 00	945400 00
3	EOO FOR PSE	Cybm	400000	149	596000 00
4	EOO FOR RE	Cybm	42000	203	85260 00
5	EOO FOR SOM	Cybm	114000	132	150480 00
6	EOCC FOR SOM	Cybm	74000	163	120620 00
7	EOPS	Cybm	400000	048	192000 00
8	EOR	Cybm	42000	048	20160 00
9	FFSP FROM OE	Cybm	114000	013	14820 00
10	FFSP FROM CCE	Cybm	74000	013	9620 00
	Total bid				1221484 00
	Total costs				1038510 30
	Total profit				182973 70
	% of profit				150

The Unbalanced Bid

It is not uncommon to unbalance a bid. The contracting officer and the owner frown on this method of bidding. Nevertheless, it is done now and then for one or both of the following reasons.

1. A desire to have large profits available during the early stages of the work. An example is to bid the excavation item high and the paving item low on a highway job because the excavation must be done and paid for some time before the paving work is done and paid for.

2. A desire to have a large profit on a bid item which is apt to overrun beyond the quantity set up in the bid schedule. An example is excavation in a slide-prone area, where the total quantity might be 50 percent more than that of the bid schedule.

In the previous example of bid preparation there is little likelihood of an appreciable overrun in any bid item. However, there are four opportunities to unbalance the bid so as to have available initial profits which are larger than estimated. These opportunities arise with: item 1, clearing and grubbing; item 2, development of water supply; item 3, excavation of overburden for plant site embankment; and item 7, embankment of plant site.

The method is to increase the unit bid prices on these four items and to decrease the unit bid prices on the remaining six items so that the total bid price for the job remains the same. The calculations are simple. An example is the change of bid prices in the summary Sheet F of the example of bid preparation. The following are changes in total bid prices for items 1, 2, 3, and 7:

Bid item no.	Unit price changes		Total bid prices: Original bid	Total bid prices: Unbalanced bid	Total bid prices: Difference in bids
1	Unit price:	$300	$ 23,070		
		$500	. . .	$ 38,450	+$ 15,380
2	Lump sum	$9,454	9,454	18,908	+9,454
3	Unit price:	$1.49	596,000		
		$1.75	. . .	700,000	+104,000
7	Unit price:	$0.48	192,000		
		$0.56	. . .	224,000	+32,000
Totals			$820,524	$981,358	+$160,834

To calculate the resultant changes in total bid prices for items 4, 5, 6, 8, 9, and 10, the total, $160,834, subtraction from the original bid price is distributed over the remaining six items in proportion to the ratio of their original total bid prices to the sum of their original total bid prices. The calculations are:

Bid item no.	Adjustment to original total bid price	Subtraction from original total bid price	Original total bid price	Unbalanced total bid price
4	21.3%(×$160,834)	$ 34,258	$ 85,260	$ 51,002
5	37.5%	60,313	150,480	90,167
6	30.1%	48,410	120,620	72,210
8	5.0%	8,042	20,160	12,118
9	3.7%	5,951	14,820	8,869
10	2.4%	3,860	9,620	5,760
Totals	100.0%	$160,834	$400,960	$240,126

When the foregoing changes are made and when the unit bid prices are rounded out to the nearest cent, the unbalanced bid prices are as given in the following tabulation.

Bid item no.	Bid item description	Unit	Quantity	Unit bid price,$	Unit bid price,$
1	Clearing and grubbing	acre	76.9	500.00	38,450.00
2	Development of water supply	lump sum	1	18,908.00	18,908.00
3	Excavation of overburden for plant site embankment	cybm	400,000	1.75	700,000.00
4	Excavation of overburden for roadway embankment	cybm	42,000	1.21	50,820.00
5	Excavation of overburden for stockpile of material	cybm	114,000	0.79	90,060.00
6	Excavation of channel change for stockpile of material	cybm	74,000	0.98	72,520.00
7	Embankment of plant site	cybm	400,000	0.56	224,000.00
8	Embankment of roadway	cybm	42,000	0.29	12,180.00
9	Fill for stockpile from overburden excavation	cybm	114,000	0.08	9,120.00
10	Fill for stockpile from channel change excavation	cybm	74,000	0.08	5,920.00
Total bid					$1,221,978.00

The total cost for the job is $1,038,510.30. The profit for the unbalanced bid is $183,467.70, or 15.0 percent. The profit for the balanced bid is $182,973.70, or also 15.0 percent.

The financial effects of the unbalanced bid are both positive and negative.

1. During the first 2 weeks of work, during which clearing and grubbing and development of water supply will be completed, the engineer's estimate for work completed will be increased by $24,834.

2. During the next 11 weeks of work, during which the overburden will be placed in the embankment of the plant site, the combined total bids for items 3 and 7 show that the unbalanced bid exceeds the balanced bid by $136,000, and the engineer's estimate will reflect the additional amount.

3. Accordingly, the contractor will have total engineer's estimates exceeding by $161,000 those payable by a balanced bid. Explained in another way, an average of $80,500 has accumulated over a period of 3 months. At an 8 percent annual interest rate, the unbalanced bid creates an approximate subsidy of $1610 for the contractor. At the same time, the unbalanced bid minimizes the borrowing of money during the first stages of the work.

4. Of course, later there will be a corresponding adverse effect on items 4, 5, 6, 8, 9, and 10, which are being bid at total prices less than those of the balanced bid. These total bid prices amount to $240,126. However, the overall job costs on the same items are $340,650. Thus, these six items must take a loss of $100,524 in order to make possible the unbalanced bid.

Owners and contracting officers sometimes provide safeguards in the contract against use of unbalanced bids. One perceives that a skilled estimator can detect the unbalanced bid because of the absence of a normal similarity between the unit bid price and the prevalent unit bid price for the same work. Ethically, the unbalanced bid is not a businesslike bid.

Summary of Bid Preparation

Bid preparation is a systematic procedure based fundamentally on the hourly productions of workers and machines and on the hourly costs of labor and the ownership and operation of machines. The ensuing additional costs of indirect job expense and of the share of home office overhead are important in the sense that they must be kept to a minimum consistent with efficient management.

After the overall job cost has been determined, the percentage of profit for the job depends on several factors to be considered by the home office management. Among these considerations are:

1. Desirability of additional work
2. Desirability of the work as typical or atypical of the company's average work
3. Availability of labor and machines for the work
4. Imponderable risks of the job, such as weather conditions, accessibility, and labor relations
5. Competition for the work

Ultimately the owners, the company management, the job management, and the estimator arrive at a well-considered profit margin, which determines the final total bid price.

SCHEDULE OF WORK

The estimator generally prepares a time schedule for the several bid items or operations of the work. This program is especially important if the actions are mutually dependent or if they are apt to interfere with each other with respect to working space. An example of dependence is the laying of conduits for electricity, street lighting, and telephones in the same trench during site preparation for homes. An instance of interference is the excavations for storm drains and sewers in the same street, where space is not sufficient for efficient use of machinery.

In the previous example of bid preparation the operations or bid items could be worked independently. It was necessary only to schedule them in continuous order to arrive at a total time in work hours or calendar days for completion of the job. The following example of a work schedule is much more complex, as it involves 35 independent and interdependent bid items, for which separate and simultaneous times must be arranged.

The time schedules for the bid items are almost always set up as horizontal bar diagrams or charts, each bar representing a bid item or operation. The abscissa at the bottom of the chart is time in calendar days. There are several forms for the chart, all of which have two features in common: the horizontal bars and the bottom time schedule. Representative of these charts are the diagrams of the Critical-Path Method (CPM), which is used in the following pages to illustrate planning, scheduling, and completing work.

The Critical-Path Method (CPM)

Principles of CPM Job Scheduling The *critical path* is that path which the important controlling kinds of work must take in order that they may follow each other in logical sequence and may not interfere with each other in the orderly prosecution of the job. The path of the critical work elements is shown in the wide lines of Figure 9-5, which gives the CPM schedules of a site development project for which a detailed case analysis will be presented. The *noncritical paths* are those for kinds of work which may be carried on concurrently with the critical work because there are no mutual interferences. These paths are shown by the narrow lines of the CPM chart. *Float* is extra time for an operation beyond the actual estimated time, made possible because no other operation may take place logically or physically during the same period of time.

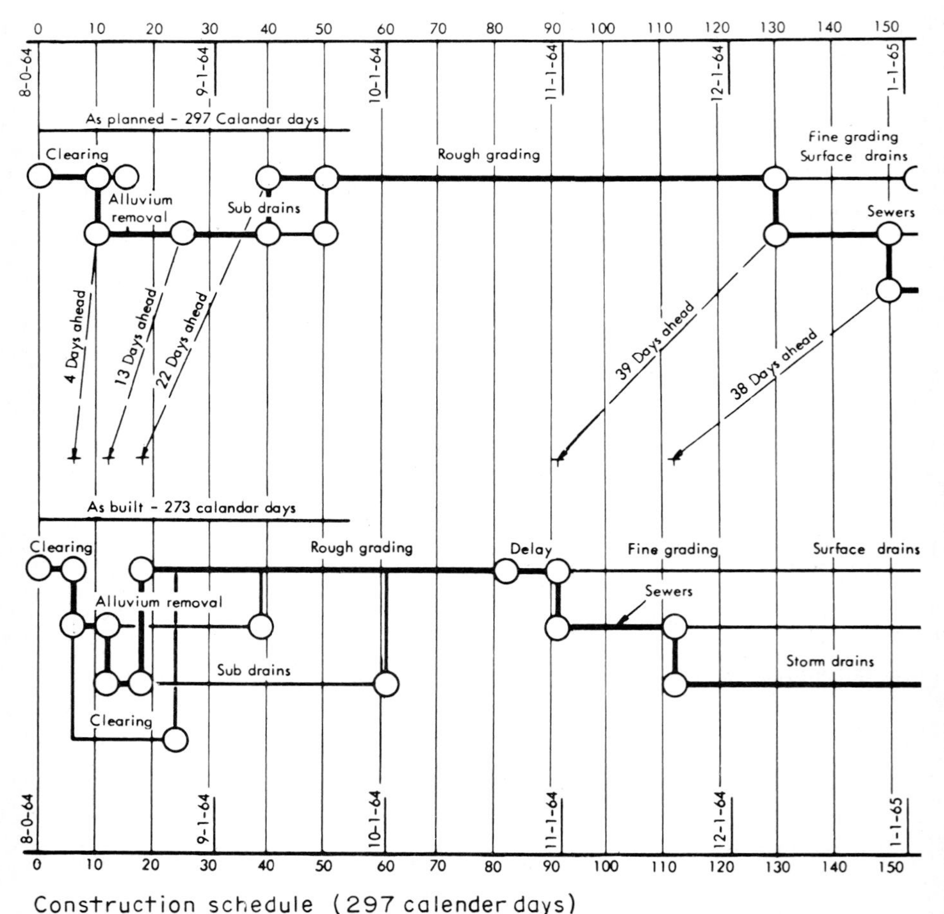

Figure 9-5 The critical-path method for scheduling construction as it was planned and as it was built. Construction was for a 22-acre mountainside development for 55 home sites. *(Horace K. Church, "How CPM Worked Out on a Grade-Sewer-Street Job,"* Roads and Streets, *August 1966, p. 84.)*

By these plottings of horizontal bars, representing times for work, against the abscissa of calendar days at the bottom of the chart, one may visualize conveniently the following important factors in the conduct of a job.

1. The individual time schedules, usually based on bid items, for the operations
2. The relationships between the critical and the noncritical work elements
3. The overall time schedule for the job

Usually corrections are made to portray the "as built" schedule in contrast to the "as planned" schedule. The correction may be made on the "as planned" chart, but a better method is to plot a new "as built" schedule below the "as planned" one, as shown in Figure 9-5. The plotting of both charts is described in the ensuing discussion of "as planned" and "as built" schedules.

Example of CPM Job Scheduling The time- and money-saving value of CPM in preplanning a project is manifest. However, a case analysis showing just how well the preset work schedule was followed and how the CPM schedule was adjusted to meet unforeseen circumstances of the job is not often available.

The example is a typical development of a mountainside site for homes. Located in the Beaudry Mountains of Glendale, California, the tract included 22 acres with 55 home sites. The roughly square area represented the first stage of Tract 29116, which eventually totaled 41 acres.

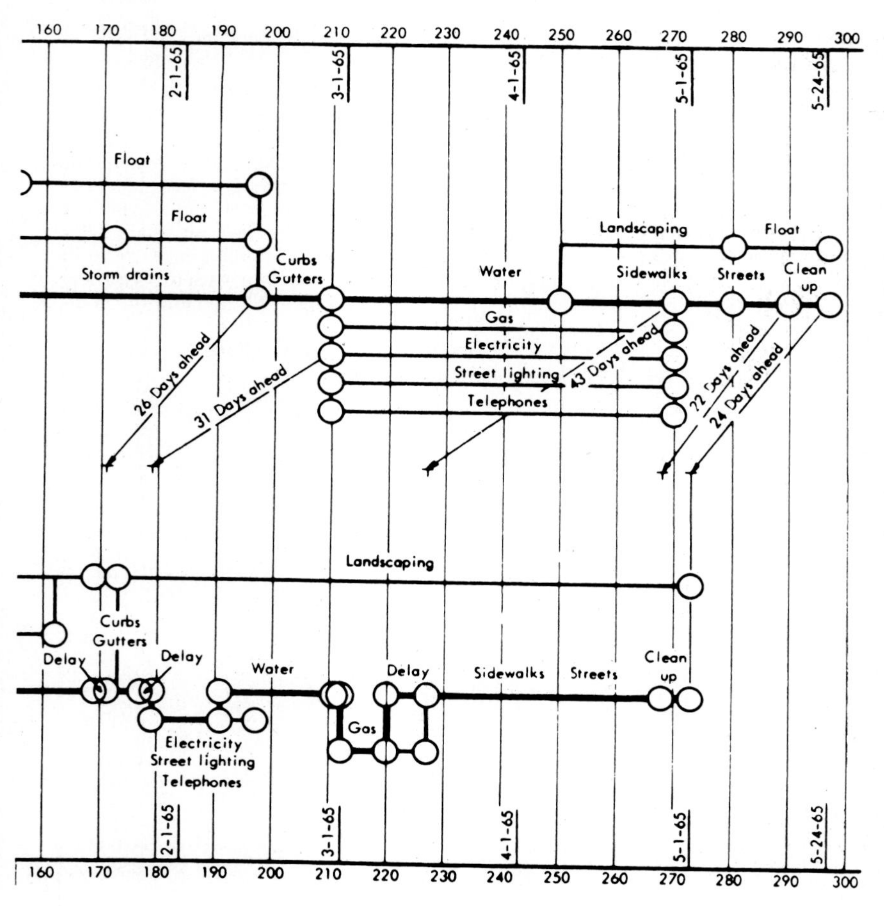

The excavation was largely weathered and semisolid granite at high elevations and terrace deposits of conglomerates at low elevations. Of the 340,000 yd³ of rough excavation, only 4500 yd³ was blasted. Some 1800 yd³ in the street cut was blasted for utility excavation, and 27,000 yd³ of alluvia in the ravines was removed and compacted in the fill, bringing the total excavation to 367,000 yd³, or 16,700 yd³ acre and 6700 yd³ per home site or lot. The range of elevations between the top of the cut and the bottom of the fill was 320 ft in a lateral distance of 1200 ft, giving a maximum gradient for haulage of −27 percent at the beginning of the excavation and an average gradient of about −13 percent.

The writer prepared preliminary estimates for the job, made up the bid schedule and contract, selected low bidders, and supervised the construction. Individual contracts for site preparation were let to the lowest responsible bidders, this method of dividing the work lending itself to lowest overall cost. These contracts are a part of the tabulated costs for the complete home site development.

The round-figure costs of the site development, ready for the building of homes, were:

Land acquisition	18.7%	$110,000
Engineering	6.8	40,000
Supervision	2.5	15,000
Construction	72.0	424,000
Totals	100.0%	$589,000
Cost per acre		$26,800
Cost per lot or home site		$10,700

All this work was done in 1964–1965, and the costs set forth in the text are far too low for 1978. For equivalent 1978 costs, all costs should be multiplied by a factor of 2.3.

During the estimating for the job prior to receiving bids, the CPM analysis for scheduling the job was prepared, as illustrated in the upper part of Figure 9-5, "as planned." The worth of this front sight is to be appraised in terms of the back sight, which is diagramed in the lower, "as built" part of Figure 9-5. Naturally, the "as built" must differ from the "as planned" because initial planning is never perfect. The differences were in terms of management and machinery and, to a minor degree, in keeping with the weather. Figure 9-6 illustrates the rock excavation work at the beginning of the job and the completed job one year later when homes were being built.

An explanation of the "as planned" CPM schedule, the foresight, follows. The bid items for the job with their overall totals by classifications were:

Grading:	$195,000
Clearing and grubbing, acres	22
Excavation, including embankment, yd³	340,000
Alluvium removal, including embankment, yd³	27,000
Terraces and surface drains, lin ft	3000
Subsurface drains, lin ft	3500
Storm drains:	$48,000
R.C.P., 18-in to 54-in, lin ft	1600
C.M.P., 18-in, lin ft	100
Inlet structures, each	2
Catch basins, each	9
Manholes, each	10
Outlet structures, each	2
Sewers:	$28,000
V.C.P., 8-in, lin ft	3000
V.C.P., 6-in, lin ft	1500
Manholes, each	15
Streets, curbs and gutters, walks	$35,000
Curbs and gutters, concrete, 18-in, lin ft	5600
Sidewalks, 3-in thickness concrete, ft²	12,700
Cross gutters and aprons, ft²	2100
Pavement, asphalt, 3-in thickness, ft²	100,000
Utilities:	$112,000
Water:	
Mains, lin ft	3200
Laterals, lin ft	1500
Gas:	
Mains, lin ft	3200

Laterals, lin ft	1500
Electricity, underground:	
Mains, lin ft	3200
Laterals, lin ft	1500
Vaults and boxes, each	8
Street lighting, underground:	
Mains and laterals, lin ft	3600
Telephones, underground:	
Mains, lin ft	3200
Laterals, lin ft	1500
Vaults and boxes, each	7
Landscaping	$6,000
Total of contracts	$424,000

(a)

(b)

Figure 9-6 Beginning and ending of development of mountainside site for homes. *(a)* Two weeks after start of granite excavation by a scraper spread. The rock is being hauled downgrade to a narrow fill in the ravine which is visible just above the push-loading tractor. The completed fill will be 80 ft high. *(b)* Twelve months later the work has been completed and home building has been started. The two homes in the upper left corner of the picture are atop the high fill which was being placed in the ravine in the upper picture. Excavation was characterized by several deep sidehill cuts to be placed in one major high fill. The site is in the Beaudry Mountains of Los Angeles County, California.

1. Clearing and grubbing were medium, involving light brush with perhaps 100 oak and eucalyptus trees with trunks of up to 24-in diameter. Ten critical days were allowed, with five extra float days for cleanup of the 22 acres. In all operations the assumed work schedule was a five-day, 45-h week, in keeping with local practice.

2. Alluvium removal and compaction were allowed 15 critical days for the 17,000 yd^3 of objectionable unconsolidated wash materials in the ravines.

3. Subdrains, 3500 lin ft, were allowed 15 critical days, at the end of which rough grading was scheduled to commence on the fortieth calendar day. The noncritical subdrains at high elevations were allowed an additional 10 days.

4. Rough grading with compaction in embankment, 340,000 yd^3, was scheduled for 90 critical days, 63 working days. The 27 idle days included Saturdays, Sundays, and holidays. It was assumed that the contractor would work a single spread of two to four scrapers with one or two push-loading tractors and one tractor-bulldozer-ripper. An average hourly production of 600 cybm was estimated.

5. Fine grading of streets and lots, part of the rough grading contract along with surface drains, 3000 lin ft, was allowed 25 noncritical days, along with 42 days of float time. The total time allowed for rough grading, including the times for other items included in the contract, was 197 calendar days.

6. Sewers, 4500 lin ft, were allowed 20 critical days, with 47 days of float time, and they were scheduled to commence on the 130th calendar day on the completion of rough grading.

7. Storm drains, 1700 lin ft, were allowed 47 critical days, commencing after 20 days of sewer work so as to allow time for the storm drain work to follow the sewer work. As a preventive measure against bank erosion, 100 lin ft of 18-in-diameter corrugated metal pipe was scheduled during the rough grading. This work was not regarded as critical to construction of other items and it was not included in the CPM schedule.

8. Curbs and gutters, 5600 lin ft, were allowed 13 critical days.

9. Underground utilities, 22,500 lin ft, including water, gas, electricity, street lighting, and telephone lines, were scheduled concurrently, with an allowance of 60 critical days before beginning of sidewalk construction.

10. Sidewalks, 12,700 ft^2, were allowed 10 critical days before street construction.

11. Streets, including cross gutters, 102,100 ft^2, were allowed 10 critical days before start of cleanup.

12. Cleanup was allowed seven critical days.

13. Landscaping, a noncritical item, was scheduled to start on the 125th day and to finish at the scheduled end of the project, the 297th calendar day.

Following is an explanation of the "as built" CPM schedule, the back sight. The "as built" part of the chart illustrates the difference between theory and practice. Expressed in another way, it is the difference between the ideas of the planner and the well-considered on-the-job changes of the contractor and the builder.

1. *Clearing and grubbing.* Instead of completing the work prior to alluvium removal, the grading contractor finished the heavy clearing in the ravines, began alluvium removal immediately thereafter, and completed light and medium clearing on the ridges later. The planned critical 10 days were reduced to 6 days, as built. The job was then four days ahead of schedule.

2. *Alluvium removal.* Instead of completing alluvium removal prior to subdrains, the contractor allowed only 6 critical days instead of 15 days as planned. The job was then 13 days ahead of schedule.

3. *Subdrains.* Instead of the planned 15 critical days the contractor scheduled only 6 days before starting rough grading. The job was then 22 days ahead of schedule.

4. *Rough grading.* Instead of using one set of two push-loading tractors, the contractor used two sets and increased the number of scrapers accordingly. Thus the 90 critical days, 63 working days, were reduced to 64 critical days, 47 working days. Instead of averaging 600 yd^3/h, production was 800 yd^3/h. Because of the unanticipated speed of rough grading, there was a delay of nine critical days before starting sewers. Nevertheless, construction was 39 days ahead of the scheduled 130 days. On November 1, 1964, rough grading was finished, ahead of the heavy winter rains.

5. *Fine grading* of streets and lots and surface drains, noncritical items, were completed 28 days ahead of schedule.

6. *Sewers.* Because of the feeling that separate contractors could schedule work

without interference on one street, sewers were allowed only 20 days of lead or critical time ahead of storm drains. This was a big mistake because of the crowded conditions. A minimum time of 21 days preceded the storm drains, and the total time for sewers was 71 days instead of the planned 42 days.

7. *Storm drains.* Forty-seven critical days after completion of sewers were allowed. Because of the aforementioned problem of interference, 57 calendar days were required. Thus, whereas construction was 39 days ahead of schedule at the completion of rough grading, it was only 26 days ahead at the completion of the storm drains.

8. *Curbs and gutters.* Starting two days after completion of the storm drains, high efficiency reduced critical time from 13 to 6 days. The work was then 31 days ahead of schedule.

9. *Utilities.* Water, gas, electricity, street lighting, and telephone lines were scheduled concurrently for 60 critical days. They were built in 41 days, so that at the end of the work the job was 50 days ahead of the planned 270 calendar days.

10. *Sidewalks and streets.* The contractor was delayed a week in moving in for the work so that sidewalks began 43 days ahead of schedule. The seasonal rains delayed sidewalks and streets so that 41 days were required instead of the planned 20 days. The job was then 22 days ahead of schedule.

11. *Cleanup.* This rather indefinite operation, along with landscaping, which had been started much ahead of schedule, required five days instead of the planned seven days. The project ended after completion of cleanup 24 days ahead of the CPM "as planned."

The overall time saving of 24 days under the planned 297 calendar days is 8 percent.

General Observations on the Value of Preplanning and Scheduling

The following observations apply specifically to the development of the mountainside site for the building of homes.

Estimating the Total Cost and the Total Time for the Work The estimator prepared the schedule for the work. In making up the chart he put in visual form the times and the sequences for the operations or the bid items. These times determined the necessary hourly productions on which cost estimates were based. Thus an easily changeable diagram was obtained which the owner and estimator could modify in order to secure the lowest total cost for the project in keeping with the desired total time for the work.

Bidding for the Work Prior to bidding, the selected bidders were able to visualize their required scheduled work and to see clearly their mutual time-motion relationships. Being practical contractors, they were able to give valuable time- and cost-saving suggestions for modifications in the schedule, which benefited both owner and contractors.

Prosecution of the Work The superintendent for the owner was able to see quickly the progress schedule and to discuss the facts with the contractors. At the same time he secured valuable advice from the contractors. The superintendent was able to keep all contractors informed concerning the best possible time sequences, and this facility was especially valuable to contractors about to come in on the job for their particular work. The contractors were able to visualize their own progress during the entire job, as well as the progress of their fellow contractors. This knowledge and understanding assisted in the solution of the problem of interference and congestion between the sewer and storm drain contractors.

Both owner and contractors were able to see mistakes in timing and to correct them on the job. And, most certainly, mistakes were recognized as avoidable on the next job. In this and in other senses, the charted schedule became a device for teaching correct procedures.

Financing the Work The owner was able to present a well-thought-out plan for the job to the lending institution. This, coupled with the well-known business acumen of the owner, contributed to a good understanding between borrower and lender. During construction the lending institution was able to see actual progress related to planned progress. This picture, as it was favorable to the owner, assisted the owner in borrowing money to finance the payroll of one of the contractors.

The project finished 24 days ahead of the planned 297 calendar days. This saving of time was attributable to the cooperation of contractors and owner, but it was also apparent that the CPM scheduling contributed to the cooperation. The resultant savings to the owner were estimated to be in accordance with the following tabulation.

Interest on money for land acquisition: $110,000, for 24 days at 5.0%, 1964–1965 interest rate	$ 361.64
Interest on money for engineering, supervision, and construction: average $239,500 borrowed, for 24 days at 5.0%.	787.40
Total saving during 24 days	$1149.04

SUMMARY

Preparation of bid and schedule for work are of equal importance. Getting the job with the lowest qualified bid must be backed up by the execution of the work so as to realize the intent of the bid insofar as costs and profit are concerned. Not only are they of equal significance but they are also mutual efforts in that bidding a job necessarily is based on a work schedule, as costs are reflected in the time allotted to the various bid items. Thus, a schedule for work is contemplated in the preparation of the bid. In a general sense the superintendent follows the schedule for work on which the bid was based but deviates from that schedule to adapt it to the problems encountered in the prosecution of the work. Obviously, mutuality of preparation of bid and schedule for work exists from the initial desire for the job to the final acceptance of the work.

Appendixes

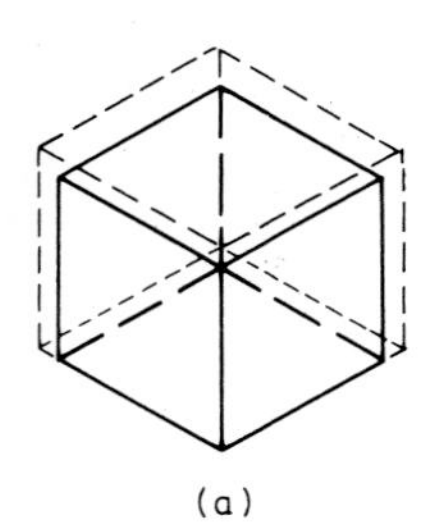

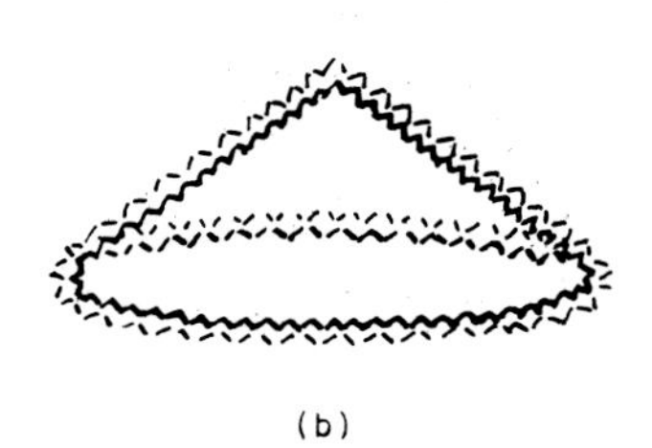

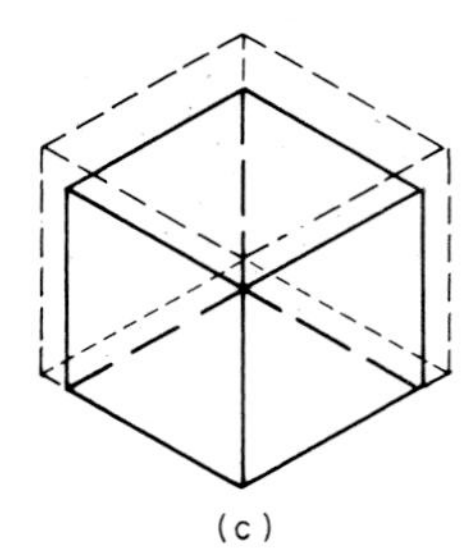

(a) (b) (c)

PLATE A-1

Relationship between cubic yard and cubic meter of semisolid and solid rock in bank, when transferred to hauler or stockpile, and when in compacted embankment. *(a)* Bank measurement: 1 $yd^3 = 0.76\ m^3$; 1 $m^3 = 1.31\ yd^3$. *(b)* Loose measurement in hauler or in stockpile; swell factor = 1.67. One unit bank measurement becomes 1.67 unit loose measurement; one unit loose measurement equals 0.60 unit bank measurement. *(c)* Compacted measurement in embankment; swell factor = 1.33. One unit bank measurement becomes 1.33 units compacted measurement; one unit compacted measurement equals 0.75 unit bank measurement. Cubic yards in solid and dashed lines; cubic meters in dashed and dotted lines.

Note: For corresponding data for specific material see Appendix 1.

APPENDIX 1

Approximate Material Characteristics*

*I am indebted to some 25 authorities for the data in this Appendix. Approximately 1500 values for specific gravities, weights, and swell and shrinkage factors were analyzed, interpreted, and weighted for good averages.

For about 100 years, commencing with Trautwine's pioneering handbook of 1882, *Civil Engineer's Pocketbook,* authoritative sources in the United States have been publishing tables of material characteristics. Generally speaking, these tables include specific gravities, weights in natural bed, swell factors from the natural bed or cut to the loose condition, weights in the loose condition, swell or shrink factors from the natural bed or cut to uncompacted fills or compacted embankments, and weights in uncompacted fills or compacted embankments. Engineers, both public and private, contractors, mining companies, machinery manufacturers, and writers of handbooks have contributed to this array of data.

The following table in this appendix is a summary of existing data, commencing with Trautwine's tables based on his own meticulous laboratory and field work, and ending with personal data gathered during the past 50 years. The table is necessarily based on properly interpreted and weighted averages. It is therefore not absolute for a specific case, and engineering experience and judgment will guide the user in its proper application. Prior to examination of the table, the reader is referred to these explanatory notes.

Materials

Rock materials are noted to be I, igneous; S, sedimentary; or M, metamorphic. Materials marked by asterisks are ores in the mineral or near mineral state, and the weights do not allow for the containing gangues of the ore body. The weight of the mineral is constant, with a set specific gravity, but the weight of the gangue, such as the associated earthy materials contained in quartz, rhyolite, schist, and feldspar, varies considerably with respect to the weight of the mineral. In mining the engineer must estimate the unit weight of the ore body and the weight of the contained mineral.

For example, hematite, the iron mineral, weighs 8560 lb/yd^3. Associated gangue, however, varies with respect to the hematite. Suppose that the mineral hematite samples 40 percent by weight of the ore and that the gangue, weighing 4000 lb/yd^3, samples 60 percent by weight of ore. Then 1000 lb of ore in the natural bed occupies a volume of

$$\frac{40\% \times 1000}{8560} + \frac{60\% \times 1000}{4000} = 0.197 \text{ yd}^3$$

The ore, then, weighs $1000/0.197 = 5080$ lb/yd^3, as contrasted to the weight of 8560 lb/yd^3 for the contained mineral hematite. At this juncture it is well to explain that miners sometimes use the word *hematite* for both the mineral and the ore.

Specific Gravity

When the value for specific gravity is in parentheses, it is an *apparent specific gravity* because the material is not in the solid state. Examples are gravel and rock-earth mixtures, which contain voids when in their natural bed.

Cubic Yard in Cut

The weight in the natural bed, or bank measurement, includes natural moisture. The average weight is subject to a maximum ± 10 percent variation. Again, it is emphasized that ore weights are for the mineral only and not for an impure ore body containing gangue.

Cubic Yard in the Loose Condition

Percent swell from the natural bed to the loose condition is an average which is subject to a maximum 33 percent variation in both rock and earthy materials. Variations are multipliers and not percentages to be added to or subtracted from the given percent of swell. The swell factor of 67 percent, given for several rocks, is an average figure obtained from existing data for solid rock, and it has been applied to solidly bedded unweathered rocks for which no swell factors are available specifically. Percent swell factors for ores are in terms of the entire ore body rather than in terms of the contained mineral. Weights in the loose condition are averages, except when calculated on the basis of the aforementioned average 67 percent swell factor. All weights are subject to any adjusted value of the swell factor.

Cubic Yard in the Fill

In the table a cubic yard in a fill is a cubic yard in a compacted embankment. No values are given for ores in a fill as they are not construction materials. When they are in a fill, they are in a stockpile, and the values for a cubic yard in the loose condition are applicable. Percent swell or shrink from cut or natural bed to fill is an average, subject to a maximum 33 percent variation in both rock and earthy materials. Percentage variation is a multiplier.

It is absolutely necessary, especially in the case of rock materials, to distinguish between two methods of fill construction:

1. Natural or gravity compaction, which was common years ago before the development of compacting machinery, is little used now except in the building of waste fills and stockpiles of materials and ores. The swell and shrink factors from the cut or natural bed vary from 10 percent shrinkage for earthy materials to 67 percent swelling for rock materials. Because of different degrees of fragmentation in the cut and because of the wide variations of fill construction methods in natural or gravity compaction, no figures are tabulated.

2. Mechanical compaction by rollers, along with wetting of the fill, is today's accepted method for fill consolidation. The tabulated swell and shrink factors and weights are for this modern method of fill compaction.

Two other influences affect swell and shrink factors and resultant weights. First, crawler-tractor-rippers produce better fragmentation and better grading of both rock and earthy formations in the cut. Second, the average so-called rock job really consists of a rock-earth mixture which in itself is pretty well graded.

These three factors, nature of materials, use of tractor-rippers, and modern compacting methods, have made possible the prevalent high densities of fills, densities not in accordance with some previously tabulated data for swell and shrink factors from cut to fill. In the case of construction materials the writer has used swell and shrink factors and weights, including moisture, resulting from average compaction methods.

It is a fact that certain friable rocks in weathered and parent rock zones have low swell factors from cut to fill. These rocks are really equivalent to rock-earth mixtures in their behavior during excavation and compaction. Rock swell factors are in terms of solid rock in the cut and do not include allowances for overlain residual and weathered rocks or for earthy and friable materials, all of which would reduce greatly the swell factor from cut to fill.

		Cubic yards,	Cubic yards loose		Cubic yards in fill	
Material	sp gr	in cut—weight, lb	Percent swell	Weight, lb	Swell or shrink, %	Weight, lb
Adobe, S	(1.91)	3230	35	2380	−10	3570
Andesite, I	2.94	4950	67	2970	33	3730
Asbestos	2.40	4040	67	2420		
Ashes, coal	(0.61)	1030	33	800	−50	2060
Asphaltum, S	1.28	2150	67	1390		
Asphalt rock, S	2.41	4050	62	2500		
Aragonite, calcium ore*	3.00	5050	67	3020		
Argentite, silver ore*	7.31	12300	67	7360		
Barite, barium ore*	4.48	7560	67	4520		
Basalt, I	2.94	4950	64	3020	36	3640
Bauxite, aluminum ore*	2.73	4420	50	2940		
Bentonite	1.60	2700	35	2000		
Biotite, mica ore*	2.88	4850	67	2900		
Borax, S	1.73	2920	75	1670		
Breccia, S	2.41	4050	33	3040	27	3190
Calcite, calcium ore*	2.67	4500	67	2700		
Caliche, S	(1.44)	2430	16	2100	−25	3200
Carnotite, uranium ore*	2.47	4150	50	2770		
Cassiterite, tin ore*	7.17	11380	67	6800		
Cement				2700		
Cerrusite, lead ore*	6.50	10970	67	6560		
Chalcocite, copper ore*	5.70	9600	67	5750		
Chalcopyrite, copper ore*	4.20	7060	67	4220		
Chalk, S	2.42	4060	50	2710	33	3050
Charcoal				1030		
Chat, mine tailings				2700		
Cinders	(0.76)	1280	33	960	−10	1420

Material	sp gr	Cubic yards, in cut—weight, lb	Cubic yards loose		Cubic yards in fill	
			Percent swell	Weight, lb	Swell or shrink, %	Weight, lb
Cinnabar, mercury ore*	8.10	13630	67	8170		
Clay, S:						
Dry	(1.91)	3220	35	2380	−10	3570
Damp	(1.99)	3350	40	2400	−10	3720
Clinker				2570		
Coal, S:						
Anthracite	1.55	2610	70	1530		
Bituminous	1.35	2280	67	1370		
Coke	(0.51)	860	0	860		
Colemanite, borax ore*	1.73	2920	75	1670		
Concrete:'						
Stone	2.35	3960	72	2310	33	2910
Cyclopean	2.48	4180	72	2430	33	3150
Cinder	1.76	2970	72	1730	33	2240
Conglomerate, S	2.21	3720	33	2800	−8	4030
Decomposed rock:						
75% R, 25% E	(2.45)	4120	25	3300	12	3700
50% R, 50% E	(2.23)	3750	29	2900	−5	3940
25% R, 75% E	(2.01)	3380	26	2660	−8	3680
Diabase, I	3.00	5050	67	3010	33	3810
Diorite, I	3.10	5220	67	3130	33	3930
Diatomite, S:						
Ditomaceous earth	(0.87)	1470	62	910		
Dolomite, S	2.88	4870	67	2910	43	3400
Earth, loam, S:						
Dry	(1.84)	3030	35	2240	−12	3520
Damp	(2.00)	3370	40	2400	−4	3520
Wet, mud	(1.75)	2940	0	2940	−20	3520
Earth-rock mixtures:						
75% E, 25% R	(2.01)	3380	26	2660	−8	3680
50% E, 50% R	(2.23)	3750	29	2900	−5	3940
25% E, 75% R	(2.45)	4120	25	3300	12	3700
Feldspar, I	2.62	4410	67	2640	33	3320
Felsite, I	2.50	4210	67	2520	33	3170
Fluorite, S	3.10	5220	67	3130		
Gabbro, I	3.10	5220	67	3130	33	3940
Galena, lead ore*	7.51	12630	67	7570		
Gneiss, M	2.71	4550	67	2720	33	3420
Gob, mining refuse	(1.75)	2940	0	2940	−20	3520
Gravel, average graduation, S:						
Dry	(1.79)	3020	15	2610	−7	3240
Wet	(2.09)	3530	5	3350	−3	3640
Granite, I	2.69	4540	72	2640	33	3410
Gumbo, S:						
Dry	(1.91)	3230	50	2150	−10	3570
Wet	(1.99)	3350	67	2020	−10	3720
Gypsum, S	2.43	4080	72	2380		
Hematite, iron ore*	5.08	8560	75	4880		
Hessite, silver ore*	8.50	14300	67	8560		
Ice	0.93	1560	67	930		
Ilmenite, titanium ore*	4.75	8000	69	4730		
Kaolinite, S:						
Dry	(1.91)	3230	50	2150		
Wet	(1.99)	3350	67	2010		
Lignite	(1.25)	2100	65	1270		
Lime				2220		
Limestone, S	2.61	4380	63	2690	36	3220
Linnaeite, cobalt ore*	4.89	8230	67	4930		
Limonite, iron ore*	3.80	6400	55	4140		
Loam, earth, S:						
Dry	(1.84)	3030	35	2240	−12	3520

Material	sp gr	Cubic yards, in cut—weight, lb	Cubic yards loose		Cubic yards in fill	
			Percent swell	Weight, lb	Swell or shrink, %	Weight, lb
Damp	(2.00)	3370	40	2400	−4	3520
Wet, Mud	(1.75)	2940	0	2940	−20	3520
Loess, S:						
Dry	(1.91)	3220	35	2380	−10	3570
Wet	(1.99)	3350	40	2400	−10	3720
Magnesite, magnesium ore*	3.00	5050	50	3360		
Magnetite, iron ore*	5.04	8470	54	5520		
Marble, M	2.68	4520	67	2700	33	3400
Marl, S	2.23	3740	67	2240	33	2820
Masonry, rubble	2.33	3920	67	2350	33	2950
Millerite, nickel ore*	5.65	9530	67	5710		
Molybdenite, molybdenum ore*	4.70	7910	67	4750		
Mud, S	(1.75)	2940	0	2940	−20	3520
Muscovite, mica ore*	2.89	4860	67	2910		
Niccolite, nickel ore*	7.49	12600	67	7550		
Orpiment, arsenic ore*	3.51	5900	50	3940		
Pavement:						
Asphalt	1.93	3240	50	1940	0	3240
Brick	2.41	4050	67	2430	33	3050
Concrete	2.35	3960	67	2370	33	2980
Macadam	1.69	2840	67	1700	0	2840
Wood block	0.97	1630	72	950	33	1220
Peat	(0.70)	1180	33	890		
Phosphorite, phosphate rock, S	3.21	5400	50	3600		
Porphyry, I	2.74	4630	67	2770	33	3480
Potash, S	2.20	3700	50	2470		
Pumice, I	0.64	1080	67	650		
Pyrites, iron ore*	5.07	8540	67	5110		
Pyrolusite, manganese ore*	4.50	7560	50	5050		
Quartz, I	2.59	4360	67	2610	33	3280
Quartzite, M	2.68	4520	67	2710	33	3400
Realgar, arsenic ore*	3.51	5900	50	3930		
Rhyolite, I	2.40	4050	67	2420	33	3040
Riprap rock, average	2.67	4500	72	2610	43	3150
Rock-earth mixtures:						
75% R, 25% E	(2.45)	4120	25	3300	12	3700
50% R, 50% E	(2.23)	3750	29	2900	−5	3940
25% R, 75% E	(2.01)	3380	26	2660	−8	3680
Salt, rock, S	2.18	3670	67	2200		
Sand, average graduation, S:						
Dry	(1.71)	2880	11	2590	−11	3240
Wet	(1.84)	3090	5	3230	−11	3460
Sandstone, S	2.42	4070	61	2520	34	3030
Scheelite, tungsten ore*	5.98	10100	67	6050		
Schist, M	2.59	4530	67	2710	33	3410
Serpentine, asbestos ore*	2.62	4440	67	2650		
Shale, S	2.64	4450	50	2970	33	3350
Silt, S	(1.93)	3240	36	2380	−17	3890
Siltstone, S	2.42	4070	61	2520	−11	4560
Slag:						
Furnace	2.87	4840	98	2690	65	2930
Sand	(0.83)	1400	11	1260	−11	1570
Slate, M	2.68	4500	77	2600	33	3380
Smaltite, cobalt ore*	6.48	10970	67	6560		
Snow:						
Dry	(0.13)	220	0	220		
Wet	(0.51)	860	0	860		
Soapstone, talc ore*	2.70	4550	67	2720		
Sodium niter, chile saltpeter	2.20	2710	50	2470		
Stibnite, antimony ore*	4.58	7710	67	4610		
Sulfur	2.00	3450	50	2310		

Material	sp gr	Cubic yards, in cut—weight, lb	Cubic yards loose		Cubic yards in fill	
			Percent swell	Weight, lb	Swell or shrink, %	Weight, lb
Syenite, I	2.64	4460	67	2670	33	3350
Taconite, iron ore*	3.18	5370	60	3360		
Talc, M	2.70	4640	67	2780	33	3490
Topsoil, S	(1.44)	2430	56	1620	−26	3280
Trachyte, I	2.40	4050	67	2420	33	3050
Trap rock, igneous rocks, I	2.79	4710	67	2820	33	3540
Trash				400	−50	800
Tuff, S	2.41	4050	50	2700	33	3050
Witherite, barium ore*	4.29	7230	67	4320		
Wolframite, tungsten ore*	7.28	12280	67	7350		
Zinc blende, zinc ore*	4.02	6780	67	4060		
Zincite, zinc ore*	5.68	9550	67	5710		

Key to table:
I—igneous rock. **S**—sedimentary rock. **M**—metamorphic rock.
*—ores in the mineral state, with no gangues. Adjust for percentage of mineral bearing gangue or rock to estimate weight of entire ore body, as explained previously in text.
( )—apparent specific gravity, as material is not solid.

Weights per cubic yard in cut are subject to average ± 10 percent variation. Swell and shrinkage factors for loose condition and embankment are subject to average ± 33 percent variation. Weights in loose condition and in embankment are subject to adjustments in accordance with modified swell and shrinkage factors.

APPENDIX 2

Rock Clauses

When an owner and a contractor enter into a contract for rock excavation, the excavation may be described as either *classified* or *unclassified* in the bid schedule. If it is unclassified, the contractor must excavate all material in the cut at the agreed-upon unit bid price. On the other hand, if the material is classified as *common* and *rock,* the contractor will be paid accordingly once the kind of excavation is established. Naturally some means must be used to distinguish between common and rock, or else there will be disagreements. It is to the owner's interest to delay the change from common to rock because rock costs more than common. On the other hand, it is to the contractor's advantage to hasten the change if rock ripping becomes difficult because it may result in higher costs than drilling and blasting.

Before the advent of efficient tractor-rippers for fragmentation there was really no middle ground between common and rock. Common was described in specifications as earth, hardpan, and loose rock, which could be excavated by hand or by horse-drawn scrapers. Rock was described as solid rock or rock in place and boulders measuring ½ yd^3 and up, the removal of which required drilling and blasting. The use of the tractor-ripper created a middle zone of weathered and semisolid rock between common and rock, which might be fragmented either by tractor-ripper or by blasting. It was then necessary to use some new impartial means to distinguish between common and rock. For some years the seismic timer described in Chapter 7 has been used to establish this line of demarcation.

The current 1978 rock clauses for surface and subsurface excavation are for the heavyweight crawler-tractor-rippers and for the heavyweight trenchers which are used regularly in rock excavation.

Rock Clauses

For Surface Excavation Normally Handled by Heavy Excavation Machinery Any surface excavation material with seismic shock-wave velocity greater than 6000 ft/s will be considered to be economically nonrippable by crawler-tractor-ripper and to require blasting prior to excavation. The contractor, when of the opinion that the velocity exceeds 6000 ft/s, will so advise the owner, and both owner and contractor will test the material to be excavated. If the velocity exceeds 6000 ft/s, the rock-earth will be classified as rock instead of common and the contractor will be paid accordingly. Both owner and contractor agree to expedite testing by seismic timer.

For Subsurface Excavation Normally Handled by Heavy Trenchers Any subsurface excavation with seismic shock-wave velocity greater than 4000 ft/s will be considered to be economically "nondiggable" by trenchers and to require blasting prior to excavation. The contractor, when of the opinion that the 4000 ft/s velocity is exceeded, will so advise the owner, and both owner and contractor will test the excavation. If the velocity exceeds 4000 ft/s, the rock will be classified as rock and the contractor will be paid accordingly. Both owner and contractor agree to expedite testing by seismic timer.

APPENDIX 3

Depreciation Schedule for Machinery and Facilities

Chapter 8 contains suggested rates of depreciation in years for machinery and facilities used in rock excavation. Table 8-3 is based on averages for the construction and mining industries.

For a long time an authoritative and useful table has been Schedule F of the U.S. Bureau of Internal Revenue. It has been used for all construction machinery and facilities since 1931, and has since been revised on several occasions. Below is the schedule, abridged to cover essentially machinery and facilities used in excavation.

Machinery and Facilities	*Life, yr*
Automobiles:	
Light	2
Medium	3
Heavy	5
Backfillers, power:	
Light	3
Medium	5
Heavy	6
Tractor-mounted	5
Barges:	
Steel	30
Wood	25
Bending machines:	
Angle	15
Pipe and rail	10
Bins:	
Steel and concrete	6
Steel	12
Wood	8
Bin frames, steel	6
Blacksmith shop, portable	4
Boilers:	
Upright	7
Locomotive	15
Stationary	20
Buckets:	
Cableway	6
Clamshell	6
Elevator	5
Orange-peel	6
Bail, pivot turnover	5
Scraper or dragline	6

A–12 Depreciation Schedule for Machinery and Facilities

Machinery and Facilites	*Life, yr*
Bulldozers:	
Grade builders	8
Tractor-mounted	4
Burner equipment, gas and oil	12
Cables, wire	4
Cableways, cable only	3
Cableway carriage	5
Capstans, electric	10
Cars:	
Ballast spreader	10
Boarding and tool	20
Dump, steel	8
Dump, wood	6
Flat, steel	12
Flat, wood	10
Skip, hoist	10
Tank	20
Carts, tool, steel	4
Chains:	
Hawsers and lines	6
Power, transmissions	5
Channelers, rock	6
Cleaning machines, steam or sand	15
Compressors:	
Belt-driven	10
Electric, portable	8
Gasoline, portable	6
Motor truck unit	5
Steam, portable	6
Controllers, motor	12
Conveyors:	
Belt elevating, portable	3
Belt elevating, stationary	6
Buckets:	
Cable-drag	6
Monorail	15
Chain, portable	6
Portable	5
Scraper	6
Cranes:	
Crawler, electric:	
2½–5 tons	5
10–15 tons	7
20 tons and over	9
Crawler, gasoline:	
2½–5 tons	5
10–15 tons	9
20 tons and over	12
Locomotive, gasoline	7
Steam:	
2½–5 tons	6
10–15 tons	10
20 tons and over	12
Locomotive	10
Dragline	10
Universal, gasoline 2½–5 tons, mounted on 10-ton truck	6
Dock or wharf, traveling	20
Crushers, rock:	
Portable	8
Stationary	10
Cutting and welding outfits	4
Diggers, clay, pneumatic	3
Draglines:	
Electric:	
½–¾ yd^3	6

Machinery and Facilites	*Life, yr*
1–1½ yd^3	8
2 yd^3 and over	10
Gasoline:	
½–¾ yd^3	5
1–1½ yd^3	9
2 yd^3 and over	12
Steam:	
½–¾ yd^3	6
1–1½ yd^3	10
2 yd^3 and over	12
Dredges:	
Clamshell	16
Dipper	8
Hydraulic	20
Pipe	10
Drill boats	12
Drill points, well	5
Drills:	
Air drifter	3
Rock, electric	3
Jackhammer	3
Steam	5
Traction, well	7
Tripod	7
Tunnel, carriage	5
Well	10
Engines:	
Gasoline	10
Marine	20
Oil	20
Steam	11
Excavators:	
Cableway, complete	4
Trench, gasoline:	
Depth, 7–12 ft	6
Depth, 18 ft	8
Trench, steam	
Depth, 7–12 ft	8
Depth, 18 ft	10
Trench, vertical boom	5
Trench, wheel or ladder type	5
Generator sets:	
Steam engine	12
Turbine, headlight or floodlight	4
Graders:	
Blade, road:	
7- to 8-ft blade	4
9- to 10-ft blade	5
10-ft blade and over	8
Elevating	8
Rooters, wheel	5
Hoists:	
Air and electric	8
Gasoline	6
Steam	12
Hose:	
Fire, linen- or rubber-lined, cotton	5
Rubber, air, steam, or water	10
Levee construction:	
Draglines	8
Shovels	8
Tower excavators	12
Light plant	12
Lighters	22
Loaders, bucket:	

A–14 Depreciation Schedule for Machinery and Facilities

Machinery and Facilites	*Life, yr*
Crawler and portable	5
Stationary	6
Locomotives, industrial:	
Diesel	10
Electric	16
Gasoline:	
Up to 10 tons	8
10–20 tons	15
20 tons and over	20
Steam:	
Up to 10 tons	8
10–20 tons	18
20 tons and over	20
Locomotives, standard gauge	30
Motors:	
Electric, small	8
Electric, medium	10
Electric, large	12
Hydraulic	5
Pneumatic	5
Pipe, black or galvanized	4
Pipelines and fittings for floating dredges	10
Pit and quarry plants	6
Ploughs:	
Furrow	3
Rooter	6
Pumping units:	
Electric	6
Gasoline	6
Highway contractor's pump	4
Piston	5
Steam centrifugal	10
Pumps:	
Air lift	10
Centrifugal	6
Hydraulic	15
Oil	10
Steam piston	6
Rails, steel	10
Razing equipment for buildings	8
Rollers, road, gasoline- or steam-powered	10
Sandblast outfits	10
Saws, hand, electric and pneumatic	3
Scales, large-track and wagon	20
Scarifiers:	
Attachments	4
Drag, all steel	4
Grader-type	4
Scrapers:	
Blade, carryall	6
Fresno	4
Slip	2
Wheel	5
Shovel attachments	6
Shovels:	
Electric or gasoline, crawler- or wheel-mounted:	
½–¾ yd^3	5
1–1½ yd^3	6
2 yd^3 and over	8
Steam, crawler- or wheel-mounted:	
½–¾ yd^3	7
1–1½ yd^3	8
2 yd^3 and over	10
Railroad, steam	10
Tunnel	4

Machinery and Facilites	*Life, yr*
Spreaders, stone:	
Hopper wagon	5
Steel box	5
Switches:	
Portable	4
Stationary	5
Tampers, backfill, pneumatic	3
Tamping machines	10
Tanks, gasoline (storage, steel)	6
Tanks, water or air (storage, steel)	10
Towers, cableway:	
Steel	6
Wood	3
Steel boom with counterweights	5
Tractors, gasoline or steam:	
3 tons	4
5 tons	6
10 tons	8
20 tons	10
Trailers:	
Dump, steel or wood	10
Platform, wood	4
Drop platform, heavy duty	5
Trucks, automobile, dump:	
$1/3$–$2/3$ yd^3	3
1–$1\frac{2}{3}$ yd^3	5
2 yd^3 and over	8
Tugs, screw-propelled, steam or gasoline	25
Wagons:	
Dump, steel or wood	6
Road oilers, steel tank	10
Tank or sprinkler:	
Steel	10
Wood	8
Welders, acetylene or electric	10
Winches, electric or pneumatic	10
Wire and cables, electric	6

NOTE 1: The depreciation schedule is not modern as far as kinds and sizes of excavation machinery are concerned. However, it is useful for the values for particular machines which are listed, and also for the selection of values for machines with characteristics similar to the machine under consideration.

NOTE 2: A supplementary pamphlet of the U.S. Internal Revenue Service is *Tax Information on Depreciation,* Publication 534, 1978. This pamphlet contains a guideline schedule of depreciation periods, not as complete as this table, and other helpful information concerning the general subject of depreciation of machinery and facilities.

APPENDIX 4

Conversion Factors for Systems of Measurement

ENGLISH UNITS TO ENGLISH UNITS

This unit	Multiplied by	Equals this unit
Feet (ft)	12	Inches (in)
Yards (yd)	3	Feet (ft)
Fathoms	6	Feet (ft)
Rods, poles, or perches	16.5	Feet (ft)
Furlongs (fur)	660	Feet (ft)
Miles, statute	5280	Feet (ft)
Miles, statute	1760	Yards (yd)
Miles, nautical	6080	Feet (ft)
Miles, nautical	2027	Yards (yd)
Square feet (ft^2)	144	Square inches (in^2)
Square yards (yd^2)	9	Square feet (ft^2)
Square miles (mi^2)	27,878,400	Square feet (ft^2)
Square miles (mi^2)	3,097,600	Square yards (yd^2)
Acres	43,560	Square feet (ft^2)
Acres	4840	Square yards (yd^2)
Gallons (gal)	231	Cubic inches (in^3)
Gallons, U.S. (gal)	0.833	Imperial gallon
Cubic feet (ft^3)	1728	Cubic inches (in^3)
Cubic feet (ft^3)	7.48	Gallons (gal)
Cubic yards (yd^3)	27	Cubic feet (ft^3)
Cubic yards (yd^3)	202	Gallons (gal)
Acre-feet (acre·ft)	43,560	Cubic feet (ft^3)
Acre-feet (acre·ft)	1613	Cubic yards (yd^3)
Acre-feet (acre·ft)	325,829	Gallons (gal)
Pounds, lb	16	Ounces (oz)
Tons, short	2000	Pounds (lb)
Tons, long	2240	Pounds (lb)
Miles per hour (mi/h)	88	Feet per minute (ft/min)
Miles per hour (mi/h)	1.47	Feet per second (ft/s)
Horsepower (hp)	33,000	Foot-pounds per minute (ft·lb/min)
Horsepower (hp)	550	Foot-pounds per second, standard (ft·lb/s)
Horsepower (hp)	0.746	Kilowatts (kW)
British thermal units (Btu)	778	Foot-pounds (ft·lb)
British thermal units (Btu)	0.0236	Horsepower (hp)
Equals this unit	Divided by	This unit

METRIC UNITS TO METRIC UNITS

This unit	Multiplied by	Equals this unit
Meters (m)	100	Centimeters (cm)
Kilometers (km)	1000	Meters (m)
Square meters (m^2)	10,000	Square centimeters (cm^2)
Hectares	10,000	Square meters (m^2)
Square kilometers (km^2)	1,000,000	Square meters (m^2)
Cubic meters (m^3)	1,000,000	Cubic centimeters (cm^3)
Cubic meters (m^3)	1000	Liters (L)
Kilograms (kg)	1000	Grams (g)
Quintals	100	Kilograms (kg)
Tons, metric	1000	Kilograms (kg)
Kilometers per hour (km/h)	16.7	Meters per minute (m/min)
Kilometers per hour (km/h)	0.278	Meters per second (m/s)
Equals this unit	Divided by	This unit

ENGLISH UNITS TO METRIC UNITS AND METRIC UNITS TO ENGLISH UNITS

This unit	Multiplied by	Equals this unit
Inches (in)	2.54	Centimeters (cm)
Feet (ft)	0.305	Meters (m)
Yards, (yd)	0.914	Meters (m)
Miles, statute	1.609	Kilometers (km)
Square inches (in^2)	6.45	Square centimeters (cm^2)
Square feet (ft^2)	0.0929	Square meters (m^2)
Square yards (yd^2)	0.836	Square meters (m^2)
Acres	0.405	Hectares
Square miles (mi^2)	2.590	Square kilometers (km^2)
Cubic inches (in^3)	16.4	Cubic centimeters (cm^3)
Cubic feet (ft^3)	0.0283	Cubic meters (m^3)
Cubic yards (yd^3)	0.765	Cubic meters (m^3)
Gallons, U.S. (gal)	3.79	Liters (L)
Ounces (oz)	28.4	Grams (g)
Pounds (lb)	0.454	Kilograms (kg)
Tons, short	0.907	Tons, metric
Tons, long	1.016	Tons, metric
Feet per minute (ft/min)	0.305	Meters per minute (m/min)
Feet per second (ft/s)	0.305	Meters per second (m/s)
Miles per hour (mi/h)	1.609	Kilometers per hour (km/h)
Pounds per square inch (lb/in^2)	0.0703	Kilograms per square centimeter (kg/cm^2)
Pounds per square foot (lb/ft^2)	4.887	Kilograms per square meter (kg/m^2)
Pounds per cubic yard (lb/yd^3)	0.593	Kilograms per cubic meter (kg/m^3)
Foot-pounds or pound-feet (ft·lb)	0.138	Kilogram-meter (kg·m)
Horsepower (hp)	0.746	Kilowatts (kW)
British thermal units (Btu)	0.252	Calories (cal)
Equals this unit	Divided by	This unit

Degrees Fahrenheit, °F, to degrees, Celsius, °C:

$$°C = \tfrac{5}{9}(°F - 32) = 0.55(°F - 32)$$

Degrees Celsius, °C, to degrees Fahrenheit, °F:

$$°F = \tfrac{9}{5}\,°C + 32 = 1.80\,°C + 32$$

APPENDIX 5

Formulas Frequently Used in Calculations for Excavation Projects

ABBREVIATIONS USED IN FORMULAS

A area
A_b area of base
a acceleration
a_o angle in degrees
a_r angle in radians, 1 radian = 57.3°
b base
b_u upper base
b_l lower base
c circumference of circle
d diameter of circle
E energy
f force
f_c centrifugal force
g acceleration of gravity, 32.2 ft/s²
h height
h_s slant height
hp horsepower
kWh kilowatthour
l length
m mass
min minute
π pi, 3.1416
r radius of circle
s distance
s_p perpendicular distance
s second
t time
V volume
v velocity
W work
w width
wt weight

FORMULAS

Circumference of circle: $c = \pi d$

Areas:

Square: $A = h^2$
Rectangle: $A = hb$
Parallelogram: $A = hb$
Trapezoid: $A = h \dfrac{b_u + b_l}{2}$
Triangle: $A = \dfrac{hb}{2}$
Trapezium: A = sum of the areas of the two integral triangles
Circle: $A = \pi r^2 = 0.785 d^2$
Sector of circle: $A = \pi r^2 \dfrac{a_0}{360}$
Segment of circle: $A = 0.5 r^2 (a_r - \sin a_0)$
Ellipse: $A = \pi \dfrac{hw}{4}$

Cone: $A=A_b+(0.5ch_s)$
Sphere: $A=4\pi r^2$
Irregular area, approximately: Divide area into a number of strips with parallel sides and lengths $l_0, l_1, l_2, \ldots l_{n-1}, l_n$, separated by equal distances, s_p:

$$A=s_p\left[\frac{l_0}{2}+l_1+l_2+\ldots+l_{n-1}+\frac{l_n}{2}\right]$$

Volumes:

Cube: $V=h^3$
Prism, rectangular: $V=hwl$
Parallelopiped: $V=As_p$
Wedge: $V=0.5(hwl)$
Cylinder: $V=\pi r^2 h$
Cone or pyramid: $V=0.333A_b h$
Sphere: $V=1.333\pi r^3=0.524d^3$
Spherical sector: $V=0.667\pi r^2 h=2.09\pi r^2 h$
Spherical segment: $V=\frac{\pi h}{6}(3r^2+h^2)$

Irregular volume, approximate: Divide volume into a number of zones with parallel sides of areas $A_0, A_1, A_2, \ldots, A_{n-1}, A_n$, separated by equal distances, s_p.

$$V=s_p\left(\frac{A_0}{2}+A_1+A_2+\ldots+A_{n-1}+\frac{A_n}{2}\right)$$

Motion:

$$s=vt$$
$$v=at$$
$$=gt$$
$$s=0.5at^2$$
$$=0.5gt^2$$
$$v^2=2as$$
$$=2gs$$

Mass and weight:

$$\text{wt}=mg$$

Force:

$$f=ma$$
$$=mg$$
$$f_c=\text{wt}\times\frac{v^2}{rg}$$

Work and energy:

$$W=fs=fvt$$
$$E=fs=fvt$$

Power:

$$\text{hp}=\frac{W/\text{s}}{550}$$
$$\text{hp}=\frac{W/\text{min}}{33{,}000}$$
$$\text{kW}=1.341\ \text{hp}$$

Efficiency:

$$\text{Eff, }\%=\frac{\text{Output hp}}{\text{Input hp}}\times 100$$

Efficiency of overall work of a machine:

$$\text{Eff}=\frac{\text{min actually worked during hour}}{60}$$

The efficiency may be expressed as a working-minute hour, such as a 40- or 50-minute hour, or as a percentage, 67 and 83%, respectively. The efficiency factor is based on the machine's ability to work provided it is not held up by any extraneous delays.

APPENDIX 6

Swell versus Voids of Materials and Hauling Machine Load Factors

Swell, %	Voids, %	$\frac{cybm}{cylm}$
5	4.8	0.952
10	9.1	0.909
15	13.0	0.870
20	16.7	0.833
25	20.0	0.800
30	23.1	0.769
35	25.9	0.741
40	28.6	0.714
45	31.0	0.690
50	33.3	0.667
55	35.5	0.645
60	37.5	0.625
65	39.4	0.606
70	41.2	0.588
75	42.9	0.571
80	44.4	0.556
85	45.9	0.541
90	47.4	0.526
95	48.7	0.513
100	50.0	0.500

SOURCE: Caterpillar Tractor Co.

APPENDIX 7

Approximate Angles of Repose of Materials*

Material	Slope ratio, horizontal:vertical	Angle of repose, °
Ashes, coal	1.0:1	45
Cinders, coal	1.0:1	45
Clay:		
Dry	1.3:1	38
Damp	2.0:1	27
Coal, broken	1.4:1	36
Earth:		
Dry	1.3:1	38
Damp	2.0:1	27
Gravel:		
Round	1.7:1	30
Angular	1.3:1	38
Rock, broken:		
Soft	1.5:1	34
Hard	1.3:1	38
Rock, weathered:		
Residuals and weathered rock	1.5:1	34

SOURCE: Caterpillar Tractor Co.

*Angle of repose is the angle between the horizontal and the slope of a heaped pile of material.

APPENDIX 8

Bearing Powers of Materials

Material	English system		Metric system	
	lb/in²	tons/ft²	kg/cm²	tons/m²
Rock, solid	350	24	24.6	240
Rock, semishattered	70	5	4.9	50
Clay:				
Dry	55	4	3.9	40
Damp	27	2	1.9	20
Wet	14	1	1.0	10
Gravel, cemented	110	8	7.7	80
Sand:				
Dry, compacted	55	4	3.9	40
Dry, clean	27	2	1.9	20
Quicksand and alluvial soil	7	0.5	0.5	5

SOURCE: Caterpillar Tractor Co.

APPENDIX 9

Abbreviations

A	ampere
A	area
a	acceleration
ac	alternating current (adj.)
acre	acre
acre·ft	acre foot
A·h	ampere-hour
atm	atmosphere
avg	average
bhp	brake horsepower
bhp·h	brake horsepower-hour
Btu	British thermal unit
°C	degree Celsius
c	candle
cal	calorie
cg	centigram
cgs	centimeter-gram-second metric system
cif	cost, insurance, and freight
cm	centimeter
cm^2	square centimeter
cm^3	cubic centimeter
coef	coefficient
const	constant
cos	cosine
cp	candlepower
c to c	center to center
cwt	hundredweight
cybm	cubic yard bank measurement
cycm	cubic yard compacted measurement
cylm	cubic yard loose measurement
cysm	cubic yard struck measurement
. . .°	degree
d	diameter of circle
dbh	diameter breast high of tree
dc	direct current (adj.)
eff	efficiency
el	elevation
eq	equation
°F	degree Fahrenheit
f	force

fbm	feet board measure or board feet
fc	footcandle
fob	free on board
ft	foot
ft^2	square foot
ft^3	cubic foot
ft·lb	foot pound
ft/min	feet per minute
ft^3/min	cubic feet per minute
ft/s	feet per second
ft^3/s	cubic feet per second
ft/s^2	foot per second per second
fur	furlong
g	acceleration of gravity, 32.2 ft/s^2
g	gram
gal	gallon
gal/min	gallons per minute
GR	grade resistance
GVW	gross vehicle weight (tare wt+payload wt)
h	hour
hp	horsepower
hp.h	horsepower hour
Hz	hertz (cycles per second)
ihp	indicated horsepower
in	inch
in^2	square inch
in^3	cubic inch
in·lb	inch-pound or pound-inch
kc	kilocycle
kg	kilogram
kg·m	kilogram-meter
kg/m^3	kilograms per cubic meter
kL	kiloliter
km	kilometer
km^2	square kilometer
kV	kilovolt
kVA	kilovolt-ampere
kW	kilowatt
kWh	kilowatt-hour
L	liter
lb	pound
lb/ft^2	pounds per square foot
lb/in^2	pounds per square inch
lin	linear
lin ft	linear feet
log	logarithm
lp	low pressure
lm	lumen
lm·h	lumen-hour
m	mass
m	meter
m^2	square meter
m^3	cubic meter
max	maximum
mg	milligram
mi	mile
mi^2	square mile
mi/h	miles per hour
mi/h/s	miles per hour per second
min	minimum
min	minute
mm	millimeter
oz	ounce

pt	pint
qt	quart
r/min	revolutions per minute
RR	rolling resistance
r/sec	revolutions per second
s	distance
s	second
scp	spherical candlepower
shp	shaft horsepower
sin	sine
sp gr	specific gravity
sta	station, 100 feet
sta yd	station cubic yard
tan	tangent
temp	temperature
ton/h	tons per hour
ton·mi	ton-mile
ton/min	tons per minute
V	volume
V	volt
v	velocity
VA	voltampere
W	work
W	watt
w/c	watts per candle
Wh	watthour
whp	wheel horsepower
wk	week
wt	weight
yd	yard
yd^2	square yard
yd^3	cubic yard
yr	year

Glossary

Ablation The formation of residual deposits by the washing away of loose or soluble materials.

Abrasion The mechanical wear of rock on rock.

Acre Unit for measuring land, equal to 43,560 ft^2 (4840 yd^2, or 160 square rod).

Acre-Foot The amount of water required to cover 1 acre to a depth of 1 ft, equal to 1613 yd^3 (1233 m^3). Also used to measure materials in place, such as coal and gravel.

Adamantine Drill Core drill using chilled shot as a cutting agent. Cores from 4 to 30 in (10 to 76 cm) in diameter are obtained.

Adit A nearly horizontal or horizontal passage from the surface into a mine. Frequently called a drift.

Adobe An impure calcareous clay which, when drying, breaks into roughly cubical blocks up to 20 in (51 cm) in dimension.

Agglomerate A coarse-grained pyroclastic rock consisting largely of bombs and blocks.

Aggradation The building up of any portion of the earth's surface toward a uniformity of grade or slope by the addition of materials.

Aggregate Mineral material such as sand, gravel, shells, slag, or broken stone to be used in concrete.

Air Receiver The air storage tank on an air compressor.

Air-Slaked Wetted by exposure to moisture in the air.

Alkali Flat A sterile plain, containing an excess of alkali, at the bottom of an undrained basin in an arid region. Sometimes called a playa.

Alluvial Fan The outspread sloping deposit of boulders, gravel, and sand, left by a stream where it spreads from a gorge upon a plain or open valley floor.

Alluvium The general name for all the sediment deposited in land environments by streams.

Altimeter An aneroid barometer graduated to show elevation instead of pressure.

Amorphous Without definite form. Applying to rocks and minerals having no definite crystalline structure.

Amortization The repayment of debt, principal and interest, usually in equal installments. The process of gradually recovering the cost or value of an asset such as machinery.

Ampere The intensity of electric current produced by 1 volt acting through a resistance of 1 ohm.

Amphibolite A coarse-grained, mafic metamorphic rock containing more than 50 percent ferromagnesian minerals. Dark color. Hard and heavy.

Amygdule A mineral that fills the vesicles or holes in igneous rock.

Andesite An igneous extrusive rock of fine-grained texture, consisting chiefly of feldspar and ferromagnesian minerals. Medium color, hard and heavy.

Angle of Repose Measured from the horizontal, it is the minimum angle of plane, along which coarse particles in the material begin to fall under the influence of gravity. Also known as *critical slope.*

Angledozer A bulldozer with a blade which can be pivoted on a vertical center pin so as to cast its load to either side.

Anthracite Coal A hard, black lustrous coal containing 85 to 90 percent carbon. It produces an intense, hot fire. A metamorphic rock.

Anticline An upfold of layered rocks in the form of an arch, with the oldest strata in the center. Reverse of a syncline.

Apron The front gate of a scraper body, which is raised and lowered.

Archeozoic The second of the six groups of rocks. Corresponds to the same geologic era. From 2000 to 1200 million yr ago.

Area The amount of surface included between certain closed boundary lines. Any particular extent of surface, region, or tract.

Areal Geology Branch of geology that pertains to the distribution, position, and form of the areas of the Earth's surface occupied by different kinds of rocks or different rock formations.

Argillaceous Containing or consisting of clay, as the cementing agency in argillaceous limestones and sandstones.

Argillite Synonym of slate.

Arkose A sandstone containing at least 25 percent feldspar as well as quartz. Light to medium color. Medium hard and medium heavy.

Ash, Volcanic Extremely fine-grained particles ejected from a volcano.

Asset An owned value such as machinery.

Atmospheric Pressure Pressure of air enveloping the earth, averaged as 14.7 lb/in^2 at sea level or 29.92 in of mercury as measured by a standard barometer.

Auger A rotating drill having a screw thread that carries cuttings away from the surface being drilled.

Availability Factor A measure of the reliability of machines as regards freedom from mechanical failures. It is the actual working time divided by the available working time, in hours. An availability factor of 95 percent is excellent, 90 percent is good, and 85 percent is acceptable.

Average-End-Area Method A means of calculating the volume of earthwork between two cross sections. The volume equals the average of the cross-sectional areas times the distance between the cross sections.

Azoic The first of the six groups of rocks. Corresponds to the geologic era. From 4.5 billion to 2.0 billion yr ago.

Backfill Rock-earth used to replace excavation previously removed.

Backhoe A hoe-type or pull-type shovel.

Badlands A system of closely spaced ridges with little or no vegetation.

Baking The hardening of rocks through heating by magma or lava. It produces *contact metamorphism.*

Ballast Material such as water, sand, or metal, which increases the weight of a machine such as a compactor or dragline.

Bank Measure Measurement of material in its original place in the ground or in situ.

Bank Yards Cubic yards of material in its original place in the ground.

Barranca Spanish term meaning a gully with somewhat steep banks.

Basalt Extrusive igneous rock of dark color and fine structure. Sometimes it contains vesicles and the formation is columnar. Hard and heavy.

Base Line The main traverse or surveying line through the site of proposed construction, from which lines and grades for the work are plotted. Not to be confused with the centerline of construction.

Basement Complex Assemblage of igneous and metamorphic rocks lying beneath the oldest sedimentary rocks of the region.

Batholith Large body of intrusive igneous rock having an exposed area of more than 40 mi^2 (104 km^2).

Bauxite The clayey ore of aluminum. Light to medium color. Soft and medium-weight.

Bed A base for machinery. The bottom of the body of a hauling machine.

Bed Stratum of rock more than ½ in (1.2 cm) in thickness.

Bedding Plane The top or bottom surface of a bed or stratum of rock. Usually a separation and weakness in the rock structure.
Bedrock Continuous solid rock that everywhere underlies the regolith and which sometimes forms the Earth's surface, the regolith having been eroded away and the bedrock having become an outcrop or shield.
Belt Loader A movable or traveling belt conveyor with cutting edge or disk.
Bench A working level or step in a cut which is made in several layers.
Bench Mark A point of known or assumed elevation used as reference in determining and recording other elevations.
Bench Terrace A more or less level step in a slope of cut.
Berm An artificial ridge of earth.
Bid To submit a contract price. Also the tender or bidding price.
Bid, Unbalanced A bid in which some of the unit prices are abnormal, either too high or too low, but in which the total bid represents the total of a balanced bid.
Bit The part of a drill or excavating machine which cuts the rock-earth.
Bit, Carbide A bit having inserts of tungsten carbide.
Bit, Chopping A bit that is worked by raising and dropping.
Bit, Coring A bit that grinds the outside ring of the hole, leaving an inner core intact for sampling.
Bit, Diamond A rotary bit having diamonds set in its cutting surfaces.
Bit, Drag A diamond or fishtail bit. A bit that cuts by rotation of fixed cutting edges or points.
Bit, Fishtail A rotary bit having cutting edges or knives.
Bit, Multiuse A bit that is sharpened for new service when worn.
Bit, Plug A diamond bit that grinds out the full width of the hole.
Bit, Roller A bit that contains cutting elements that are rotated inside as it turns.
Bit, Throwaway A bit that is discarded when worn.
Bituminous Coal Black and firm soft coal that breaks into blocks and contains alternating layers having dull and bright luster. Its carbon content is between 75 and 85 percent. Lightweight sedimentary rock.
Black Powder Gunpowder used in blasting. Mixture of carbon, sodium or potassium nitrate, and sulfur.
Blade Usually, a part of an excavator which digs and pushes rock-earth but does not carry it.
Blast To loosen or move rock-earth by means of explosives or an explosion.
Blasthole A drill hole used for a charge of explosives.
Blasthole Drill A drill of any type used in the drilling of blastholes, usually air percussion, cable, rotary, or fusion-type.
Blasting Agent An inert explosive detonated by a high-strength primer.
Blasting Cap A copper shell closed at one end and containing a charge of detonating agent which is ignited from the spark of the fuse. Also, a fully enclosed copper shell containing a detonating agent which is ignited by an electrical current.
Blasting Fuse A slow-burning fuse used for igniting blasting charges.
Blasting Gelatin A jellylike high explosive made by dissolving nitrocotton in nitroglycerin.
Blasting Machine A battery- or hand-operated generator used to supply current to blasting circuits.
Blasting Mat A steel blanket composed of woven cable or interlocked steel rings to prevent the throw of blasted material.
Blockholing Blasting boulders by means of drilled holes and explosives.
Blowout A deflation basin excavated in shifting sand or other easily eroded regolith by the winds.
Blue Tops Grade stakes, the tops of which indicate finish grade level.
BM Bench mark.
Body The load-carrying part of a hauler or scraper.
Bog Soft spongy ground, usually wet and composed of more or less vegetable matter.
Bog Iron Ore A spongy variety of limonite. Found in layers or lumps on level sandy soils which have been covered by swamp or bog. Found also in existing swamps and bogs.
Bomb Rounded mass of lava expelled from a volcano by steam explosions.
Bonanza Mining term for a rich body of ore.

Booster Small amount of high explosive attached to a detonator for the purpose of increasing the rate of detonating the charge.
Borehole Hole drilled into rock for exploration or for blasting.
Boring Rotary drilling.
Borrow Pit An excavation from which material is taken for a nearby job.
Boulder Detrital material greater than 8 in (21 cm) in diameter.
Boulder Clay Stiff, hard, and usually unstratified clay of the glacial periods which contains boulders. Also called *till* or *hardpan*.
Bowl The body of a carrying scraper.
Brake Horsepower The horsepower output of an engine, motor, or mechanical device. Measured at the flywheel or belt.
Breccia General term for rock of any origin containing angular particles.
British Imperial Gallon The metric gallon, containing 277 in^3, equivalent to 1.20 U.S. gallons.
British Thermal Unit The quantity of heat required to raise the temperature of 1 lb of water 1°F at or near the temperature of maximum density, 39.1°F. Equivalent to 0.252 cal.
Brush Trees and shrubs of less than 4-in (10-cm) stump diameter. The Spanish term *chaparral* is a synonym.
Bucket The part of an excavator which digs, lifts, loads, and carries material.
Bucket Loader Usually a chain-bucket loader, but also a tractor-loader.
Buckshot A tough, tenacious earthen material which, when dry, shatters into irregularly shaped particles about the size of buckshot or larger, about ⅛ in (3 mm) or more.
Buffer Blasting Secondary fragmentation by crushing action, produced by setting off a row of holes nearest the face first and the other parallel rows behind it in succession. Successive crushing actions result, as material from each successive blast is thrown against that of the preceding one.
Bulldozer A term for the complete machine, tractor with bulldozer.
Bulking Increase in volume of fine materials due to moisture content.
Bumboat A small boat equipped with hoist for handling the pipes, lines, and anchors of a dredge.
Burden The distance of a blasthole from the face, or the volume of rock to be moved by the explosive in each blasthole.
Burn Cut A narrow section of rock pulverized by exploding heavy charges in parallel holes, as in tunneling.
Buttress Fill A compacted fill placed so as to prevent movement or sliding of an existing earthwork or foundation.
Cable Excavator A long-range cable-operated machine which works between a mast and an anchor or between a head tower and a foot tower.
Calcareous Composed essentially of calcium carbonate or cemented by calcium carbonate, as a calcareous sandstone.
Calcareous Tufa An open cellular deposit of calcium carbonate. Light color, medium hard, and medium weight.
Caliche Material composed essentially of soft limestone with varying percentages of clay. Generally soft to medium hard, although sometimes requiring blasting. Light in color. Found chiefly in deserts, where it occurs as a sedimentary rock in a shallow stratum.
Calorie The amount of heat required to raise the temperature of 1 gram of water 1°C at or near the temperature of maximum density, 4°C. Equivalent to 3.97 Btu.
Cambrian The first or oldest of the five rock systems of the Paleozoic group. Also the corresponding geologic period, from 570 to 500 million years ago.
Candle Unit of light intensity. At a distance of 1 ft, a candle produces 1 footcandle (fc) equivalent to 1 lumen per square foot on a surface normal to the beam.
Candlepower A light unit or that amount of light given by a sperm candle burning 120 grams per hour.
Cap A detonator set off by electric current or burning fuse.
Cap, Delay An electric blasting cap that explodes at a set interval after passage of the current.
Cap Rock The more impervious and harder rock found above the more weathered and softer rock.

Carbonaceous Coaly, containing carbon or coal. Used especially for shale or rock containing small particles of carbon distributed throughout the whole mass.
Carboniferous The last of the five systems of rocks of the Paleozoic group of rocks. Also the corresponding geologic period. Subdivided into the Mississippian, Pennsylvanian, and Permian series of rocks. From 340 to 230 million years ago.
Carbonite An explosive mixture of nitrobenzene, potassium nitrate, sulfur, and kieselguhr.
Carriage Cable The cable for traveling the carriage on a cable. Also the traction cable.
Cartridge A wrapped stick of dynamite or other explosive.
Casing A pipe lining for a drilled hole to prevent raveling of the surface.
Cassiterite An important ore of tin.
Cat A trademarked designation for any machine made by the Caterpillar Tractor Co. Widely used to indicate any crawler-tractor.
Catskinner The operator of a crawler-tractor.
Cave Collapse of an unstable bank.
Cellular Porous texture and fairly large cavities of certain volcanic rocks such as basalt and pumice.
Celsius Metric scale of temperature, sometimes called centigrade.
Cement Materials from solution which bind together the particles of a sedimentary rock.
Cementation The binding together of the particles of a sedimentary rock by such mineral cement as calcite, silica, and iron oxide.
Cenozoic The last of the six groups of rocks. Also the corresponding geologic era, from 63 million yr ago to the present.
Center of Gravity The point in a body about which all the weights of all the various parts balance each other. The center of mass of a cut or fill.
Center of Mass In a cut or fill, a cross-sectional line that divides the volume into its halves.
Center of Moments The point at which a body tends to rotate. A point arbitrarily selected for determining the resultant moment of a series of forces.
Centerpin In revolving machinery the fixed vertical shaft around which the machinery deck turns.
Center Stakes Stakes indicating the centerline of construction.
Centigrade See *Celsius.*
Centrifugal Force Outward force exerted by a body moving in a curved line.
Centripetal Force Inward force exerted by a body to keep it moving in a curved line. Force exerted as a result of a superelevated curve on haulage road.
Chain Bucket Loader A mobile loader using a series of buckets mounted on a chain to excavate and load material.
Chaining A term which originally meant measuring with a chain, but as now used denotes measuring with either a chain or tape.
Chalcocite An important copper ore, the mineral containing 80 percent copper.
Chalcopyrite An important copper ore, the mineral containing 35 percent copper.
Chalk A limestone that is weakly cohesive of white color and medium weight.
Chaparral In a general sense, brush that is stiff and thorny. Spanish term.
Chats The gangue which is found intimately mixed with some lead-zinc ores. Light to medium color. Medium weight.
Cherry Picker A small derrick or hoist, usually mounted on a truck or tractor.
Chert A sedimentary rock composed of silica which is either a precipitate or a replacement of calcium carbonate minerals. Light to medium color and medium weight.
Cheval-Vapeur The French term for horsepower.
Chip Blasting Blasting of shallow-depth ledge rock.
Chipping Loosening of shallow-depth rock by blasting or by air hammers.
Chips Small angular fragments of stone containing no dust.
Choker A short length of cable or chain with a noose, used to pull out stumps and for other similar purposes.
Chord The line joining two points on a curve.
Churn Drill A machine that drills a hole by raising and dropping a string of drilling tools suspended by a reciprocating cable. A cable drill.
Cinder Scoriaceous lava from a volcano. The residue from the burning of coal or other carbonaceous material.

Cinder Cone Small- to moderate-size cone built up from the cinders of a volcano.
Cinnabar The vermilion-colored, medium-weight common ore of mercury.
Circuit In electricity, the complete path of a current. In blasting, the path of the detonating electric circuit or detonating fuse such as Primacord.
Cirque A steep-walled amphitheatric recess in a mountainside, generally ascribed to glacial erosion.
Clamshell A bucket with two jaws, which clamp together to load by their own weight when lifted by the closing line.
Clastic Composed of broken fragments or grains cemented together. Sedimentary sandstone is an example.
Clay Fine argillaceous material which is more or less plastic when wet. It has grain size less than 0.0002 in (0.005 mm), light to medium color, and medium weight.
Claypan Stratum of stiff, compact, and relatively impervious clay. Not cemented.
Claystone Clastic rock consisting predominantly of clay-size particles.
Clearing Removing natural and artificial obstructions to excavation. Cutting down and removing trees and brush.
Cleavage Tendency of rock to split along definite, parallel, and closely spaced planes, which may be highly inclined to the bedding planes. Slate is an example.
Climate The total of all atmospheric and meteorological influences which combine to characterize a region and give it individuality by influencing the nature of its vegetation, soils, rocks, and landforms.
Clinometer An instrument for measuring the inclination of a line with respect to the horizontal. Used also for measuring the dip of a rock stratum or formation.
Clod Buster A drag that follows an excavating machine to break up lumps.
Close-Jointed Term applied to rock containing joints that are near together. Example is columnar basalt.
Coal Black sedimentary and metamorphic rock consisting chiefly of decomposed plant matter and containing less than 40 percent inorganic matter.
Coalescence The capacity for growing together, fusing, and binding.
Coarse and Fine Aggregates The term *coarse* is applied to graded mineral aggregate in which the largest particles have a diameter greater than ¼ in (6 mm). The term *fine* is applied to one in which the largest particles have a diameter less than ¼ in (6 mm).
Cobble Sediment particle having a diameter between 3 in (8 cm) and 8 in (21 cm).
Coefficient of Friction The ratio of the force required to move an object resting on a horizontal plane to the weight of the object. Expressed as a percentage as related to the natures of the object and the horizontal plane. It is also the tangent of the angle of inclination to the horizontal at which the object slides.
Cofferdam A barrier built so as to form an enclosure to keep water or soil from entering an area of excavation.
Cohesion The quality of a soil which makes it stick together.
Collar In drilling, the open end of the hole.
Colluvium Body of sediment consisting of alluvium in part and also containing angular fragments of the original rock.
Columnar Structure The contraction phenomenon in which prismatic columns commonly with six uniform faces form on the cooling of the magma. Examples are tabular bodies of igneous rocks such as andesites and basalts.
Common Excavation General term applied to soft excavation such as earth and residuals, as contrasted to hard excavation such as weathered rock and semisolid and solid rock.
Compacted Cubic Yard Measurement of material after it has been placed and compacted in the fill or embankment. Equivalent to 0.765 m^2.
Compaction Reduction in volume of fill or embankment by rolling, tamping, and wetting the material.
Compressor A machine for densifying air or gas from an initial intake pressure to a higher discharge pressure.
Conchoidal Shell-shaped. Compact rocks such as flint, argillite, and felsite break with concave or convex surfaces or conchoidal fracture.
Concretion A spheroidal aggregate formed by the segregation and precipitation of some soluble material such as quartz or calcite around a nucleus.
Concussion Shock waves or sharp air waves caused by an explosion or heavy blow.

Conglomerate Clastic sedimentary rock containing rounded pebbles or larger particles.
Contact Metamorphism Metamorphism in the vicinity of igneous rock resulting from the intense heating effect of the magma. Slate near granite is an example.
Continental Deposit Sedimentary rock deposit laid down within a general land area and in lakes and streams, as contrasted to marine deposits laid down in the sea.
Continental Shield An extensive area in which the Precambrian foundation rocks of a continental mass are exposed.
Contour Line An imaginary line on the surface of the ground, every point of which is at the same elevation.
Conveyor A machine that transports or hauls material by belts, cables, or chains.
Coquina An aggregation of shells and large shell fragments cemented by calcium carbonate. A form of limestone, with light color and light weight.
Core Cylindrical sample of rock recovered from an exploratory drill hole.
Core Drill A rotary drill, usually a diamond or shot drill, equipped with a hollow bit and core lifter.
Corrasion The mechanical detachment or wearing away of rock material by running water, glaciers, winds, waves, or mass movements.
Correlation In rocks and rock formations this term is used to mean determination of equivalence in geologic age and position in the sequence of strata in different areas.
Cost of Excavation The sum of direct job costs, indirect job expenses, and indirect home office expenses.
Cotton Rock Local name for soft, fine-grained siliceous magnesian limestone of the Silurian system.
Counterweight A nonworking load attached to one side or to the end of a machine to balance a working weight on the other side or end of the machine. An example is the counterweight on the back of a dragline in order to balance the weights of the long boom and the loaded bucket.
Country Rock A relative term used for the older rock into which the younger magma has intruded. The term is extended to mean the oldest rocks encountered.
Coyote Holes Horizontal tunnels into which explosives are placed for blasting a high rock face.
Crane A machine used for lifting and moving loads and for the handling of buckets for excavation.
Crawler One of a set of roller-chain tracks used to support and propel a machine, or a machine mounted on such tracks, known also as a *track layer.*
Creep The almost imperceptible slow downslope movement of the regolith of the Earth's surface.
Cretaceous Literally, of the nature of chalk. Also, the third and last system of Mesozoic rocks and the corresponding geologic period, 135 to 63 million years ago.
Crocus A term used in quarries to denote gneiss or other rock in contact with granite.
Crop Coal Coal of inferior quality near the surface, known also as *grass roots coal.*
Cross Bedding Laminations in sedimentary rocks confined to single beds and inclined to the general stratification, caused by actions of wind and water.
Cross Section A vertical section of the earthwork at right angles to the centerline. Used along with centerline distances to calculate earthwork quantities.
Crowd Forcing the bucket or dipper into the digging or the mechanism which does the forcing.
Crusher A machine which reduces rock to smaller and more uniform sizes.
Crust The topmost zone of the Earth, also called the *regolith.*
Crystalline Rock texture resulting from simultaneous growth of crystals.
Crystallization The process of development of crystals by condensation of materials in a gaseous state, by precipitation of materials in solution, or by solidification of materials in a melt.
Cubic Yard Unit of rock-earth excavation, equivalent to 0.765 m^3.
Cubic Yard Bank Measurement (cybm) Unit of excavation in cut or natural bed.
Cubic Yard Compacted Measurement (cycm) Unit of excavation in fill or embankment after compaction.
Cubic Yard Loose Measurement (cylm) Unit of excavation in machine, in stockpile, or in uncompacted fill or embankment.
Cubic Yard Struck Measurement (cysm) Unit of capacity of bucket, body, bowl, or

dipper of a machine. Measured by striking off the ends and sides of the container by a straight edge, excluding the teeth. It is sometimes called *water level measurement.*

Cuprite An important copper ore, the mineral containing 89 percent copper.

Cut Depth to which material is to be excavated. Also the volume of excavation for a given cut.

Cut and Cover A means of building a sanitary fill or dump. A cut is made, alternate compacted layers of trash and excavated material from a stockpile are placed in the area of the cut, and the entire area of the resulting fill is covered.

Cut and Fill A process of building earthworks by excavating the cuts and using the excavated material for the adjacent fills. In a balanced cut and fill, the volume of the excavated material equals the volume of the fill, with allowance for swell or shrink from cut to fill.

Cutterhead On a hydraulic or suction dredge, the set of revolving blades at the beginning of the suction line for fragmenting the hard materials.

Cutting An English term for excavating.

Cuttings The rock fragments from a drilling operation.

Cycle A complete set of individual operations which a machine performs before repetition, as the cycle for a shovel, which consists of load, swing, dump, and return. Frequent moving along the face of the cut is not part of the cycle.

Cycle of Erosion The sequence of landforms, essentially highlands and lowlands, through which a land mass evolves from the time it begins to be eroded until it is reduced to base level.

Dacite Igneous extrusive rock of the andesite family, containing quartz. Occurs in dikes as well as in lava flows. Medium to dark color. Heavy.

Data Facts, particularly those that can be expressed numerically.

Datum Any level surface taken as a plane of reference from which to measure elevations.

Deadheading Traveling without load, except from the dumping area to the loading point.

Dead Load The tare weight or unladen weight of a hauling machine.

Deadman The anchorage for a guy line or cable, usually embedded in the ground.

Decking Separating charges of explosives in the blasthole by inert material which prevents passing of concussion, each charge having a primer.

Decomposition The chemical alteration of rock minerals.

Deflagration Burning with sudden and startling combustion, as the explosion of black powder, contrasted to the more rapid detonation of dynamite.

Deflation The picking up and removal of loose rock particles by the wind.

Deformation Change of volume or shape or both in a rock body or a change in the original position of the rock body within the Earth's crust.

Degradation The lowering of any portion of the Earth's surface by erosion.

Delta A body of sediment deposited by a stream flowing into the standing or slow-current water of a lake, bay, or ocean.

Density *Weight per unit volume.* Generally used for rock-earth in terms of pounds per cubic yard or kilograms per cubic meter.

Denudation The wearing down and disintegration of rock masses by rain, frost, wind, running water, and other surficial agencies.

Deposit Anything laid down, such as rocks, minerals, and ores. Also the process of laying down.

Depreciation Loss of useful life for any cause. Specifically, in the case of machinery, loss resulting from wear, obsolescence, inadequacy, and other causes deemed to lessen usefulness. **Rate of depreciation** is an appraisal term defined as the percentage rate at which the value is being exhausted.

Desert Arid land where rainfall is less than 10 in (25 cm) annually. Deserts may have all kinds of topography and landforms at all elevations.

Detonation Practically instantaneous decomposition or combustion of an unstable compound with tremendous increase in volume.

Detonator Device to explode a charge of dynamite or other high explosive.

Detritus Collective term referring to broken pieces of older rocks, of minerals, or of skeletal remains of organisms.

Devonian The fourth system of rocks in the Paleozoic group. Also, the corresponding geologic period from 400 to 340 yr ago.
Dewatering Removing water by pumping, drainage, or evaporation.
Diameter Breast High Measurement of trees by diameter at the chest height of a man, about 5 ft (1.5 m).
Diamond Drill A rotary drill using a coring bit studded with black diamonds that is used chiefly in exploratory drilling.
Diastrophism The processes by which the crust of the Earth is deformed, producing continents and oceans, mountains and valleys, and flexures, folds, and faults in the rocks.
Differential Weathering The results of variations in the rate of weathering on different parts of a rock body.
Dike A mass of igneous rock which has intruded through a narrow fissure of existing country rock.
Diorite A coarse-grained igneous intrusive rock lacking quartz. Generally of medium color and medium weight.
Dip The angle of inclination of the plane of stratification of rock with a horizontal plane.
Dipper A digging bucket rigidly attached to the dipper stick, arm, or handle.
Disconformity See *Unconformity.*
Disintegration The mechanical breakup of rock.
Ditch Generally a long narrow trench or excavation.
Ditcher A machine used to excavate a ditch.
Dolerite A medium-grained, heavyweight, dark igneous rock of the basalt family.
Dolomite A heavyweight magnesian limestone, hard and generally of light bluish color.
Dolomitic Limestone A limestone containing dolomite, in which calcium carbonate predominates over magnesium carbonate.
Dome An uplift of rocks in which the beds dip outwards in all directions from the center.
Donkey A winch the drums of which are controlled separately by clutches and brakes.
Dozer Abbreviation for bulldozer or angledozer.
Dragscraper A digging and hauling device consisting of a bottomless bucket working on the cable of a mast and anchor, corresponding to headworks and tailworks. Also, a towed bottomless scraper used for leveling land and maintaining haul roads.
Drainage Basin The total area that contributes water to a stream.
Draw A topographical term referring to a natural depression or swale. Also a small watercourse.
Drawbar In a tractor the fixed or hinged bar extending to the rear and used as a fastening for lines and towed loads or machines.
Drawbar Horsepower The horsepower of the tractor engine, minus friction and slippage losses in the drive mechanisms and in the tracks or tires.
Drawbar Pull The force a tractor can exert on a load attached to the drawbar. The force depends on drawbar horsepower and speed if there is no slippage between tracks and ground. If there is a slippage the drawbar force is limited by the weight of the tractor and the coefficient of traction.
Dredge A machine for excavating material at the bottom or at the banks of a body of water. The material may be discharged on the bank or it may be emptied into a scow for transport to a distant point. Dredges are classified as: mechanical, of the grapple, dipper, and bucket types; and hydraulic of the self-propelled hopper, plain suction, and suction cutterhead types.
Drift In mining, a horizontal or nearly horizontal tunnel, generally following the vein of material.
Drifter An air percussion drill mounted on a column or crossbar and used for drilling underground.
Drillings The cuttings arising from the process of drilling.
Drilling Log In exploratory holes and blastholes, the detailed record of the rocks passed through in the drilling operation.
Drumlin Rounded oval hill of boulder clay formed by glacier deposition.
Dune Mound or ridge of sand deposited by wind action.
Dunite A hard and heavy intrusive igneous rock of medium color with granitic texture.
Dynamite A high explosive, made in three basic types. *Straight dynamite* contains

nitroglycerin as the principal or only explosive. *Extra dynamite* is, grade for grade, less sensitive to shock and friction than straight dynamite. *Gelatin dynamite* has an explosive base of nitrocotton-nitroglycerin gel, which is insoluble in water, for blasting under wet conditions.

Earth In this text the term is used both for the planet Earth, or the complete globe, and for the softer material making up part of the Earth's regolith, as contrasted to the harder material, rock.

Earth Drill An auger-type drill with bucket used for exploratory drilling in connection with rock-earth excavation.

Earthquake Powder An explosive mixture containing 79 percent nitrate and 21 percent charcoal.

Efficiency, Machine The ratio of time rate of energy output, horsepower, to time rate of energy input, horsepower.

Efficiency, Machine Operating The ratio of actual work time to available work time, expressed as minutes worked per hour, for example, the 50-min hour, or as a percentage, 83 percent.

Elevating Grader A mobile belt loader.

Elevation Vertical distance above or below a plane of reference.

Embankment A fill the top of which is higher than the surrounding surface. An embankment is usually compacted but it may be an uncompacted waste area such as overburden removed from a quarry.

Embayment A deep depression in a shoreline forming a large bay.

Emergence Rise of the land relative to the level of the sea. Generally a land of emergence is gentle and characterized by sandy soils, barrier beaches, and swamps. An example is southern New Jersey.

Energy The capacity to do work. Kinetic energy is that due to motion. Potential energy is that due to position or condition. Total energy is the sum of kinetic and potential energies.

Eocene The second series of rocks in the Tertiary system and the corresponding geologic epoch from 58 to 36 million yr ago.

Epoch That part of geologic time, a subdivision of a period, in which a series of rocks was deposited.

Era The largest subdivision of geologic time corresponding to the group of rocks deposited during the era. Six eras are recognized.

Erosion A general term describing mechanical disintegration, chemical decomposition, and movement of rock-earth materials from place to place on the Earth's surface.

Erratic A transported rock fragment, different from the bedrock beneath it. Agents of transport are water, ice, and wind.

Escarpment A steep slope or cliff. Sometimes called a *scarp*, especially in relation to a fault.

Esker A narrow long ridge, commonly sinuous, of sands and gravels formed by glacial meltwater.

Essential Minerals Those minerals which are necessary to the formation and identification of a rock.

Estimate A statement showing probable quantities and probable costs for proposed work.

Evaporate Nonclastic sedimentary rock precipitated from water solution as a result of evaporation. Rock salt is an example.

Excavation The cutting down of the natural ground surface, the material taken from the excavation, or the space formed by the removal of the material. Also known as *cutting*.

Excavation, Classified Excavation paid for at a unit price for common excavation and a unit price for rock excavation.

Excavation, Common Excavation in earthy materials not requiring blasting.

Excavation, Rock Excavation in rocky materials requiring blasting.

Excavation, Unclassified Excavation paid for at one unit price, whether common or rock excavation.

Exfoliation The peeling off of a rock formation in sheets, generally concentric, due to changes in temperature or to other causes.

Explosives Solid, liquid, or gaseous mixtures or chemical compounds which by chemical action suddenly generate large volumes of heated gas. The energetic action of the explosives used in blasting rock depends largely on their chemical reactions.

Exposure A body of bedrock not covered by the regolith and forming part of the Earth's surface.

Extrusive Igneous Rock Rock that originated from solidification of molten lava. Also called *volcanic rock.*

Face The more or less vertical surface of rock exposed by excavation or blasting.

Facies Distinctive group of characteristics within a rock unit that differs as a group from those found elsewhere in the same unit.

Fan A fan-shaped body of alluvia built at the base of a steep slope by the action of water. Also known as an *alluvial fan.*

Fanglomerate The materials of a fan which are only slightly worn by the action of water. They are usually made up of coarse particles and characterized by a steep gradient down the slope.

Fault An abrupt break in the continuity of strata or formation of rock by the elevation or depression of the strata or formation on one side of the plane of the fault.

Fault Breccia A breccia consisting of irregular pieces of rock broken as the result of faulting.

Feeder A pushing device or short belt that supplies material to a crusher or belt conveyor.

Feldspar One of the most important groups of rock-forming minerals. Crystalline and light to medium color. Ultimately weathers to clay.

Felsite Extremely fine-grained medium- to dark-colored igneous intrusive rock with granitic texture. Heavy and hard.

Fill The height to which material is to be placed. Also the resulting earthwork. Also called *embankment,* especially if compacted.

Fill, Net In sidehill excavation, the yardage of fill required at any station, less the yardage of cut obtained at the same station.

Fill, Net Corrected Net fill after making allowance for swell or shrink from cut to fill.

Fine Grading The finish grading after rough grading in order to bring the grade to the necessary close elevation.

Fine-Grained Refers to those rocks in which the minerals are less than 0.040 in (1 mm) in diameter.

Fines The finer-grained particles of the soil, clays, silts, and sands.

Finish Grade The final grade called for by the specifications.

Fishing The operation of recovering an object left, lost, or dropped in a drill hole.

Fissile Capable of being split, as schist, slate, and shale.

Fissure A fracture of rocks along which the opposite walls have been pulled apart.

Flint A popular name for chert.

Float The loose fragments or particles of rock, as distinguished from the outcrop or bedrock.

Float Time In the time schedule for a job, the available time in addition to the necessary time for an operation due to possible interference with a simultaneous critical operation.

Floodplain That part of any stream valley which is inundated during floods.

Flow Rate of movement of water in a conduit, in terms of gallons or cubic feet per second or minute, or liters or cubic meters per second or minute.

Flow Breccia An extrusive or volcanic breccia caused by the breaking up of a hardened crust of extrusive rock as a result of a further flow of liquid lava.

Fluvial Of or pertaining to rivers, or produced by river action, as a *fluvial plain.*

Fold A pronounced bend in layers of rock.

Foliation Parallel or nearly parallel structure in metamorphic rocks along which the rock tends to split into flakes or thin slabs.

Fool's Gold Pyrites, a sulfide of iron. Often mistaken for gold because of its resemblance in color.

Foot In compactors, one of a number of projections from the drum which resemble the feet of sheep.

Footcandle (fc) The unit of illumination defined as the illumination on a surface which receives from all light sources, directly or by reflection, 1 lumen per square foot.

Foot Pound The unit of work done by a force of 1 pound when its point of application moves 1 foot in the direction of the force.

Force That which changes the state of rest or motion in matter, measured by the rate of change of momentum.

Formation The ordinary unit of geologic mapping consisting of a large and persistent stratum or strata of some kind of rock. Generally named after the locality where it was first identified.

Formula Any general equation. A rule or principle expressed in algebraic symbols. A method of reasoning stated in the form of an equation. An empirical formula is one derived from experience or experiments and often used in rock-earth excavation.

Fossil The naturally preserved remains or traces of an animal or a plant, generally associated with sedimentary rocks because igneous and metamorphic rocks destroy living things and their remains.

Fossiliferous Containing organic remains or fossils, said of rock-earth.

Foundation The material which supports a structure, cut or fill, whether strengthened or not by piles, mats, or other means to secure adequate bearing.

Fractures Cracks in rocks large enough to be distinctly visible to the naked eye, without regard to definite direction.

Fracture Zone A great linear system of breaks in the Earth's crust, sometimes called a *rift zone,* formed in connection with earthquake activity.

Fragment A rock or mineral particle larger than a grain (1/8 in, or 3 mm, in diameter).

Free Air Air at normal atmospheric conditions; the pressure exerted by a column of 76 cm (29.9 in) of mercury at sea level and 32.0°F (0°C), at standard acceleration of gravity, 32.2 ft/s^2 (9.80 m/s^2).

Freehaul The distance within which material is moved without extra compensation, usually 1000 ft or less.

Freeway A limited-access highway devoted exclusively to the use of motor vehicles, in which all intersections are separated so that no traffic crosses at grade and opposing streams of traffic are separated. Distinguished by no stopping and free-flowing traffic.

Friction Resistance to motion when one body is sliding or tending to slide over another.

Fringe Benefits Those amounts which are added to a basic wage for work done, to take care of such costs as health and welfare, pensions, vacations, and education. As of 1978 they amount to an average 40 percent of basic wage.

Front The working attachment of an excavator, such as the boom of a dragline or the boom, handle, and dipper of a shovel.

Front-End Loader A bucket-type loader, in which the bucket operates at the front of the machine.

Frontal Apron The deposits of alluvia spread out in front of a glacier.

Frost Heaving The lifting of rock waste or human-built structures by the expansion of material during the freezing of the contained water.

Frost Wedging The pushing up or apart of rock particles by the action of ice.

Fumarole A volcano discharging gas nonexplosively. Indicative of latent geothermal energy.

Fusion Drill A drill which burns out a blasthole by means of fuel oil, oxygen, and cooling water delivered to a blowpipe within the hole. Also called a jet piercing drill.

Function A mathematical quantity which has value depending on the values of other quantities, called *independent variables* of the function.

Fuse A thin core of black powder surrounded by wrappings which, when ignited at one end, burns to the other firing end at a fixed speed.

Fuse, Detonating A stringlike core of PETN, a high explosive, contained within a waterproof reinforcing sheath. Primacord is an example.

Gabbro A medium- to dark-colored, generally coarse-grained intrusive igneous rock. Medium to hard and heavy.

Galena An important lead ore. The mineral contains 87 percent lead.

Gallon Unit of liquid volume. The U.S. standard gallon contains 231 in^3 and holds 8.34 lb of water. The Imperial standard gallon contains 277 in^3 (4.55 L) and 10.02 lb (4.55 kg) of water.

Gangue The nonvaluable minerals of an ore, or the waste materials.

Gantry An overhead structure of an excavator, which supports operating parts.

Gauge Thickness of wire or sheet metal; spacing of tracks, wheels, or tires; or spacing of rails.
Gauge Size The width of a drill bit along the cutting edge.
Geode A hollow nodule or concretion, the cavity of which is lined with crystals.
Geologic Column A composite diagram combining in a single column the succession of all known strata, fitted together on the basis of their fossils or other evidences or determinations of relative ages.
Geologic Cross Section A diagram showing the arrangement of rocks in a vertical plane.
Geologic Cycle The total of all internal and external processes working on the materials of the Earth's crust.
Geologic Map A map showing the distribution at the Earth's surface of rocks and rock formations of various kinds and various ages.
Geologic Record The archive of the Earth's history represented by regolith, bedrock, and the Earth's morphology.
Geologic Time Scale The time relationships established for the geologic column.
Geology The science of the Earth. *Historical geology:* The chronological order and events of the Earth's history, with special attention to organisms and fossils. *Physical geology:* The study of the composition and configuration of the Earth's physical features and rock masses, their relationships to each other, and the surficial and subsurface processes that operate on the Earth. Physical geology is sometimes divided into *structural* and *dynamic geology.*
Geomorphology Physical geology which deals with the form of the Earth, the general configuration of its surface, the distribution of land and water, and the changes that take place in the evolution of the landforms.
Geophysical Exploration Exploration to determine subsurface conditions based on the distribution within rocks of some physical property such as specific gravity, magnetic susceptibility, electrical conductivity, elasticity, or seismic characteristics.
Geyser An orifice in the ground that erupts steam, boiling water, and mud intermittently.
Giant A nozzle, manually or mechanically controlled, for directing a jet of water for hydraulicking. Also known as a *monitor.*
Gilsonite A hard, brittle native asphalt which is mined like coal.
Glacial Drift Any rock material, such as alluvia, transported by a glacier and deposited by the ice or by the meltwater from the glacier.
Glaciation The alteration of the land surface by the massive movement of glacial ice. The plucking, polishing, and striation of rocks by glacial action.
Glacier A body of ice, consisting mainly of crystallized snow, flowing on a land surface.
Glance A term used to describe various minerals with a splendent luster, such as silver glance and lead glance.
Glassy Texture The texture of some extrusive igneous rocks such as lavas which have cooled so quickly that they have not undergone crystallization. An example is obsidian.
Glory Hole A kind of open-pit mine in a large, vertical-sided pit. The material is fed by gravity through a funnel-shaped hole in the floor of the pit to hauling units located in an adit beneath the pit.
Gneiss A metamorphic coarse-grained rock breaking along irregular surfaces, generally with alternating layers of light- and dark-colored minerals. Hard and heavy. An example is granite gneiss, an altered granite.
Gorge A small canyon or narrow passage between hills.
Gossan A term referring to the weathered, iron-stained outcrop of a mineral deposit. An important indication of the presence of the ore.
Gouge Finely pulverized rock flour, sometimes claylike, caused by the grinding action between the walls of a fault.
Graben The depressed land between two faults, generally a long narrow debasement and typical of a basin range province.
Grade (1) The profile of the center of a roadway, canal, or like structure. (2) To prepare the ground by cutting and filling according to a definite plan. (3) The gradient, that is, the percentage of rise or fall to the horizontal distance.

Grade Assistance The assistance to a machine or body when descending a grade caused by the force of the inclined component of the weight of the machine or body.
Grade Resistance The resistance to a machine or body when ascending a grade caused by the force of the inclined component of the weight of the machine or body.
Grade Stake A stake indicating the amount of cut or fill in feet at a stated point.
Grain A mineral or rock particle having a diameter less than ⅛ in, or 3 mm.
Granite A coarse-grained light to dark intrusive igneous rock. Soft to hard and medium- to heavyweight.
Granite, Decomposed (Sometimes called DG.) Well-weathered granite in which some of the minerals have altered and softened, weakening the rock.
Granodiorite Coarse-grained light- to medium-colored rock intermediate between intrusive igneous granite and diorite. Soft to hard and medium- to heavyweight.
Granular Textural term referring to uniform size of grains or crystals of a rock, particles ranging in size from 0.002 in (0.05 mm) to 0.25 in (0.6 cm).
Grapple A clamshell-type bucket, sometimes with heavy tongs, having three or more jaws and generally used to handle rock such as riprap.
Gravel Inclusive term for materials resulting from the natural erosion of rock and ranging in size from coarse sand to cobbles, that is, from ⅛ in (3 mm) to 8 in (20 cm) in diameter.
Gravel, Pit Run The mixture of alluvia and foreign materials as they occur in any natural deposit. The source of prepared gravels.
Graywacke An old term applied to metamorphosed shaly sandstones that yield a tough, irregularly breaking rock different from both slate and quartzite.
Grid A set of surveyor's closely spaced reference lines laid out at right angles with elevations taken at intersections of the lines. The framework is used to calculate quantities of excavation.
Grit A coarse sand or sandstone formed mostly of hard angular quartz grains.
Grizzly A coarse screen or set of rails used to remove oversize pieces from excavation.
Ground Moraine The irregular sheet of till deposited partly beneath an advancing glacier and partly from the ice when it melts away.
Ground Pressure The weight of a machine divided by the area of the ground directly supporting the machine.
Group One of the six major divisions of rocks, corresponding to a geologic era.
Grouser A heavy lug attached to crawler tread plates and tractor wheels to obtain better traction in soft ground.
Grubbing Removing the stumps and roots from an area to be excavated.
Gypsite A dirty variety of gypsum with some impurities. A sedimentary rock, grayish in color and of medium hardness and medium weight.
Gypsum Hydrous calcium sulfate evaporate. Massive light-colored sedimentary rock. Medium to hard and heavy.
Gyratory Crusher A primary crusher having a central conical breaker with an eccentric motion contained in a chamber of circular cross section tapering from a wide top opening.
Half Track A heavy-duty truck with a high-speed crawler-track drive in the rear and steering wheels in the front.
Hammer, Air A machine hammer driven by compressed air, as a jackhammer.
Hand Level A small instrument consisting of a telescope and a leveling bubble tube for sighting a horizontal line.
Handle In a shovel or backhoe, the arm connecting the boom to the dipper. Same as *stick*.
Hanging Valley A valley the floor of which is notably higher than that of the valley or shore to which it leads. Hanging valleys are usually tributary to a glaciated main valley.
Hardness, Mineral A measure of the cohesion of the surface particles as determined by the capacity of the mineral to scratch another mineral. Minerals are referred to a scale of hardness of 10 minerals in ascending order of hardness. Each mineral will scratch the one below it on the scale, diamond being the hardest. The hardness ratings of this Mohs scale are: talc 1, gypsum 2, calcite 3, fluorite 4, apatite 5, orthoclase 6, quartz 7, topaz 8, corundum 9, diamond 10.
Hardness, Rock An arbitrary, relative term to define the resistance to fragmentation, either by ripping by heavyweight tractor-ripper or by blasting. The 1978 standards in

terms of ripping by tractor-ripper in the 100,000-lb (45,000-kg) to the 160,000-lb (73,000-kg) weight class and in the 380- to 530-flywheel hp class are based on the range of seismic shock-wave velocities in the rock.

	Range of velocities					
	feet/second			meters/second		
No ripping	To		1500	To		460
Soft ripping	1500	to	4000	460	to	1260
Medium ripping	4000	to	5000	1220	to	1520
Hard ripping	5000	to	6000	1520	to	1830
Extremely hard ripping or blasting	6000	to	7000	1830	to	2130
Blasting	7000		Up	2130		Up

Hardpan A stratum of material accumulation that has been thoroughly cemented into a hard rocklike layer which will not soften when wet.

Hardwood Wood from a tree which is sturdy, with a texture of growth not given to breakage. Typical are oak, hard maple, and hickory.

Haul The distance material is moved from the loading point to the point of disposition.

Haul, Average The average distance material is hauled from the cut to the fill. This distance is sometimes calculated by means of a mass diagram.

Haul, Free See *Freehaul.*

Haul, Over See *Overhaul.*

Head The height of water above a specified level, consisting of the sum of static head, velocity head, and friction head.

Heading A collection of joints in a rock formation.

Heap The load of rock-earth carried above the sides of the hauler or the bucket of an excavator.

Hectare A metric measure of area, equivalent to 10,000 m², or 2.47 acres.

Hematite A common iron ore. The mineral contains 70 percent iron.

High Explosive An explosive which decomposes with extreme rapidity.

Highway A main road or thoroughfare with a mixture of traffic and without grade separations and free-flowing traffic as in the case of a freeway.

Hogback Ridge formed by the outcropping edge of tilted strata of rock.

Hoist A machine for lifting weights and loads, or the part of a machine which performs the hoisting operation, such as the hoist for the bucket of a dragline.

Holocene The second and latest series of rocks of the Quaternary system and also the corresponding geologic period, which commenced with the Neolithic age about 12,000 yr ago and extends to the present.

Hopper Body The body of a hauler which permits dumping the load through the bottom hopper doors or gates. Sometimes known as a bottom dump.

Hornfels A fine-grained, compact, highly metamorphosed rock, the original texture of which has been completely obliterated by contact metamorphism with nearby intruded magma. Derived from shale or other sedimentary rock; light to dark in color, heavy and hard.

Horse A large fragment of rock broken from one block and caught between the walls of the fault.

Horsepower A unit of power equal to a rate of 33,000 ft·lb/min (4560 m·kg/min) or 550 ft·lb/s (76 m·kg/s). The unit is actually about 1.5 times the output of a continuously working 1000-lb horse.

Horsepower, Engine Rated The average horsepower developed by an engine when operating at full throttle at rated speed and at specified altitude and temperature. Working horsepower is that delivered at the flywheel with full engine accessories for the working conditions.

Horsepower, Wheel The average horsepower delivered at the points of contact between the driving wheels and the ground surface. This horsepower equals the flywheel horsepower less the power lost through the power transmission assemblies. In a wheels-tires machine, such as a rear-dump truck, wheel horsepower equals approximately 0.76 times flywheel horsepower.

Horsepower, Drawbar The average horsepower delivered to the drawbar of a crawler-tractor. This horsepower equals the flywheel horsepower less power lost through the power transmission assemblies and to the rolling resistance of the tractor when tested under standard conditions of rolling resistance of 110 lb per ton of tractor weight. Drawbar horsepower equals approximately 0.69 times flywheel horsepower.

Horst The elevated land between two faults, generally a long narrow block and typical of a basin range province.

Humus The dark decomposed residue of plant and animal tissues.

Hydration The reaction of a compound with water to form a hydrate.

Hydraulic Fill A fill built by transporting the material by water. The material may be moved in an open natural channel or it may be pumped by a dredge.

Hydraulicking Use of water under high pressure directed against the face of the material to be moved. The dislodged material is carried away by the water, as in the case of overburden from deposits of rock and ore.

Ice Age The Pleistocene epoch, during which the four glacial periods occurred.

Igneous Rock Rock formed by the solidification of molten silicate materials.

Ilmenite An important iron and titanium ore. The mineral contains 36 percent iron and 32 percent titanium.

Impeller The member of a centrifugal pump using centrifugal force to discharge a fluid into the outlet passages.

Impervious Resistant to the percolation or absorption of water.

Incipient Erosion The early stages of erosion, especially with reference to gullying.

Inclined Plane A slope used to change the direction and the speed/power ratio of a force.

Indicated Horsepower The mechanical power developed in a steam cylinder by the steam working against the piston.

Indurated Hardened, as a rock hardened by heat, pressure, or the addition of some ingredient not commonly contained in the rock.

Inertia The property of matter by which it will remain at rest or in uniform motion in a straight line unless acted on by an external force.

Indigenous Rocks Rocks formed by magmas, the source of which has been the Earth's interior. Granite is an example. Same as intrusive igneous rocks.

Inhaul The line or mechanism by which a cable excavator bucket is pulled toward the dumping point.

Inlier An outcrop of rock surrounded on all sides by geologically younger rocks.

In situ As applied to rock or earth, in the natural position in which it was originally formed or deposited.

Intake That portion of a pipe, pump, or structure through which water or air enters from the source of supply.

Interface The boundary between two rocks or formations with different physical characteristics.

Intermontane Lying between two mountains.

Internal Friction The resistance of rock or earth particles to sliding over each other.

Intrusion A mass of igneous rock which, while molten, was forced between other rocks.

Intrusive Rocks Term applied to a rock which has been forced into other rocks. Sometimes called *plutonic*. Granodiorite is an example.

Invert The inside bottom of a pipe, trench, or tunnel.

Itacolumite A flexible sandstone which is composed of quartz grains, muscovite, talc, and a few other minerals. Peculiar to North Carolina and a few other places.

Jackhammer An air drill which can be operated by one person.

Jars A tool in the string of tools of a cable drill which contains slack to allow upward hammering to free a stuck drill bit.

Jaw Crusher A primary crusher with a fixed and an oscillating jaw widely spaced at the top and closely spaced at the bottom and a mechanism to move the oscillating jaw to and from the fixed jaw.

Jetty A long fill or structure extending into the water from the shore which serves to change the direction or velocity of flow of water.

Joint A fracture on which no appreciable movement parallel with the fracture has occurred. Common rock joints are caused by flexure in sedimentary rocks and by cooling in igneous rocks.

Joint System A combination of two or more intersecting sets of parallel joints.

Jurassic The middle of the three rock systems of the Mesozoic group and the corresponding geologic period, from 180 to 135 million years ago.

Kame A body of stratified drift material due to ice contact and shaped as a short, steep-sided knoll. Glacier-formed.

Kaolin A white nonplastic sedimentary rock formed from the residual minerals of granite. Soft and medium-weight.

Karst Topography Topography developed in a region of easily soluble limestone bedrock, characterized by sinkholes, caverns, and underground streams.

Kelly A square or fluted pipe which is turned by a rotary drill table while it is free to move up and down through the table. It carries the rotary bit.

Kettle A closed depression in drift material created by the melting of a mass of underlying ice.

Kieselguhr German name for *diatomaceous earth,* a filler or carrying agent in high explosives.

Kilogram A metric unit of weight, 1000 g, equivalent to 2.205 lb.

Kilometer A metric unit of length, 1000 m, equivalent to 0.621 mi.

Kilovolt-Ampere An electrical unit of power, equivalent to about 0.89 kW.

Kilowatt Metric and English unit of power, 1000 W, equivalent to 1.34 hp.

Kinetic Energy Energy due to motion.

Knife The cutting edge of some excavating machines.

Kyrock Bituminous silica sandstone from Kentucky, containing up to 12 percent natural bitumen. Dark, medium-hard, and heavy.

Laccolith Lenticular intrusion of igneous rock into a sedimentary rock formation, generally with a plane bottom and a domed top.

Ladder The digging-boom assembly of a hydraulic dredge or a chain-bucket-type excavator.

Lamina A stratum of less than ⅜ in (1 cm) thickness.

Laminations The banding of rocks caused by variations in different minerals, usually distinguishable by different colors. Sometimes laminations result in weakened planes between them.

Landforms Physical features of the surface of the land, such as hills and valleys, developed by the processes of aggradation and degradation.

Lapilli Particles of dust and small rock fragments ejected from volcanoes, with diameters ranging from ⅛ in (3 mm) to 1¼ in (32 mm).

Lapilli Tuff Medium-grained pyroclastic rock consisting of fused lapilli. Light to medium color, soft to hard, and medium-weight.

Laterite Reddish residual, product of weathering of soft rocks rich in iron and aluminum.

Latite A rock that is intermediate in properties among the igneous extrusive rocks such as rhyolite, andesite, and basalt. Medium to dark color, medium to hard, and medium-weight.

Lava General name for the magma outpourings of volcanoes; an igneous extrusive rock.

Leaching The continuous removal by water of soluble matter from regolith or bedrock.

Lead Wire In blasting circuits the heavy wire that connects the blasting machine or source of current with the connecting or cap wires.

Ledge A bedding or several beddings of rock, as in a quarry. An outcropping of horizontal or nearly horizontal rock.

Levee A rock or earth embankment to prevent inundation or erosion.

Leverman One who operates the controlling levers of a machine; an operator.

Life An appraisal term defined as the mean or expected duration of the life of a property.

Lift The depth of a cut taken out or of a fill placed during a cycle of excavation.

Lignite A brownish black soft coal, intermediate in composition between peat and bituminous coal. A sedimentary rock. Lightweight.

Lime Rock A natural sedimentary rock composed essentially of calcium carbonate with varying percentages of silica. Light to medium color, medium soft, and medium-weight.

Limestone A sedimentary rock consisting predominantly of calcium carbonate. Light to medium color, hard, and heavy.

Limonite An iron ore of hydrous oxide of iron. Brownish rust color when powdered. Soft and heavy. Mineral contains 86% iron.

Lip The cutting edge of a bucket or dipper. Applied chiefly to edges, including tooth sockets.

Liter Metric unit of capacity, equivalent to 1000 cm^3, 61.0 in^3, or 0.908 U.S. quart.

Lithification Rock formation. The conversion of sediments to sedimentary rock.

Load The placement of explosives in a blasthole.

Load Factor Average load carried by an engine, machine, or plant, expressed as a percentage of its maximum capacity.

Loam Earth having a relatively even mixture of clay, silt, and sand and generally containing a considerable portion of organic matter. Also called *topsoil.*

Local Metamorphism Contact metamorphism as distinguished from regional metamorphism.

Location The centerline and grade line of an engineering structure such as a highway, preparatory to its construction.

Loess Wind-deposited silt, usually accompanied by some clay and fine sand. Somewhat consolidated so that it maintains a steep bank of cut.

Log In geology and construction, the detailed record of the rocks passed through during a drilling operation.

Loose Yards Cubic yards loose measurement, a unit of volume of excavation.

Low Explosive An explosive which decomposes more slowly than high explosives and by deflagration rather than detonation.

Lumen (lm) The unit of light flux in terms of which the output of light sources is expressed, 1 lumen per square foot equaling 1 footcandle.

Luster The quality and the intensity of light reflected from a mineral or rock.

Machine An apparatus for applying mechanical power which has several parts, each with a different function.

Macrostructure A structural feature of rocks that can be discerned by the unassisted eye or with the help of a simple magnifier.

Mafic Rock A rock in which ferromagnesian minerals exceed 50 percent. Gabbro is an example.

Magazine A structure or container in which explosives are stored.

Magma Liquid molten rock, consisting of hot silicic solutions containing water and gases, from which igneous rocks are formed.

Magnetite An important magnetic iron ore. Black, hard, and heavy. The mineral contains 72 percent iron.

Malachite An important ore of copper. Greenish in color, medium hard, and mediumweight. The mineral contains 72 percent cupric oxide.

Mantle Rock The loose and more or less consolidated weathered rock resting on the bedrock.

Map Representation to a definite scale on a horizontal plane of the physical features of a portion of the Earth's surface.

Marble A hard, light-colored metamorphic rock derived from limestone or dolomite. Light color, hard, and heavy.

Marine Sediment Sediment deposited in the sea. Also the sedimentary rocks of the sea.

Marl An earthy mixture of calcium carbonate and clay. Light color, soft, and medium weight. A common lake deposit.

Mass Quantity of matter. Equal to weight divided by the acceleration of gravity.

Mass Diagram In earthwork calculations, a graphical representation of the algebraic cumulative quantities of cut and fill along the centerline, where cut is positive and fill is negative. Used to calculate haul in terms of station yards.

Massif A single mountainous mass of rock, which may be considered a unit.

Massive As applied to rock, homogeneous in structure, without stratification, flow-banding, foliation, schistosity, or the like.

Mass Shooting Simultaneous exploding of charges in all of a large number of blastholes, as contrasted with sequential firing with delay caps.

Mass Wasting The gravitative movement of rock debris downslope without the help of another agency such as water, ice, or wind.

Mean A quantity having an intermediate value among several others of which it expresses the average, obtained by dividing the sum of the quantities by their number.

Meander A looplike bend in the channel of a stream.

Mechanical Efficiency The ratio of the energy or work of the output of a machine to the energy of the input.

Mechanics The science of force and its effect on matter.

Medium-Grained Rock Rocks whose crystals range in diameter from 1/32 in (1 mm) to 3/16 in (5 mm).

Mesa A high, broad, flat tableland, bounded on at least one side by a steep cliff rising from the lower land.

Mesozoic The fifth of the groups of rocks preceding the present Cenozoic group. Also the corresponding geologic era, from 230 to 63 million yr ago.

Metamorphic Rock A rock which has been changed by temperature, pressure, and/or the chemical action of fluids into a new form which is more stable under the new conditions. Examples are gneiss from granite, quartzite from sandstone, and slate from shale.

Meteorite A particle of solid matter from outer space that has fallen to the ground from the atmosphere.

Meter A metric measure of length, equivalent to 3.28 ft.

Metric Ton A unit equivalent to 1000 kg, 2205 lb, or 1.102 short tons.

Mile Yard The movement of 1 yd^3 of excavation through a horizontal distance of 1 mi. A measure of payment for such excavation.

Millimeter A metric measure of length, equivalent to 0.0394 in.

Millisecond Delay A type of delay cap with a definite but extremely short interval between passage of current and detonation, usually in increments between 12 and 700 ms (0.012 and 0.700 s).

Mineral A naturally occurring inorganic substance of definite chemical composition and physical properties which is a constituent of rocks.

Mineralogy The science or study of minerals.

Miner's Inch An old measure of the flow of water as related to mining. The accepted value of one hundred years ago is the quantity of water flowing through an orifice 1 in square under a head of 6.5 in in 1 min. This flow is equivalent to 1.53 ft^3/min or 11.4 gal/min.

Mining The removal of rock or earth having value because of its chemical composition.

Miocene The fourth series of rocks in the Tertiary system. Also the corresponding geologic epoch, from 25 to 13 million yr ago.

Misfire Failure of all or part of an explosive charge to detonate.

Mississippian The first series of rocks of the Carboniferous system. Also the corresponding geologic epoch from 340 to 310 million yr ago.

Moil Point A short length of drill steel sharpened to a conical point used with a jackhammer for breaking rock or concrete or for hole-punching work.

Moisture Content The percentage weight of water in a given weight of rock, earth, or rock-earth.

Moldboard A curved surface of a plow, bulldozer blade, motorgrader blade, or other excavator which gives the material moving over it a rotary, spiral, or twisting movement.

Monadnock A conspicuous residual hill on a peneplane or leveled plain.

Monitor In hydraulicking, a high-pressure nozzle mounted in a swivel frame. Also called *giant.*

Monzonite A light- to medium-colored extrusive igneous rock intermediate between syenite and diorite. Medium-hard and medium-weight.

Moraine An accumulation of drift deposited by a glacier. If deposited along the side or sides of the glacier, it is a *lateral moraine.* If deposited at the end, it is a *terminal moraine.*

Motor In this text, a rotating machine which transforms electrical energy into mechanical energy. Not to be confused with an internal combustion engine, which transforms fossil fuel into mechanical energy.

Motor-Generator A generator propelled by an electric motor, known as an MG set. An example is the MG set of the Ward Leonard system for transforming ac power to dc power to mechanical power by means of a dc motor.

Mountain In a general sense, any land mass or landform which stands conspicuously above its surroundings. Geologically, it refers to parts of the Earth's crust having thick, crumpled strata, metamorphic rocks, and granitic batholiths.

Muck Soft mud containing vegetable matter. Also a general term referring to all kinds of excavation, including rock, with particular application to the excavation for a tunnel.

Mud Generally, any earthy material containing enough water to make it soft.

Mudflow Flow of a torrent so heavily charged with earth and debris that the mass is thick and viscous. Blocks of rock many feet in dimension can be transported vast distances.
Mudcapping Blasting boulders by means of explosives placed on the surface and covered with mud, the inertia of which intensifies the action of the explosives.
Net Cut In sidehill excavation, the cut required less the fill required at a particular station or part of the centerline.
Net Fill In sidehill excavation, the fill required less the cut required at a particular station or part of the centerline.
New Construction Term used in the economics of construction which applies to projects on entirely new locations or to reconstruction projects in which there is no salvage value for the existing construction.
Nitroglycerin A powerful liquid explosive, which is dangerously unstable unless combined with other materials such as kieselguhr (diatomaceous earth).
Normal Fault A generally steeply inclined fault in which the hanging wall block appears to have moved relatively downward.
Normal Haul A haul the cost of which has been included in the cost of excavation, so that no additional charge is made for it.
Obsidian An extrusive igneous rock that has cooled rapidly so as to produce a glasslike surface. Generally black. Hard and heavy.
Obsolescence The factor in depreciation of machinery resulting from changes in methods or design which makes the machinery less desirable or valuable for continuation of work.
Oil Shale A compact sedimentary rock containing organic matter, which yields oil when destructively distilled. Dark, medium-hard, and heavy.
Oligocene The third series of rocks in the Tertiary system. Also the corresponding geologic epoch, from 36 to 25 million yr ago.
Oolitic Limestone Limestone consisting largely of minute spherical grains of calcium carbonate resembling fish roe. Light in color and of medium hardness and medium weight.
Open-Pit Quarry or Mine An excavation in which the working area is kept open to the sky. Used to distinguish the method of quarrying or mining from underground work.
Operating Costs Machinery costs such as repairs and replacements, including labor, and fuel, oil, and lubricants. Sometimes operator's wages and fringe benefits are included in the cost but usually they are kept separate, as in this text.
Operator One whose work is to operate a machine.
Optimum Moisture Content The moisture content in percent by weight of dry rock-earth which results in the least voids or greatest density when the rock-earth is compacted.
Ordovician The second of the five systems of rocks in the Paleozoic group. Also the corresponding geologic period, from 500 to 430 million yr ago.
Ore A mineral or association of minerals that may under favorable circumstances be worked commercially for extraction of one or more of the minerals.
Original Horizontality The principle which states that most strata are nearly horizontal when originally deposited.
Orogeny Another term for mountain making. The deformation of the Earth's crust in the development of mountains.
Outcrop Underlying bedrock which comes to the surface of the ground and is exposed to view.
Outlet That portion of a pipe, pump, or structure through which water or air leaves.
Outlier An outcrop of rock surrounded on all sides by geologically older rocks.
Outwash Stratified drift deposited by streams of meltwater as they flow away from a glacier.
Overbreak Moving or loosening of rock beyond the intended line or plane of cut as a result of blasting.
Overburden Rock-earth mantle, waste material, or other matter found directly above the deposit of material to be excavated.
Overhang Projecting parts of a bank, face, or high wall.
Overhaul Transportation of excavation beyond certain specified limits known as *free-haul distances.*

Overhead Those indirect job and home-office expenses which cannot be charged to individual costs or bid items except by proration, as they are incurred in the maintenance of the business in general.

Overhead Shovel A bucket loader which loads at one end, swings the bucket overhead, and dumps at the other end. Used in closely confined areas, as in tunnels.

Overthrust The lateral thrusting of a mass of rock over or upon other rocks along a thrust fault.

Ownership Costs Machinery costs which are more or less fixed, such as depreciation, interest, taxes, insurance, storage, and replacement cost escalation.

Packsand A very fine-grained sandstone so loosely cemented by a slight calcareous cement as to be readily cut by a spade. A light- to medium-colored, soft, medium-weight sedimentary rock.

Pad Ground contact part of a crawler-type track. Also called a *shoe* or *plate.*

Paleocene The first of the five series of rocks of the Tertiary system. Also the corresponding geologic epoch, from 63 to 58 million yr ago.

Paleozoic The fourth of the six major groups of rocks. Also the corresponding geologic era, from 570 to 230 million yr ago.

Pan Name for a carrying scraper.

Parallel An arrangement of electric blasting caps in which the firing current passes through all of them at the same time.

Parallel Series Two or more series of electric blasting caps arranged in parallel.

Particle A general term referring to any size from microscopic mineral grains to huge rocks. Equivalent to an entity.

Parting A surface of separation within a rock body.

Pass A working trip or passage of an excavating machine.

Pass A defile or passageway through rough terrain.

Pay Formation A body of rock, earth, or ore the value of which is enough to justify excavation.

Pay Items Units in a contract which are covered by the specifications as bid items and listed as separate units for payment.

Payload The load of excavating machinery for which payment is made. In terms of cubic yards it may be bank measurement (cybm), loose measurement (cylm) or compacted measurement in the embankment (cycm). The load may be also in terms of short tons as in the case of borrow-pit excavation or in short or long tons as in the case of quarrying or mining.

Pea Gravel Clean gravel the particles of which equal peas in size, about 3/16 in (5 mm) in diameter.

Peat A brownish lightweight mixture of decomposed plant tissues in which parts of the plant are easily recognized. The beginning of formation of coal.

Pebble Sediment particles having diameters between 1/16 in (2 mm) and 2½ in (64 mm).

Pediment A sloping surface, cut across bedrock, adjacent to the base of a highland in an arid country.

Pegmatite A very coarse-grained granite occurring in irregular dikes or lenses in granites and some other rocks. Light to dark in color, hard, and heavy.

Peneplane Almost a plain. A land surface worn down to very low relief by streams and mass wasting.

Penetration The rate of drilling at 100 percent efficiency with the drill bit working continuously. Expressed in feet per minute for fast drilling and inches per minute for slow drilling.

Pennsylvanian The second of the three series of rocks in the Carboniferous system. Also the corresponding geologic epoch, from 310 to 280 million yr ago.

Perched Water Table The upper surface of a body of free groundwater in a zone of saturation separated by unsaturated material from an underlying body of groundwater in a differing zone of saturation.

Peridotite An ultramafic dark intrusive igneous rock. Hard and heavy.

Perlite Glassy extrusive or volcanic igneous rock of rhyolitic composition. Light to medium color, hard, and lightweight.

Permian The third and last series of rocks in the Carboniferous system. Also the corresponding geologic epoch, from 280 to 230 million yr ago.

Petrology The study of rocks.

Phosphate Rock A sedimentary rock composed chiefly of calcium phosphate, along with impurities such as clay and lime. Medium color, medium hardness, and medium weight.
Phyllite A group of rocks associated with slate and sometimes called *leaf stone.* Of metamorphic origin. Medium to dark color, hard, and heavy.
Pi, π A number, approximately 3.1416, used in the determination of the properties of the circle and the sphere.
Pile A steel or wooden member usually driven or jetted into the ground and deriving its support from the underlying rock-earth and by the friction of the rock-earth on its surface. In earthwork its usual function is to form a wall, as in a cofferdam, to exclude water and soft material from the area of excavation.
Pillow Lava Ellipsoidal mass of extrusive igneous rock formed by the extrusion of lava under water. Dark, hard, and heavy.
Pioneering The first working over of rough or overgrown areas prior to excavation.
Pioneer Road A semipermanent road built along the job to provide means for moving workers and machinery.
Pit An open excavation, deep with respect to area.
Placer A deposit of heavy minerals concentrated mechanically, as by stream flow, wave action, or wind.
Placer Diggings Areas where placer mining has overturned or removed the earth and left a rough, eroded, and scarred surface.
Planation The bringing of a surface of the Earth to level or plane condition by natural or artificial means, as the complex process by which a stream develops its floodplain.
Planimeter An instrument for measuring the area of a plane figure by passing a tracer around the boundary of the area.
Plastic Limit The minimum amount of water, measured in percent based on the dry weight of material, that will make the material plastic.
Plastic Material A material that can be rolled into strings ⅛ in (3 mm) in diameter without crumbling.
Plateau An extensive upland underlain by essentially horizontal strata and having large, nearly flat areas.
Playa The dry bed of a lake on the nearly level floor of an intermontane basin.
Pneumatic Powered or inflated by compressed air.
Pleistocene The first of two series of rocks in the Quaternary system. Also the corresponding geologic epoch, from 2 million to 12,000 yr ago.
Pliocene The fifth and last series of rocks in the Tertiary system. Also the corresponding geologic epoch, from 13 to 2 million yr ago.
Plucking The process by which a glacier pulls away or quarries blocks of rock of considerable size from its channel walls.
Pluton Any body of intrusive igneous rock.
Point A general term to describe the teeth and the bits of excavating machinery.
Pontoon A float supporting part of a structure, as that which supports the pipe of a dredge from dredge to shore.
Poorly Sorted Sediments A sediment consisting of particles of many sizes.
Pore Spaces The spaces or voids in a body of rock or sediment which are unoccupied by solid materials.
Porphyritic Texture A texture of igneous rock in which some particles are conspicuously larger than the rest of the particles.
Porphyry An igneous rock with phenocrysts, the largest particles, making up more than 25 percent of the volume.
Potato Dirt A general term for any rock-earth which is easy to dig.
Pound-Foot The same as *foot-pound.*
Powder A general term for any low explosive such as black powder and earthquake powder.
Power The time rate at which work is done, expressed by horsepower or kilowatts.
Power Control Unit One or more winches mounted on the tractor and used to manipulate the working parts of bulldozers, loaders, scrapers, and other machines.
Power Train All moving parts connecting a prime mover with the point or points where work is performed.
Precambrian Rocks Rocks which are older than those of the Cambrian system. These rocks include the Azoic, Archeozoic, and Proterozoic groups. The demarcation marks

the beginning of real fossil life, important in the classification and the correlation of rocks. From 4500 to 570 million yr ago.

Primacord Trademarked name for a a detonating fuse.

Prime Mover A machine to pull other machines. An engine or motor to drive a machine or machines.

Primer A high explosive and blasting cap used to initiate the explosion of a blasting agent. A primer is used in conjunction with ammonium nitrate explosives, which are in themselves inert.

Profile The intersection of a vertical plane through the centerline with surface of the ground and the plane or planes of the finished earthwork, or a drawing indicating the same, so as to indicate grades and distances, depth of cut, and height of fill for the earthwork.

Proterozoic The third of the six major groups of rocks. Also the corresponding geologic era, from 1200 to 570 million yr ago.

Province An area throughout which geologic history has been essentially the same over a long period of time or which is characterized by particular structural or physiographical features.

Puddingstone A sedimentary conglomerate rock in which the pebbles are rounded. Of light to medium color, hard, and medium-weight.

Puff Blowing Blowing chips out of a drill hole by means of exhaust air from the drill.

Pumice Extremely vesicular frothy natural glass. An extrusive igneous rock. Of light to medium color, sharp, and lightweight.

Push-Tractor A crawler-tractor equipped with a push block which loads push-loaded scrapers.

Pyroclastic Rock Fragmental extrusive igneous rocks, such as tuffs and breccias, produced by volcanic explosions. Sometimes classified as sedimentary rock. Light to dark in color, hard, and medium- to heavyweight.

Quantity The amount of material to be excavated and handled, expressed in the prescribed unit, usually cubic yard or ton.

Quarry A deposit of rock from which material is excavated for broken stone, crushed stone, or dimension stone.

Quartzite A metamorphic rock, derived from quartz sandstone. Light in color, hard, and heavyweight.

Quartz Monzonite A kind of granite in which the light-colored minerals such as the feldspars predominate. Light to medium in color, hard, and heavy.

Quaternary The last of the two systems of rocks in the Cenozoic group. Also the corresponding geologic period, from 2 million years ago to the present.

Quicksand A mass of silt and fine sand, thoroughly saturated with water, forming a semifluid.

Rake, Brush A rake blade for a crawler-tractor having a high top and light construction.

Rake, Rock A heavy-duty rock blade for a crawler-tractor equipped with teeth along the cutting edge.

Ramp An incline connecting two levels, such as a ramp connecting two working benches in a quarry.

Raveling The loosening and falling of materials from a bank or face.

Ravine A deep, more or less linear, depression or hollow worn by running water.

Receiver The air tank or reservoir on an air compressor.

Reclaiming Removing material from a stockpile.

Reef A rocky ledge or bar of sand under water in a stream channel.

Regional Metamorphism Extended metamorphism of rocks that, as contrasted to contact metamorphism, extends over a large area.

Regolith The noncemented rock fragments and materials which overlie the bedrock in most places. Regolith is of two kinds, residual and deposited.

Relief The difference in elevation between the high and the low parts of a land surface.

Relief Holes Holes drilled closely along a line which are not loaded but which serve to weaken the rock so that it breaks along the line when blasted.

Relocation A new alignment varying from the original location, generally of highways and railroads.

Residuals Earth and rock fragments formed in situ by the weathering of rocks and left as an overburden upon the country rock.

Reverse Fault A generally steeply inclined fault along which the hanging wall has moved relatively upward.

Revolving Shovel A shovel in which the upper works or deck revolves independently of the lower works about a centerpin.

Rhyolite A fine-grained extrusive igneous rock of light to medium color. Medium to hard and medium-weight.

Rift Zone The fractured and brecciated zone between the walls of a fault.

Right Bank That bank of a stream, reservoir, or dam which is on the right when one looks in the direction in which the current flows.

Right-of-Way The land or water rights necessary for a construction project.

Ripping The fragmentation of rock by a crawler-tractor equipped with ripper shanks and points.

Riprap Heavy rocks placed to form a revetment to prevent stream erosion, a seawall to prevent wave erosion, or a jetty to protect the entrance to a harbor or for some similar purpose. The rocks vary generally in size from 2 ft^3 to several cubic yards and in weight from about 300 lb to 10 tons.

Riverwash Alluvial deposits in streambeds and flood channels, subject to erosion and deposition during recurring flood periods.

Roadbed For a railroad, the finished surface of the roadway upon which the ballast rests. For a highway, the finished surface of the roadway between shoulder lines.

Roadway For a railroad, that part of the right-of-way prepared to receive construction of ditches, shoulders, and roadbed. For a highway, the entire construction area for the highway and its appurtenances.

Rocks A mass of material, loose or solid, which makes up an integral part of the Earth.

Rock Avalanche The extremely rapid downslope flow of a mass of dry rock particles.

Rock Cleavage Closely spaced partings in rocks controlled by platy particles that have been aligned in response to pressures within the Earth's crust.

Rock Cycle That part of the geologic cycle concerned with the creation, destruction, and alteration of rocks during erosion, transport, deposition, metamorphism, plutonism, and volcanism.

Rock Excavation The fragmentation, loading or casting, hauling, and depositing of rock or rock-earth.

Rock Fall The rapid descent of a rock mass, vertically from a cliff or by leaps down a slope.

Rock Slide The rapid descent of a rock mass down a slope.

Rolling Resistance A measure of the force which must be overcome to pull or roll tracks or wheels-tires over the ground. It is affected by ground conditions and the total weight of the machine, and is expressed for different ground conditions in terms of pounds per ton of weight of machine and in percentage of weight of machine.

Rotary Drill A drill using a rotary bit such as a simple fishtail type or a complex rotary-cone type.

Rotational Firing Blasting a block of rock nearest the face with a first explosion and afterwards timing other blasts successively so as to throw their burdens toward the space created by the preceding blast. Equivalent to *row shooting* or *buffer blasting.*

Rough Grading First stage of excavation, when the grade in the cut and fill is held to about ± 0.1 ft prior to the finish grading by a motor grader or other finishing machine.

Round A blast, including a succession of delay shots.

Row Shooting In a large blast, setting off the row of holes nearest the face first and the other rows behind it in succession. Same as *rotational firing.*

Rubble Rough rocks of irregular shapes and different sizes, broken from larger masses either naturally or artificially, as by geologic action or by blasting.

Rule-of-Thumb A statement or formula that is not accurate, but which is sufficiently precise for figuring an approximate value.

Run of Pit, Quarry, or Mine Material as it comes directly from the pit, quarry, or mine before any processing.

Runner The operator of some machines, such as a shovel or dragline.

Running An operating or producing machine, such as a drill.

Saddleback A hill or ridge having a concave outline along the top.

Sand Particles of sediment having diameters of more than 1/300 in (0.08 mm) and less than 1/12 in (2.0 mm).

Sandstone A clastic sedimentary rock consisting predominantly of sand-size particles. Light to medium in color, medium-weight, and soft to hard.

Sandy Loam Material containing much sand and silt, but having enough clay to make it somewhat coherent.

Scaling Prying loose pieces of rock off the face of a cut or roof of a tunnel to avoid danger of their falling unexpectedly.

Scalping The process of removing residuals and weathered rock of a cut prior to excavation of the hard rock.

Scarifier An accessory or tool of a motor grader, tractor, or other machine used chiefly for the loosening of materials of shallow depth.

Schist A well-foliated metamorphic rock having a tendency to split into thin layers. Medium to dark in color, hard, and heavy. An example is mica schist.

Scoria Lava in which the gas cavities are numerous and irregular in shape, producing a sharp rock. Light to dark in color, medium-weight, and hard.

Seams Parting planes in rocks, such as the cooling joints in igneous rocks and the separations between the strata of sedimentary rocks.

Secretions Rock materials which have been deposited from solution by infiltration into the rock cavities. The crystals of a geode are examples.

Sediment Regolith that has been transported and deposited by water, air, or ice; predecessor of sedimentary rocks.

Sedimentary Breccia A clastic sedimentary rock containing numerous angular particles of pebble size and larger. Of various colors and hardnesses and of medium weight.

Sedimentary Rock A rock formed by the cementation of sediment or by other processes acting at ordinary temperatures at or close beneath the surface.

Seismic Studies The analysis and appraisal of rock-earth excavation by seismic means by which shock-wave velocities at different depths are determined. The degree of consolidation of the material and, indirectly, the cost of the excavation are proportional to the shock-wave velocities.

Semiarid A term applied to a country neither entirely arid nor strictly humid but intermediate. Annual rainfall varies from about 10 in (25 cm) to about 30 in (76 cm).

Sequential Firing In a large blast, setting off the row of holes nearest the face first and the other rows behind it in succession. Same as *rotational firing* or *row shooting.*

Series An arrangement of electric blasting caps in which the firing current passes through them successively.

Series The third division of rocks, a part of the second division, or *system,* of rocks. Corresponds to a geologic *epoch.*

Series Parallel Two or more parallel circuits of blasting caps arranged in series.

Serpentine A metamorphic rock composed chiefly or wholly of the mineral serpentine. Greenish and massive, sometimes fibrous and lamellar. Medium to hard and medium-weight.

Shale A fine-grained sedimentary rock composed of clay-size and silt-size particles, generally with thinly laminated structure. Medium to dark in color, medium to hard, and medium- to heavyweight.

Shear Zone A zone of rock and in which shearing has occurred on a large scale so that the rock is crushed and brecciated.

Shoot Same as to *blast.*

Shooting Rock Material or rock which requires blasting.

Shot Rock Rock which has been blasted.

Shrinkage The diminution in dimensions and volume of a material. In excavation, the percentage loss in volume when 1 yd^3 in the cut is placed in the fill or embankment.

Sidehill Cut An excavation in a hill involving only one cut slope and, usually, one fill slope. It is sometimes considered to be any cut in a hillside.

Sill A relatively thin tabular sheet of magma which has penetrated a rock formation along approximately horizontal bedding planes.

Silt A sediment of particle size between 0.0002 in (0.005 mm) and 0.002 in (0.05 mm) in diameter.

Siltstone A clastic sedimentary rock consisting predominantly of silt-size particles.

Silurian The third system of rocks in the Paleozoic group and the corresponding geologic period, from 430 to 400 million yr ago.

Sink A large solution cavity open to the sky. A large shallow bowllike valley from which drain water escapes by evaporation or percolation.

Slab A large and flat but relatively thin mass of rock.

Slag Fused or partly fused compounds resulting in secondary products from the reduction of metallic ores.

Slate A fine-grained metamorphic rock derived from shale, in which pressure has produced very perfect cleavage planes. Sometimes called *argillite.* Medium to dark in color, hard, and heavy.

Slickensides Striated and polished surfaces of rocks abraded by movement along a fault.

Slide The movement of a part of the Earth's surface under the force of gravity.

Slide Rock Any loose fragmental rock lying on a slope. *Talus* is a term applied to slide rock.

Slip A fault in rock or a smooth joint or crack where the rocks have moved relative to each other.

Slope An incline or gradient as measured from the horizontal. Measured in degrees with the horizontal or as the ratio of horizontal distance to vertical distance, as 34° and 1.5:1 for the same inclination.

Slope Stake A surveying stake set at the point where the finished side slope of a cut or fill intersects the surface of the ground. Usually set along a line at right angles to the centerline and sometimes set a few feet from the actual point as a reference stake not to be lost in the prosecution of the work.

Sloughing Sliding of overlying material such as overburden upon rock.

Slump The downward slipping of a coherent body of earthy material along a surface of rupture.

Soapstone A metamorphic rock of interlocking fibrous texture and soft soapy feel, such as talc. Light to medium in color, soft, and medium- to heavyweight.

Softwood Wood which is light in texture, nonresistant, and easily worked. Typical are pines and cottonwoods.

Soil A mixture of earthy materials intermingled with organic matter so as to support rooted plants.

Sorting Selection by natural processes during transport of rock particles according to size, specific gravity, shape, durability, or other characteristics.

Specific Gravity A number stating a ratio of the weight of solid rock to the weight of an equal volume of pure water.

Specific Gravity, Apparent A number stating the ratio of the weight of the total volume of permeable rock with its included voids to the weight of an equal volume of pure water. Thus, the apparent specific gravity depends on the specific gravity of the rock and on the amount or volume of the contained voids.

Speed Indicators A measure of the travel speeds of a hauling machine. When laden it equals the engine(s) bhp divided by the GVW in 1000s of lb. When unladen it equals the engine(s) bhp divided by the tare weight in 1000s of lb.

Spoil Rock-earth wasted because it has no use. Overburden in a copper mine is an example.

Spoil Bank Bank or pile of spoil or wasted materials.

Spread The excavating machinery necessary to do the complete work for the whole or a part of the operation or job.

Square A unit of area equal to 100 ft^2 (equivalent to 9.29 m^2).

Stability The resistance of the material in a cut or fill to movement downslope due to inherent characteristics or to the weight of a superposed load.

Stack A small prominent island of bedrock, the remnant of a former narrow promontory eroded by wave action.

Station A distance of 100 ft measured along the centerline and designated by a stake bearing its number.

Station Yard A unit of quantity times distance, corresponding to 1 yd^3 moved horizontally through a distance of 100 ft.

Stemming Inert material such as cuttings placed in a blasthole instead of explosives, as in deck loading.

Stock A body of igneous rock intruded upward into an older rock formation.

Stockpile Material excavated and piled for future use.

Stone Any natural rock deposit or formation of igneous, sedimentary, or metamorphic origin, either in its original or altered form.

Stratification The deposition of sediment beds, layers, or strata. The arrangement of the rocks in such beds, layers, and strata. The parallel structure resulting from such deposition and arrangement.

Stratigraphy The systematic study of stratified rocks.

Stratum A bed or layer of rock.

Stream Terrace A bench along the side of a valley, the upper surface of which was formerly the alluvial floor of the valley.

Strength The ability of a rock body to resist stresses tending to change both shape and volume.

Striations Scratches and grooves on bedrock surfaces, caused by grinding of rock against rock.

Strike The direction of the line of intersection of the plane of rock stratification or bedding with the horizontal plane.

Stripping Removing the undesirable material from a deposit intended to be be used.

Strip Mining The open-pit mining of materials by removal of overburden.

Structural Geology The study of rock deformation and the delineation of geologic structural features.

Submergence Fall of the land relative to the level of the sea. A land of submergence has generally rough topography. The coast of Maine is an example.

Superelevation Rise of the outside curve above the inside curve of a haul road to accommodate the centrifugal force of the hauling units and thus prevent skidding and overturning.

Surficial A common word describing the unmoved surface of the earth.

Surge Bin A compartment for temporary storage of materials which allows converting a variable rate of supply to an even rate of discharge of the same average amount. An example is the receiving hopper of a conveyor belt.

Swell The increase in dimensions and volume of a material. In excavation, the percentage gain in volume when a cubic yard in the cut is placed in the fill or embankment.

Swing In revolving shovels and like machines, to rotate the upper works or deck with respect to the lower works. In cable drills, to operate a string of drilling tools.

Swing-Return Angle The angle for swing-return through which a revolving excavator must pass in a cycle of load-swing-dump-return.

Switchback A hairpin turn of about 180° in a haul road.

Syenite A granular intrusive igneous rock of generally dark color, containing little or no quartz. Medium to hard, and heavy.

Syncline A downfold of rocks with troughlike form and having the youngest rocks in the center.

System The second division of rocks within a major *group.* Corresponds to a geologic *period.*

Taconite An important iron ore of ferruginous chert. Medium to dark, hard, and heavy.

Talc A metamorphic rock with a soapy feel. Light to medium in color, soft, and medium-to heavyweight. Same as *soapstone.*

Talus An apron of rock waste sloping outward from the cliff which supplies it.

Tamping Compacting loose material by means of weights or weighted machines such as sheepsfoot or tamping-foot compactors.

Tamping Rollers One or more steel drums fitted with projecting feet in a machine to be towed or self-propelled.

Tectonic Pertaining to the rock structures and external forms resulting from the deformation of the Earth's crust.

Tephra A collective term designating all particles ejected from volcanoes, irrespective of size, shape, and composition.

Terminal Moraine The end moraine deposited by a glacier along its line of greatest advance.

Terrace A relatively flat elongated surface, bounded by a steep ascending slope on one side and a less steep descending slope on the other side.

Tertiary The first of the two systems of rocks in the Cenozoic group. Also the corresponding geologic period, from 63 to 2 million years ago.

Texture The sizes and shapes of the particles in a rock or sediment and the mutual relationships among them.
Throw Scattering of rock fragments from blasting.
Through Cut A cut with slopes on both sides. If there is little cut on one side, it is sometimes called a sidehill cut.
Thrust Fault A low-angle reverse fault with dip generally less than 45°.
Tight A term applied to rock-earth formations without weaknesses, which may require hard ripping or light blasting before excavation.
Till Nonsorted or ungraded glacial drift made up of all sizes of particles.
Tillite Sedimentary rock consisting of cemented till. Light to dark in color, medium to hard, and medium- to heavyweight.
Toe The lowest edge of a cut where the slope intersects the grade and the lowest edge of a fill where the slope intersects the ground.
Ton, Short A unit of weight equal to 2000 lb, 907 kg, or 0.907 metric ton.
Ton, Long A unit of weight used in mining, equal to 2240 lb.
Topographic Map A map that delineates surface forms, usually by contours.
Topography The relief and the form of a land surface.
Top Soil The good or productive soil as contrasted to the underlying soil. Ideally, a mixture of clay, silt, sand, and humus matter.
Track Drill A large percussion drill powered by compressed air or hydraulic oil. Mounted on crawler tracks.
Traction The friction of a machine on the surface on which it moves. Traction for excavating machinery is a measure of the weight of the machine on its driving tracks or wheels-tires and the coefficient of friction between the driving tracks or wheels-tires and the surface of the ground.
Trap Rock The dark-colored, fine-grained and dense igneous rocks with little or no quartz, such as basalt, diabase, and gabbro.
Travertine A collective variety of limestone, including dripstone, flowstone, and calcareous tufa.
Triassic The first system of rocks in the Mesozoic group. Also the corresponding geologic period, from 230 to 180 million yr ago.
Tuff A fine-grained pyroclastic rock consisting of ash and dust. Light to dark in color, soft to hard, and light- to medium-weight.
Ultramafic Rock Granular igneous rock consisting almost entirely of ferromagnesian minerals. Dark, hard, and heavy.
Unclassified Excavation Excavation not subdivided for bidding purposes into common, or earth, and rock. Regardless of nature, the excavation is bid at one unit bid price.
Unconformity A lack of continuity between units of rock in contact, corresponding to a gap in the geologic record, the gap being a period of erosion.
Uniformity Principle The concept that relationships established between processes and materials in the modern world can be applied as a basis for interpreting the geologic record and for reconstructing the history of the Earth.
Unit Cost In earthwork, the cost of producing a unit of work, such as a cubic yard of excavation or a cubic yard of compacted embankment.
Unwatering Same as *dewatering.*
Vein A tabular deposit of minerals, occupying a fracture, in which the particles have grown away from the walls toward the middle.
Vesicle A small cavity in an igneous rock formed by the expansion of a bubble of gas or steam during the solidification of the rock.
Voids The spaces between particles of a rock or rock-earth, expressed as a percentage of the entire volume.
Volcanic Ash The smallest of the particles ejected by the explosion of a volcano.
Volcanic Bomb The largest of the particles ejected by the explosion of a volcano.
Volcanic Mudflow A mudflow of water-saturated, predominantly fine-grained tephra.
Volcanism A term designating the aggregate of processes associated with the transfer of materials from the Earth's interior to the surface.
Volcano A vent or fissure through which molten and solid materials and hot gases pass upward to the Earth's surface.
Volume In excavation, the volume of cut or fill in cubic yards or cubic meters, as well

as the volume, loose measurement, in waste areas or stockpiles and in haulers measured in the same units.

Wacke Residual sand, silt, and clay formed by the weathering of igneous rocks.

Walking Dragline A huge dragline which moves itself by means of side-mounted shoes or pontoons actuated by overhead cams, the motion being akin to walking.

Wash Boring An exploratory drill hole from which samples are brought up mixed with water.

Waste Digging, hauling, and dumping valueless material to get it out of the way, as in the removal of overburden from an ore deposit. Also the material itself.

Water Gap A pass in a ridge or mountain range through which a stream flows.

Watershed Area which drains into a stream or other water passage.

Water Table A surface of underground gravity-controlled water.

Watt A standard unit of electrical energy, equivalent to 0.00134 hp.

Weathering The mechanical disintegration and the chemical decomposition of rock materials during exposure to nature's destructive forces.

Weight, Gross Vehicle (GVW) The combined tare weight of the hauling machine and payload weight.

Weight, Payload The weight of the load for which payment is made.

Weight, Tare The working weight of the hauling machine.

Welded Tuff A fine-grained volcanic rock the particles of which were so hot when deposited that they fused together. Light to dark in color, hard, and medium- to heavyweight.

Well Drill Name for the percussion cable drill.

Wellpoint A pipe fitted with a driving point and a fine mesh screen, used to remove underground water.

Wellpoint System Machinery used for the removal of underground water, consisting chiefly of centrifugal pump, header pipe, lateral pipes, riser pipes, valves, and wellpoints.

Well-Sorted Sediment A sediment consisting of particles of about the same size.

Wind Gap A former water gap through which the stream no longer flows.

Windrow A ridge of loose material thrown up by a machine.

Work-Cost Indicator A measure of the efficiency of a hauling machine. It equals the sum of gross vehicle weight plus tare weight divided by payload weight. It is an approximate expression of the ratio of total work done to pay work done during the cycle of a hauling machine.

Working Cycle The complete set of productive operations of a machine. For a hauler of rock, the cycle includes loading, hauling, turning, dumping, returning, and turning.

Xenolith A fragment of another rock, or earlier solidified portion of the same rock, in an igneous rock.

Yard In excavation, the area in which the owner stores and repairs the machinery and usually maintains his offices. Also, an expression for a cubic yard.

Yard-Mile Same as *Mile-Yard.*

Yard-Station Same as *Station-Yard.*

Zone of Aeration The zone of the regolith in which the open spaces or voids are filled mainly by air. Generally the weathered zone.

Zone of Fracture The upper portion of the Earth's crust in which rocks are deformed mainly by fracture.

Zone of Rift That volume of rock-earth between the walls of a fault which has been fractured, broken, brecciated, and pulverized by the long-term action of the fault. The zone may be 2 mi in width and several miles in depth.

Zone of Saturation The subsurface zone below the water table in which all voids are filled with water.

Bibliography

Associated General Contractors of America: *Contractors' Equipment Manual,* Washington, D.C., 1974.

Bajpai, A. C., et al.: *Mathematics for Engineers and Scientists,* Wiley, New York, 1973.

Baumeister, Avallone, and Baumeister: *Marks' Standard Handbook for Mechanical Engineers,* 8th ed., McGraw-Hill, New York, 1978.

Beiser, Arthur, et al.: *The Earth,* Time-Life Books, New York, 1970.

California Division of Highways, *Standard Specifications,* Sacramento, 1960

Caterpillar Tractor Co.: *Handbook of Ripping,* Peoria, Ill., 1972.

Caterpillar Tractor Co.: *Performance Handbook,* Peoria, Ill., 1979.

Daugherty, Robert L., and Franzini, Joseph B.: *Fluid Mechanics with Engineering Applications,* 7th ed., McGraw-Hill, New York, 1977.

Dickenson, E. H., and Slager., T., Jr.: *Rock Drill Data,* Ingersoll-Rand Co., New York, 1960.

Dobrin, Milton D.: *Introduction to Geophysical Prospecting,* 3d ed., McGraw-Hill, New York, 1976.

E. I. du Pont de Nemours & Co.: *Blasters' Handbook,* Wilmington, Del., 1966.

Eshbach, O. W., and Souders., M.: *Handbook of Engineering Fundamentals,* 3d ed., Wiley, New York, 1975.

Fenton, Carroll L., and Fenton, Mildred. A.: *Rock Book,* Doubleday, Garden City, N.Y., 1970

Fink, Donald G., and Beaty., H. Wayne: *Standard Handbook for Electrical Engineers,* 11th ed., McGraw-Hill, New York, 1978.

Hyster Co.: *Compaction Handbook,* Kewanee, Ill., 1972.

Kent., R. T., et al.: *Mechanical Engineers' Handbook:* Wiley, New York, 1950.

Kissam, Philip: *Surveying for Civil Engineers,* 2d ed., McGraw-Hill, New York, 1976.

Kummel, Bernhard: *History of the Earth,* W. H. Freeman, San Francisco, 1970.

Longwell, C. R., Flint., R. F., and Skinner, B. J.: *Physical Geology,* 2d ed., Wiley, New York, 1977.

McGraw-Hill Encyclopedia of Science and Technology, Daniel N. Lapedes (ed.), McGraw-Hill, New York, 1979.

Nichols, Herbert L., Jr.: *Moving the Earth,* 3d ed., North Castle Books, Greenwich, Conn., 1976.

Peele, Robert: *Mining Engineers' Handbook,* Wiley, New York, 1951.

Peurifoy, R. L.: *Construction Planning, Equipment, and Methods,* 3d ed., McGraw-Hill, New York, 1979.

Pough, Frederick H.: *A Field Guide to Rocks and Minerals,* 4th ed., Houghton Mifflin, Boston, 1976.

Schultz, John R., and Cleaves, A. B.: *Geology in Engineering,* Wiley, New York, 1955.

Shelton, John S.: *Geology Illustrated,* W. H. Freeman, San Francisco, 1966.
Steel, Ernest W., and McGhee, Terrence: *Water Supply and Sewerage,* 5th ed., McGraw-Hill, New York, 1979.
Thornbury, William D.: *Principles of Geomorphology,* Wiley, New York, 1969.
Trauffer, Walter E. (ed.): *Pit and Quarry Handbook,* Pit and Quarry Publications, Chicago, Ill., 1975.
United States Mineral Resources, U.S. Bureau of Mines, Washington, 1973.
Urquhart, Leonard C.: *Civil Engineering Handbook,* 4th ed., McGraw-Hill, New York, 1959.
Wyckoff, Jerome: *Rock, Time, and Landforms,* Harper and Row, New York, 1966.

Index

ANDERSONIAN LIBRARY
WITHDRAWN
FROM
LIBRARY
STOCK
UNIVERSITY OF STRATHCLYDE